T0215316

MAKING GEOGRAPHIES OF PEACE AND CONFLICT

This book illustrates the diversity of current geographies, ontologies, engagements, and epistemologies of peace and conflict. It emphasizes how agencies of peace and conflict occur in geographic settings, and how those settings shape processes of peace and conflict.

The essence of the book's logic is that war and peace are manifestations of the intertwined construction of geographies and politics. Indeed, peace is never completely distinct from war. Each chapter in the book will demonstrate understandings of how the myriad spaces of war and peace are forged by multiple agencies, some possibly contradictory. The goals of these agents vary as peace and war are relational, place-specific processes. The reader will understand the mutual construction of spaces and processes of peace and conflict through engagement with the concepts of agency, the mutual construction of politics and space, geographic scales, multiple geographies, the twin dynamics of empathy/othering and inclusivity/partitioning, and resistance/militarism. The book discusses the intertwined nature of peace and conflict, including reference to the environment, global climate change, borders, technology, and postcolonialism.

This book is valuable for instructors teaching a variety of senior level human geography courses, including graduate-level classes. It will appeal to those working in political geography, historical geography, sociology of geographic knowledge, feminist geography, cultural and economic geography, political science, and international relations.

Colin Flint, a geographer by training, is Distinguished Professor of Political Science at Utah State University. His research interests include geopolitics and world-systems analysis. He is the author of *Introduction to Geopolitics* (Routledge, 2022), *Geopolitical Constructs* (2016), and co-author, with Peter J. Taylor, of *Political Geography: World-Economy, Nation-State and Locality* (Routledge, 7th edition, 2018). He is editor of *The Geography of War and Peace* (2004) and co-editor (with Scott Kirsch) of *Reconstructing Conflict: Integrating War and Post-War Geographies* (2011). His books have been translated into Spanish, Polish, Korean, Mandarin, Japanese and Farsi.

Kara E. Dempsey is Associate Professor of Geography at Appalachian State University. She studies ethnonational conflicts, consolidation of state and regional power, international forced migration, and peace-building processes. She is the author of *The Geopolitics of Conflict, Nationalism, and Reconciliation in Ireland* (Routledge, 2022), and co-editor (with Orhon Myadar) of *Making and Unmaking Refugees: Geopolitics of Social Ordering and Struggle with the Global Refugee Regime* (Routledge 2023). She currently is serving as the president of the Political Geography Specialty Group, American Association of Geographers (AAG) and the AAG Honors Committee.

MAKING GEOGRAPHIES OF PEACE AND CONFLICT

Edited by Colin Flint and Kara E. Dempsey

Routledge
Taylor & Francis Group

LONDON AND NEW YORK

Designed cover image: Peter Zelei Images

First published 2024
by Routledge
4 Park Square, Milton Park, Abingdon, Oxon OX14 4RN

and by Routledge
605 Third Avenue, New York, NY 10158

Routledge is an imprint of the Taylor & Francis Group, an informa business

British Library Cataloguing-in-Publication Data
A catalogue record for this book is available from the British Library

ISBN: 978-1-032-38595-2 (hbk)
ISBN: 978-1-032-38598-3 (pbk)
ISBN: 978-1-003-34579-4 (ebk)

DOI: 10.4324/9781003345794

Typeset in Times New Roman
by SPi Technologies India Pvt Ltd (Straive)

CONTENTS

FIGURES

ACKNOWLEDGMENTS

We thank Prachi Priyanka and Faye Leerink for their support of the idea of this book and support through the whole process. Thank-you to the authors for their ideas and commitment to completing the project in a timely manner. The final thank-you is to the reader, and instructors who may assign chapters in their classes, for the commitment to teaching about the violent nature of the world and, especially, the ability to find pathways to peace and to make the choice to follow them. Teaching, learning, and practicing peace is the brave and bold choice.

Thank-you to Taylor & Francis Informa UK Ltd. for permission to reproduce Figure 7.1, and to Devon Maloney, Department of Geography, University of North Carolina at Chapel Hill, for producing Figure 8.1 and giving permission for its use in this volume.

CONTRIBUTORS

Annika Björkdahl is Professor of Political Science at Lund University. Her research includes peace and peacebuilding, politics of memory, gender, and transitional justice. Recent books include *Peacebuilding and Spatial Transformation* (2017) with S. Kappler, and *Spatializing Peace and Conflict* (2016) and *Space for Peace* (2022) with S. Buckley-Zistel a special issue of the *Journal of Intervention and Statebuilding*. She has published in journals such as *Political Geography*, *Peacebuilding*, and the *International Journal of Transitional Justice*. She is editor of *Cooperation and Conflict*, and co-editor of the Book Series *Rethinking Peace and Conflict Studies*.

Tony Colella teaches geography classes at the University of Arizona in the desert southwest, where he earned his PhD and MFA. His longest-term research is about queer experiences in university science. He is also a photographer and visual artist, and he hosted the queer sff podcast The Imaginaries for many years.

Kara E. Dempsey is Associate Professor of Geography at Appalachian State University. She studies ethnonational conflicts, consolidation of state and regional power, international forced migration, and peace-building processes. She is the author of *The Geopolitics of Conflict, Nationalism, and Reconciliation in Ireland* (Routledge, 2022), and co-editor (with Orhon Myadar) of *Making and Unmaking Refugees: Geopolitics of Social Ordering and Struggle with the Global Refugee Regime* (Routledge, 2023). She currently is serving as the president of the Political Geography Specialty Group, American Association of Geographers (AAG) and the AAG Honors Committee.

Lorraine Dowler is Professor of Geography and Women, Gender, and Sexuality Studies at Penn State University. Her scholarship is rooted in feminist approaches to geopolitics that enables more fluid conceptualizations of compassion, identity, and individuality related to understanding everyday life, private spaces, and the lives of women and other vulnerable groups. She expands on the initial questions posed by feminist geopolitics by drawing from the literatures of feminist care ethics to examine the potential for a caring geopolitics. This research examines the violent processes at the heart of the life of any nation while exploring the role of hypermasculine state practices in daily life. Her research focuses on how individual vulnerabilities to violence are rendered invisible through spatial processes such as border-making, cultural privilege, militarization, and nationalism.

Md Azmeary Ferdoush is an Academy of Finland postdoctoral researcher based at the Karelian Institute, University of Eastern Finland. As a political geographer, his interest lies particularly in the study of borders, migration, state sovereignty, identity, and critical geopolitics with a geographical focus on Bangladesh and Finland. Azmeary has published extensively in journals, including *Antipode*, *Political Geography*, *Geopolitics*, *Geoforum*, *Area*, *Citizenship Studies*, and *Ethnography*. He is co-editor of *Borders and Mobility in South Asia and Beyond* (2018). Azmeary's monograph, *Sovereign Atonement*, is expected in 2024.

Colin Flint, a geographer by training, is Distinguished Professor in the Department of Political Science at Utah State University. His research interests include geopolitics and world-systems analysis. He is the author of *Introduction to Geopolitics* (Routledge, 4th ed. 2022), *Geopolitical Constructs: The Mulberry Harbours, World War Two, and the Making of a Militarized Transatlantic* (2016), and co-author, with Peter Taylor of *Political Geography: World-Economy, Nation-State and Locality* (Routledge, 7th edition, 2018). He is editor emeritus of the journal *Geopolitics*. His books have been translated into Spanish, Polish, Korean, Mandarin, Japanese and Farsi.

Joshua Inwood is a human geographer whose work focuses on broad questions of race, racism and the making of space and place. His work is primarily centered in the American South and explores broad questions related to the geographic history of race within the United States. He draws from a range of scholars and theorists to understand the broad currents of white supremacy and resistance that have shaped and continue to shape the making of space and place in the United States. This includes focusing on a range of economic, political, and social structures, including the realities of whiteness and the realities of settler colonialism, and how race and racism intersects with the political economy. He also has a long-standing research interest in memory and landscape studies and questions about U.S.-based truth commissions. These interests often mean he collaborates with community groups and interested local activists who are involved in struggles to tell stories and remake their communities in more just ways.

Scott Kirsch is Professor of Geography at the University of North Carolina at Chapel Hill. He is author of *American Colonial Spaces in the Philippines: Insular Empire* (Routledge) and *Proving Grounds: Project Plowshare and the Unrealized Dream of Nuclear Earthmoving*, and editor, with Colin Flint, of *Reconstructing Conflict: Integrating War and Post-War Geographies* (Routledge).

Sara Koopman teaches at the School of Peace and Conflict Studies at Kent State, which is a living memorial to the students killed by the National Guard there in 1970 for protesting for peace. She is a feminist political geographer interested in the socio-spatial aspects of peace and peacebuilding and how interlocking systems of unequal power and privilege shape those. Sara looks at solidarity efforts that work across both distance and difference to build peace and justice with a particular focus on global North–South international solidarity, how it can fall into colonial patterns, and how to avoid these.

Christian C. Lentz is Associate Professor of Geography at the University of North Carolina at Chapel Hill. He is author of *Contested Territory: Điện Biên Phủ and the Making of Northwest Vietnam*, winner of the 2021 Harry J. Benda Prize for outstanding first book in Southeast Asian Studies. His articles have appeared in *The Journal of Peasant Studies*, *Political Geography*, *Modern Asian Studies*, and other journals.

Andrew Linke is Assistant Professor in the University of Utah Department of Geography and an external research affiliate of the Peace Research Institute Oslo. His research links the fields of conflict studies, political geography, and human-environment interactions. He has published research in *Political Geography*, *Global Environmental Change*, *Annals of the American Association of Geographers*, the *Journal of Peace Research*, and *American Sociological Review*, among other journals.

Nerve V. Macaspac is Assistant Professor at the City University of New York, where he is a faculty at the Department of Political Science and Global Affairs at the College of Staten Island and a graduate faculty at the doctoral programs in Earth and Environmental Sciences and Environmental Psychology. His current research examines the phenomenon of community-led peace zones amid active violent conflict to better understand the spatialities of peace.

Virginie Mamadouh is Associate Professor of Political and Cultural Geography at the University of Amsterdam (the Netherlands). She has published about geopolitics, diplomacy, and European integration and was an editor of *The Wiley Blackwell Companion to Political Geography* (2015).

Adam Moore is Professor at the University of California, Los Angeles, with a joint appointment in the Department of Geography and the International Institute. His research concerns the political and geographical dynamics of war, violence, and peace. Topics of particular interest include geopolitics, ethnic conflict, nationalism, intrastate war, postwar peacebuilding, Southeast European politics, military labor, military contracting, and the militarization of U.S. foreign policy.

Orhon Myadar is Associate Professor at the University of Hawai'i at Mānoa. As a political geographer, Myadar studies the intersection of geography and politics at various scales, specifically how borders of belonging or exclusion shape everyday struggles of underserved and marginalized individuals and communities. Her current research examines forced mobility and displacement especially in the context of political turmoil. She leads a collaborative project funded by the National Institute for Transportation and Communities on connections between mobility, transportation, and quality of life in refugee communities in Tucson, Arizona. She is the author of *Mobility and Displacement* (Routledge, 2022) and co-editor of *Making and Unmaking Refugees* (Routledge, 2023).

María Belén Noroña is a human geographer working at the intersection of higher education and grassroots-grounded research. Her research interest includes feminist political ecology, extractive industries, Amazon rain forest, Indigenous territory, Indigenous epistemologies, race and gender intersectionality, networked space and place, nonrepresentational cartography, and decolonization. Her current research centers on Indigenous ontologies that can aid our understanding of human-environmental interactions. She is currently an assistant professor of geography at the Pennsylvania State University.

Shannon O'Lear is Professor of Geography in the Geography and Atmospheric Science Department and Director of the Environmental Studies Program at the University of Kansas in Lawrence. She has published work on climate science, geography and STS, geopolitics in Azerbaijan and Armenia, genocide, and other forms of violence. Her recent books include *Environmental Geopolitics* (2018), *A Research Agenda for Environmental Geopolitics* (2019), and *A Research Agenda for Geographies of Slow Violence: Making Social and Environmental Injustice Visible* (2021). Her undergraduate students' research project podcasts are available online at https://opentext.ku.edu/environmentalgeopolitics/.

Mark Ortiz (he/him) is a scholar-activist and Presidential Postdoctoral Fellow in the Department of Geography at Penn State. He completed a PhD in Geography at the University of North Carolina, Chapel Hill, in 2022 and a BA in Environmental and Religious Studies at the University of Alabama in 2015. Broadly he is interested in transnational youth movements, the global politics of climate change, and youth popular and social media cultures. His doctoral research, funded by the National Science Foundation, characterized the pathways to institutional influence of emerging global youth climate movements such as the School Strikes for Climate and YOUNGO Youth network. His current work develops collaborative digital storytelling methods to represent the geographical diversity of contemporary youth climate activism and elevate the stories of young people on the frontlines of climate change erased from dominant narratives. He is a Leadership Team Member of the North Carolina Climate Justice Collective and has worked with numerous youth and intergenerational climate justice organizations including CliMates and the Climate Reality Project. He has also served as an expert panelist and consultant on topics related to youth empowerment for IDEO and the U.N. Foundation and a delegate to U.N. climate meetings.

Clionadh Raleigh is Professor of Geography at the University of Sussex. Her research investigates African politics and patterns of violent conflict. She is the founder and director of the Armed Conflict Location Event Data project, a global non-profit that records geographic data for thousands of political instability events worldwide. Her articles have appeared in journals including *Political Geography, Annals of the American Association of Geographers, African Affairs*, and *Global Environmental Change*, among others.

Ian Slesinger is an interdisciplinary researcher whose work explores the nexus between technology, security, and politics at a range of scales, from the elite to everyday. He is particularly engaged in questions concerning the agency of nonhumans in the politics of technology. His current post is Postdoctoral Research Assistant in the Information Security Group at Royal Holloway, University of London. He holds a PhD in political geography from the University of Birmingham, an MSc in Modernity, Space and Place from University College London, and a BA in anthropology from Reed College. His areas of specialist knowledge include geopolitics, critical security studies, science and technology studies, borders, war and conflict, and the Middle East.

James A. Tyner is Professor of Geography at Kent State University and a Fellow of the American Association of Geographers. He received his PhD in Geography from the University of Southern California. Professor Tyner is the author of more than twenty books, including *War, Violence, and Population*, which received the AAG Meridian Book Award for Outstanding Scholarly Contribution to Geography. In addition, Professor Tyner is the recipient of the AAG Glenda Laws Award for outstanding contributions to geographic research on social issues. His research coalesces around violence, genocide, militarism, and political economy.

1

INTRODUCTION

Making Geographies of Peace and Conflict

Colin Flint and Kara E. Dempsey

Questions of war and peace dominate the headlines. They are also pressing questions for students in university classrooms and scholars forming research agendas. The war between Russia and Ukraine raised the specter of another European war, or even escalation into a wider conflagration. Post–Cold War optimism now seems naïve. The persistence of terrorism, including attacks perpetuated by groups proclaiming a brand of Islamic fundamentalism, and the chaotic U.S. withdrawal from Afghanistan, indicate that the violence within the War on Terror persists. These dramatic events came on top of long-term tensions, especially the Western focus on China's growing power through the twin strategies of the Belt and Road Initiative and creating a blue-water navy. On the other hand, we see that steps toward peace are possible. Violence has remained in abeyance in Northern Ireland, despite post-Brexit tensions. Diplomacy and dialog have come to the fore within Western Europe to sustain what had appeared to be a creaky trans-Atlanticism. The Black Lives Matter movement has been successful in making racial justice and postcolonialism a central concern in many aspects of life. In sum, intertwined processes of peace and conflict are ongoing. In 2003, in the wake of the terrorist attacks of September 11, 2001, then Vice President Cheney argued, "9/11 changed everything." Perhaps so, but not in the ways expected and not forever. Geopolitics remains a process, a state of flux, rather than a state of stasis. The flux is manifest in a tapestry of geographies of war and peace that can only be understood by identifying interlinked spaces and the simultaneity of processes of peacebuilding and conflict making.

It was not just the practices of war that were changed by "9/11." The discipline of geography, especially the subfield of political geography, changed too.

DOI: 10.4324/9781003345794-1

Surprising as it may seem now, war was not a topic of interest to most political geographers in 2001. There was a dearth of literature for academics to turn to for teaching materials or intellectual debate. But now war and conflict are a strong focus for political geographers (Mamadouh 2004 and this volume). Most significantly, the topic of peace has emerged as of equal importance as a way of understanding the processes, scalar experiences, and engagements of political geography (Megoran 2011; Williams and McConnell 2011).

Geographic scholarship and its critical framing of conflict and peace are particularly relevant, rich, and provocative. It has produced many different approaches and insights. This book offers educators and students a single volume that illustrates the diversity of current geographies ontologies, engagements, and epistemologies of peace and conflict. In addition, the book will explore interactions with scholars in other disciplines who have discovered and implemented a "spatial turn" to peacebuilding (Björkdahl and Buckley-Zistal 2016; Björkdahl and Kappler 2021) and those who are exploring innovative approaches to understanding peace but would benefit from further consideration of the geographic perspective (MacGinty 2021). In sum, we hope the volume will highlight the value of the geographic perspective in understanding the processes of peacebuilding and conflict, and strengthen bridges with related disciplines.

The world has experienced changes in the practices and geographies of war-making and peacebuilding, and the way geographers understand and engage these practices. The purpose of the book is to showcase how current geographic thought informs the new geopolitical context.

Geographic Framings of Peace and War

Though certainly not the focus of his comment, Vice President Dick Cheney's comment about the change brought about by "9/11" certainly rang true for the discipline of geography. Prior to the proclamation of the War on Terror just two edited volumes on the geography of war were available: *The Geography of War and Peace* edited by David Pepper and Alan Jenkins (1985) and *The Political Geography of Conflict and Peace* edited by Nurit Kliot and Stanley Waterman (1991). The foci of these books were the Cold War and ethno-nationalist conflicts, particularly Israel-Palestine. The world had changed, and geographers had not reacted.

However, the reaction was swift and comprehensive. Colin Flint's *The Geography of War and Peace* (2004) was an attempt to collect a set of essays on war and peace that addressed the new geopolitical context. Though in some ways well received, it was rightly criticized for being too focused on war rather than peace. Soon to follow were other monographs and edited volumes on war and violence that were centered upon the dynamics of the War on Terror. Key edited volumes included *Violent Geographies: Fear, Terror, and Political Violence* edited by Derek Gregory and Allan Pred (2007) and *War, Citizenship, and Territory* edited by Deborah Cowen and Emily Gilbert (2008). John Morrissey's (2017) monograph *The Long War* described the growth of U.S. military presence in the Middle East.

The renewed interest in war soon developed into processes of conflict and violence that were not directly tied to the War on Terror. The world as a violent place came to the fore of geographic inquiry. The pervasiveness of the military in the construction of society and geographic landscapes was brought in to focus by Rachel Woodward's *Military Geographies* (2004) and *Military Legacies* by James Tyner (2009). The violent nature of globalization was illustrated by Deborah Cowen's (2014) *The Deadly Life of Logistics*. The violent and competitive nature of development was discussed by Marcus Power (2019) in *Geopolitics and Development*. The overarching processes of global warming were connected to topics of war and peace (see Simon Dalby's (2020) *Anthropocene Geopolitics*). The context of an age of refugee crises and struggles over the control of borders catalyzed key books such as *Border Wars* by Klaus Dodds (2021), and the work of Reece Jones on *Violent Borders* (2016).

The emphasis upon intersecting forms of conflict and agencies of peacebuilding has recently come to the fore. Nicole Laliberté (2016) investigated how gender, race, and class perspectives shape discriminatory politics and visions of peacebuilding work among international, national, and regional actors. Also, Dresse et al. (2019) explored how environmental peacebuilding is fostered by the assumption that global environmental change may provide incentives for global cooperation and peace. Kara E. Dempsey's *Geopolitics of Conflict, Nationalism and Reconciliation* (2022) traced the production of spaces of peace that emerge as peacebuilders navigate violent conflicts, especially when states actors fail to do so.

The renewed focus on war sparked a new field of geographic inquiry; geographies of peace and pacific geopolitics. Important scholarly articles and chapters have established geographies of peace as a topic in and of itself (Williams and McConnell 2011; Williams 2013; Megoran 2010 and 2011; Koopman 2011a, 2011b, 2018). Philippa Williams' *Everyday Peace* (2015) elucidated ways in which ordinary people forge peace through citizenship practices, tolerance, and civility. Coherence and energy were given to the rally toward geographies of peace through the edited volume *Geographies of Peace* edited by Fiona McConnell, Nick Megoran, and Philippa Williams (2014).

This brief discussion is by no means exhaustive. It merely shows that the topics of war and peace were largely, and surprisingly, ignored by political geographers. Yet the last twenty years have seen a resurgence of interest, and the important call to give equal focus to processes of peace and not let conversations about war be dominant. For a full discussion of the discipline of geography's torrid engagement with war and empire, and also peace, see the work of Virginie Mamadouh (2004 and this volume). Political geography and related disciplines are now very focused on peace and conflict. Numerous monographs and journal articles advance our conceptual understanding of peace and war. This volume builds upon this new literature to illustrate the breadth and depth of contemporary work on the geography of peace and conflict for educators and students. We give equal prominence to geographies of peace and war, and the intersections between peace and conflict, while reflecting the current geopolitical context that involves much more than the War on Terror.

Peace, Conflict, and the Making of Geographies

The essence of the book's logic is that war and peace are manifestations of the *intertwined* construction of geographies and politics (Boulding 2000; Kirsch and Flint 2011). Indeed, peace is never completely distinct from war. Each chapter in the book will demonstrate understandings of how the myriad spaces of war and peace are forged by multiple agencies, some possibly contradictory. The goals of these agents vary as peace and war are relational, place-specific processes. The temporal scope and legacies of the agency are multiple, from simultaneity to the *longue durée*.

As a heuristic device, we can initially think of two continuums: place-specific processes of peace and manifestations of war (e.g., absolute) and simultaneity to the *longue durée*, and the intersection of the two continuums. The politics of the interaction is driven by multiple forms of agents (from individuals to multinational organizations) and multiple forms of agency (from active and conscious peacebuilding to purposeful war-making). The multiplicity of agents and their agency form (and are framed by) multiple spaces and scales.

Although there is a continuum in the *expressions* of time, the *operation* of time is not linear or unidirectional. Instead, time is a combination of simultaneity of actions, with short- to long-term implications, within relative structural constants of the *longue durée*. Some actions are banal and serve to maintain existing power relations and geographic circumstances. Other actions are transformative and can move the needle from war to peacebuilding, or vice versa, and in the process rearrange spatial relations and contexts. The multitude of processes of peace and conflict moves in different temporal directions. More importantly, conditions of peace and war often exist simultaneously in the same geographic setting. Some conditions may progress from war to peace, while others may regress from peace to war; though it is best not to think of peace and conflict as binary and separate processes but as ongoing and intertwined forms of agency (Flint 2011). In sum, peace and war are not thought of as a dichotomy but as an ongoing dynamic that continually makes and remakes multiple geographies.

From our entry point of continuums, we can interrogate through their interaction via multiple agencies a set of spatial-temporal contexts that develop, regress, twist and turn, reinforce, and challenge each other. We come to a kaleidoscope of spaces/arenas/scales displaying different forms of peacebuilding and war-making forms of agency.

To advance the debates and make sense of the dynamic kaleidoscopic patterns of spaces of peace and war, the chapters will focus on a range of engagements, mechanisms, agents, and forms of agency that occur within, and rearrange different geographic contexts or articulations. Agencies of war may be anything from enacting global war to internment. Agencies of peace are (re)-produced in everyday dimensions, multi-scalar politics of power, and as tools of statecraft. They may span from decolonization to the creation of twin-cities. They also include processes that "peaceweave" elements of a just society to foster social justice and equity, community engagement, and shared governance (e.g., Shields and Soeters 2017). But agency is never straightforward in its intentions and outcomes. An act of agency by an individual, social group, country, or alliance may invoke actions of war (such as dropping nuclear bombs on Hiroshima and Nagasaki) to end World War Two and create a negative peace. Similarly, an act of agency to promote peace (such as free movement of people within the European Union) has created a violent external border.

All forms of agency occur within temporal and geographic contexts. Agency may maintain the dominant political relations of a given time, or they may be transformative. It may implement geographic understanding as a tool to achieve peace. Similarly, agency may utilize geographic arrangements that establish a set of power relations, or create new territorial and scalar settings with different power relations. The times and spaces that are made may tend toward stable peace or absolute war.

Through critical reflections on the power relations and the spatialities of war and peace, the book illustrates multiple agencies (some possible contradictory), in a myriad of spaces, with goals that range in a conceptual continuum from stable peace to global war, with varied temporal-relations from simultaneity to the *longue durée* and a combination of peace and conflict making forms of agency. The value of an edited volume is the perspective offered by different scholars on the manifold possibilities of the intersections of agency, space, and time requires multiple ontologies. The many forms of agency and the spaces that are made require multiple ontologies and epistemologies, or ways we perceive the way the world works and how to investigate it. Bringing together a range of recent engagements and authors helps the reader explore the complexity and dynamism of the many contemporary geographies of peace and war.

Consistent Themes through the Book

Current approaches to the topics of peace and war in the discipline of geography, and related disciplines, are as eclectic as they are vibrant. This is a positive development. The pursuit of peace and concern about conflict is shared by scholars, whatever their approach to studying the world. This volume allows those different approaches to be showcased. However, we also recognize that students and scholars (sometimes in their role as teachers) would like to identify threads of ideas throughout our discussion of peace and war to help with the learning process. With that concern in mind, coherence through the individual chapters may be found through the identification of consistent themes. Not every chapter will address each of these themes, but in sum we hope the volume will allow you to explore their value in understanding contemporary circumstances of peace and war.

The key themes are:

Agency: Peacebuilding and conflict-making are the outcomes of intersecting social processes at multiple scales initiated and conducted by various actors.

Mutual construction of politics and space: The agency of peacebuilding and conflict-making is situated in, and simultaneously re-creates and re-arranges, geographic settings. The settings provide opportunities and constraints for agents.

Multiple scales: All geographical settings are multi-scalar in that the global and the local, and all intervening scales, are mutually constructed through processes that operate primarily within, but also transcend, any particular scale.

Multiple geographies: There are many forms of geographical settings, but the prominent ones are places (arenas of activity and identity), territories, networks, and scales.

The twin dynamics of empathy/othering and inclusivity/partitioning: Identities based on geographical identities and attachments (including but not limited to countries and regions) and membership in social groups (class, race, religion, gender, and sexuality) can foster a sense of difference and separation that may fuel conflict or a sense of shared experience or concern that can enable empathy and peacebuilding.

Resistance/militarism: The essentially militaristic nature of capitalism and states provokes actions of resistance. Some of these actions may be anti-systemic and engender fundamental social change. Other actions may re-create the same forms of violence and militarism but in new ways; such as many cases of national separatism or the criminal actions of terrorist and fundamentalist groups.

Geographic Scale and Forms of Peace and Violence

As British women's rights activist Emmeline Pethick Lawrence argued, it is war that is negative, while peace "is the highest effort of the human brain applied to the organization of the life and being of the peoples of the world on the basis of cooperation" (Pethick Lawrence 1972 [1915], 143; see also Addams 1907/2007). In other words, peace is uneven and multifaceted. This statement can be seen as an initial search for a concept and form of politics that has become known as positive peace (Shields and Soeters 2017). Johan Galtung's distinctions between negative and positive peace, and the concepts of structural and cultural violence, are seminal ideas that underlie the geographic investigation of peace and war.

The absence of direct violence, either interpersonal or at any scale up to global war, is known as negative peace (Galtung 1965, 233). An increase in wealth and well-being that is implicitly understood as nurturing peaceful circumstances, social justice, and a growing sense of security, is known as positive peace (Galtung 1965, 233). Structural violence is the harm imposed upon people through living in situations of poverty and exploitation (Galtung 1969, 170). The difference in life expectancy and life chances from wealthy to poor areas are visible manifestations of structural violence. These gross disparities are ugly and uncomfortable for those lucky enough to live in relative prosperity and security. Hence, the disparities are justified by cultural violence, or the use of cultural representations used to justify structural or direct violence (Galtung 1990).

Our geographic inquiry into intertwined processes of war and peace is an engagement with structural and cultural violence, and positive and negative peace. We can consider the connections between these forms of peace and violence through a focus on geographic scale. For example, conceptualizing a pyramid of peace helps us explore the interaction of scales, time-periods, and forms of peace (Adolf 2009, 236–238). The scale of the body is the site of corporeal peace, and household and community settings are the scales of sanctuarial peace. The former is access to adequate nutrition, shelter and sanitation, healthcare, and education. These are the basic needs that allow an individual to survive and are the foundation for a person to fulfill themselves in a complete lifepath. Hence, it is the individual's ability to engage in the provision and benefits of social peace. However, it is only a viable route for an individual's lifepath if they also experience sanctuarial peace, or minimal harm from other people, the state, political and economic inequality and deprivation, and a nontoxic environment.

Absence of interpersonal harm is a form of negative peace, whether it be from armed groups (gangs or state forces) or a relative. The other forms of sanctuarial peace are more closely related to positive peace, the operation of society that provides for people and enables them to achieve their goals. These forms of peace are a function of relations within the home, the neighborhood, and village, town, or city. Only when these two forms of peace are being experienced can a person's inner peace, their spiritual and intellectual sense of self and calm, be attained.

Inner peace and corporeal peace, or their absence, are experienced on a daily, or even hourly basis and within the scales of the home and place of settlement. They imply, or even demand, that a person has a place of settlement to experience these types of peace. A refugee or someone made homeless by economic deprivation will struggle to find these forms of peace. To fully understand their presence and absence requires us to consider the geographic scale and temporal scope of two other forms of peace: socioeconomic and world peace. We usually consider socioeconomic peace at the national scale. Disparities of wealth, all types of discrimination, and access to employment that is not enforced, are commonly associated with national economies and the way a country is governed. Our emphasis on geographic scale means that national circumstances, and those at subnational scales, must be considered within a global context.

Galtung and Adolf point to the intertwined nature of forms of peace, in different sites and scales, involving different agents and structures. Our framework and themes hope to shed light on the complexity of the duality of peace-violence in various forms and settings.

Chapter Summaries

The first three chapters of the book provide reviews of the considerable amount of writing on issues of war and peace that have been published in the last twenty years or so. Virginie Mamdouh's chapter builds on the nineteenth- and early-twentieth-century historical foundations of the discipline of geography as a means of informing states in the endeavors of building empire and making war. In contrast, the recent academic literature has been largely critical of state practices of war and has developed an explicit focus on peace. Nerve V. Macaspac and Adam Moore develop the discussion on the literature addressing the geography of peace. Specifically, they identify four themes: peace as a set of place-specific processes; political practices and ideologies that animate peace projects; the development of a holistic peace agenda; and everyday peace. The development of transdisciplinary research is highlighted by Annika Björkdahl in Chapter 4. She discusses the "spatial turn" in peace and conflict research and how this has promoted new understandings of peace such as everyday peace, mobile peace, urban peace, and trans-scalar peace.

The following two chapters provide overviews of the changing dynamics of the geographies of conflict and peacebuilding, respectively. Chapter 5, by James A. Tyner, begins with the deceptively simple definitions of negative peace and positive peace to explore the complex entanglement of overlapping geographies of war and peace. Chapter 6, by Kara E. Dempsey, focuses on the spatial practices of peacebuilding. The chapter uses the example of Northern Ireland and the construction of "shared" spaces to illustrate how building a politics of cooperation, respect, and deep listening necessarily requires the making of new geographies.

Chapter 7, by Colin Flint, takes a macro-view to illustrate how foundational relations between what is commonly referred to as the Global North and Global South are structural context that has fostered forms of direct and indirect violence. This pervasive context has also provided opportunities for promoting new visions of peace and effective campaigns. Chapter 8, by Christian C. Lentz and Scott Kirsch, develop the idea of Global North–Global South relations through the examination of postcolonial conflict in Southeast Asia by interrogating the idea of the "shatterbelt" through the idea of colonial rupture.

The following two chapters discuss the visions and practices that may build peace. In Chapter 9, Orhon Myadar and Tony Colella use a discussion of cultural representation through movies to explore the construction of empathy, or the lack of it. Geographies of victimization are often created in a way that justifies power and militarism. These imagined geographies must be challenged to create new visions of peace. In Chapter 10, Sara Koopman uses examples of peace activism in Colombia and Ukraine to show how everyday practice can create spaces of negative peace and pathways toward positive peace – even within contexts of ongoing violence.

Chapters 11 and 12 may be read together to illustrate the interconnected geographies of territory and networks in the practices of peace and conflict. Md Azmeary Ferdoush uses the example of borders in postcolonial South Asia to show how borders are a violent practice of compartmentalization and separation. However, everyday practices of "peacework" are challenging the way borders inhibit the pursuit of positive peace. In Chapter 12, Ian Slesinger discusses the use of digital technologies in practices of war and peace. Especially, the ambiguity of digital violence shows the fluidity of practices of conflict and peacemaking, and the complexity of the spaces they make.

Chapters 13 and 14 may also be read together to consider the role of the environment and global climate change in geographies of peace and conflict. In Chapter 13, Shannon O'Lear highlights the term "slow violence" to show how particular geographies of the environment, specifically geographies of enclosure, are a form of violence. However, in some contexts new geographies of environmental practices may foster peaceful collaboration. In Chapter 14, Andrew Linke and Clionadh Raleigh emphasize the value of the geographic approach by challenging the value of generalizations that see climate change as national security threat stemming from a direct connection between weather patterns and violence. Instead, they take a human security approach that illustrates the relative stability of social systems in particular places as interacting with the impacts of climate change.

Finally, Chapter 15 by Mark Ortiz, María Belén Noroña, Lorraine Dowler, and Joshua Inwood draw our attention to the importance of teaching peace. They identify spatial-pedagogical practices that allow students and instructors to engage positive peace. These practices include recognizing that progress toward positive peace requires questioning and challenging ways institutions reinforce structural violence.

Conclusion

A conclusion to an introduction to a set of essays authored by different scholars is an invitation to read ahead rather than stop reading with a new set of ideas. We hope you enjoy your pathway through the following essays. Of course, they were written at a certain moment of world history. Yet, we hope they endure as a catalyst to think about processes of peace and conflict from a geographic perspective. Building peace and making war are complex processes. Any claim to a simple framework should be addressed with a jaundiced eye. However, the job of social science is to offer a way to think about complexity, and we hope the themes in this book, and the way they are employed differently by the authors, are a means for you to tackle the causes of whatever conflict may be raging at the time you are reading, and to consider possibilities of peacebuilding. Understanding conflict is a necessary component of working toward pathways to a just and sustainable peace. We hope we offer at least some guidance for you along that pathway.

References

Addams, J. 1907/2007. *Newer Ideals of Peace*. University of Illinois Press.

Adolf, A. 2009. *Peace: A World History*. Polity Press.

Björkdahl, A. and Buckley-Zistal, S. 2016. *Spatializing Peace and Conflict*. Palgrave Macmillan.

Björkdahl, A. and Kappler, S. 2021. Spaces of peace. In *The Oxford Handbook of Peacebuilding, Statebuilding, and Peace Formation*, eds. O.P. Richmond and G. Visoka, 139–151. Oxford University Press.

Boulding, E. 2000. *Cultures of Peace*. Syracuse University Press.

Cowen, D. 2014. *The Deadly Life of Logistics*. University of Minnesota Press.

Cowen, D. and E. Gilbert, eds. 2008. *War, Citizenship, Territory*. Routledge.

Dalby, S. 2020. *Anthropocene Geopolitics*. University of Ottawa Press.

Dempsey, K.E. 2022. *An Introduction to the Geopolitics of Conflict, Nationalism and Reconciliation in Ireland*. Routledge.

Dodds, K. 2021. *Border Wars*. Penguin.

Dresse, A., I. Fischhendler, J. Ø. Nielsen and D. Zikos. 2019. Environmental peacebuilding: Towards a theoretical framework. *Cooperation and Conflict* 54: 99–119.

Flint, C., ed. 2004. *The Geography of War and Peace*. Oxford University Press.

Flint, C. 2011. Intertwined spaces of peace and war: The perpetual dynamism of geopolitical landscapes. In *Reconstructing Conflict: Integrating War and Post-War Geographies*, eds. S. Kirsch and C. Flint, 31–48. Ashgate.

Galtung, J. 1965. On the meaning of nonviolence. *Journal of Peace Research* 2: 228–257.

Galtung, J. 1969. Violence, peace, and peace research. *Journal of Peace Research* 6: 167–191.

Galtung, J. 1990. Cultural violence. *Journal of Peace Research* 27: 291–305.

Gregory, D. and A. Pred, eds. 2007. *Violent Geographies*. Routledge.

Jones, R. 2016. *Violent Borders*. Verso.

Kirsch, S. and Flint, C. 2011. Introduction: Reconstruction and the worlds that war makes. In *Reconstructing Conflict: Integrating War and Post-War Geographies*, eds. S. Kirsch and C. Flint, 3–28. Ashgate.

Kliot, N. and S. Waterman, eds. 1991. *The Political Geography of Conflict and Peace*. Belhaven.

Koopman, S. 2011a. Altergeopolitics: Other securities are happening. *Geoforum* 42: 274–284.

Koopman, S. 2011b. Let's take peace to pieces. *Political Geography* 30: 193–194.

Koopman, S. 2018. 'Territorial peace': The emergence of a concept in Colombia's peace negotiations. *Geopolitics* 23: 464–488.

Laliberté, N. 2016. 'Peace begins at home': Geographic Imaginaries of Violence and Peacebuilding in Northern Uganda. *Political Geography* 52: 24–33.

MacGinty, R. 2021. *Everyday Peace: How So-called Ordinary People Can Disrupt Violent Conflict*. Oxford University Press.

Mamadouh, V. 2004. Geography and war, geographers and peace. In *The Geography of War and Peace*, ed. C. Flint, 26–60. Oxford University Press.

McConnell, F., N. Megoran, and P. Williams, eds. 2014. *Geographies of Peace*. I.B. Tauris.

Megoran, N. 2010. Towards a geography of peace: Pacific geopolitics and evangelical Christian Crusade apologies. *Transactions of the Institute of British Geographers* 35: 382–398.

Megoran, N. 2011. War and peace? An agenda for peace research and practice in geography. *Political Geography* 1: 1–12.

Morrissey, J. 2017. *The Long War: CENTCOM, Grand Strategy, and Global Security*. University of Georgia Press.

Pepper, D. and A. Jenkins, eds. 1985. *The Geography of Peace and War*. Blackwell.

Pethick Lawrence, E. 1972 [1915]. Opinions of the Congress. In *Women at the Hague: The International Congress of Women and Its Results*, eds. J. Addams, E.G. Balch, and A. Hamilton, 143–144. Garland publishing [Macmillan].

Power, M. 2019. *Geopolitics and Development*. Routledge.

Shields, P.M. and Soeters, J. 2017. Peaceweaving: Jane Addams, positive peace, and public administration. *The American Review of Public Administration* 47: 323–339.

Tyner, J. 2009. *Military Legacies*. Routledge.

Williams, P. 2013. Reproducing everyday peace in north India: Process, politics and power. *Annals of the Association of American Geographers* 103: 230–250.

Williams, P. 2015. *Everyday Peace? Politics, Citizenship, and Muslim Lives in India*. John Wiley and Sons, Inc.

Williams, P. and McConnell, F. 2011. Critical geographies of peace. *Antipode* 43: 927–931.

Woodward, R. 2004. *Military Geographies*. Blackwell.

2

GEOGRAPHY AND WAR, GEOGRAPHERS AND PEACE

Expanding research and political agendas

Virginie Mamadouh

Introduction

The intimate relations of the modern discipline of geography and military violence through nationalist and imperialist projects are well established. Nonetheless until recently war was rarely a prominent topic of geographical inquiries. Peace even less, although there have always been geographers trying to use geography to promote peace and/or to delegitimate war and violence. This chapter sketches the many research and political agendas recently developed by geographers regarding war and peace.

It is an addendum to a previous essay that reviewed the ways geographers have dealt with war and peace since the establishment of modern Western academic geography (Mamadouh 2005). In that chapter, I argued that two periods could be distinguished, with 1945 being the watershed, marked by the bombing of Hiroshima and Nagasaki in early August, the end of World War Two, and the establishment of the United Nations Organization. A second watershed was the attacks of September 11, 2001, on New York and Washington, D.C. and the subsequent U.S.-led Global War on Terror. This marked the beginning of a third period, the one addressed in the present chapter.

The chapter is organized in five sections. The first two deal with the general context: changes in war and peace, and changes in academic geographies. The following three sections discuss the engagement of geographers with the war and peace geographies, as researchers and as activists: geographers writing *about* war (and peace), geographers writing *about* peace (and conflict) and geographers *in* war and peace. This literature review is not a meta-study based on quantitative, bibliographic materials aimed at describing the structure of the field.[1] The chapter is biased toward review articles, agenda-setting pieces and

DOI: 10.4324/9781003345794-2

edited volumes at the detriment of empirical case studies and monographs. Moreover, contributions to edited volumes and special issues are rarely named individually, and the review is limited to English language publications (with very few exceptions).[2]

Key developments in war and peace in the 21st century

The 20th century was marked by fundamental changes regarding war and peace. Its first half featured large-scale colonial wars, national wars, interstate wars, and the major escalation aptly known as World Wars One and Two. The century also witnessed the development of military technologies with the aerial bombing of civilians, and the invention and the proliferation of nuclear weapons. At the same time, international law and international politics have been geared toward the proactive peaceful settlement of disputes. The establishment of the United Nations in 1945 came with the decisive outlawing of military might as a normal way of intervening abroad. War became an exceptional, condemnable state behavior, justifiable only in the last resort to defend its own population and territory. A mighty symbol of this change is the renaming of most ministries of war around the world into ministries of defense. Not that war disappeared, but it was seriously delegitimized.

The September 11, 2001, attacks can be seen as a tipping point. By their mere geographies (location, scope, and scale) they were an impressive example of new form of war where neither the stake nor the means were territorialized. Networked non-state actors proved to be able to carry out a large-scale bombing in the heart of the homeland of the most powerful state. The response of the United States with the invasion of Afghanistan reterritorialized the war, blaming the Afghan government for failing to control al Qaeda operating from its territory. By mobilizing a "coalition of the willing" in a Global War on Terror the United States reorganized the world order that emerged in the post–Cold War period marked by globalization and optimism about global trade and world peace. Instead, militarization and securitization increasingly pervade social life, most notably with the control of cross-border migration and in the policing of cities and protests.

By contrast the escalation of the Russian–Ukrainian war with the full-scale invasion of Ukraine on February 24, 2022, was a mighty reminder that more "traditional" wars were not obsolete: not in their logic and rhetoric, neither in their material expression. The destruction of cities like Mariupol was reminiscent of the aerial bombings that marked World War Two, while the trench war in places like Bakhmut resonates with the horrors of the World War One.

Peace institutions have proven, once again, less effective than hoped for. The early 21st century witnessed a legitimacy crisis of key international institutions, marked as Western, and increasingly contested by larger non-Western states with increasing economic, financial, military, and/or cultural power: China,

India, Iran, Saudi Arabia, and Brazil, for example.[3] This growing gap between the West and its challengers undermines the ability of existing international institutions to foster peaceful (in the meaning of nonbelligerent) settlements of disputes and hamper negotiations in (localized and/or escalating) conflicts.

Key developments in academic geography in the 21st century

By contrast, academic geography did not radically change since the turn of the century. Late-20th-century trends accelerated, including changing academic practices regarding research and publishing (Paasi 2015) and the expansion of an ever more diverse array of approaches, including a new material turn.

All these perspectives dramatically enlarged the research agendas advanced by geographers, expanding the focus well beyond states, state actors, and inter-state relations. Different ontologies and epistemologies also come with different takes on ethics. Feminist geographies have mainstreamed an ethic of care that warrants the reconceptualization of the relations between the researcher and research participants as more collaborative than extractive. Research itself is problematized, requiring more self-reflection about the positionality of researchers/authors and the situatedness of the knowledge produced. Feminist and critical geographers have made political agendas visible, both the un-marked ones (maintaining the status quo, geography as an aid to statecraft, informing policies of the powers that be) and those aiming at advancing eman-cipation and social (and climate) justice and/or at supporting marginalized groups in society.

Combined with the broadening of the research and political agendas of geographers regarding peace and conflict studies, geographical aspects are paid more attention in cognate disciplines, in other words a spatial turn has taken place in political science, international relations, humanities, and more specifi-cally in conflict and peace studies (Macaspac and Moore 2022).

Geographers about war (and peace)

Geographies of war and peace have been mostly about war – and peace as the absence of war (the desirable postwar transition). Mamadouh (2005) reviewed the limited interest of academic geographers for the topic in the first half of the 20th century and could present the rare volumes dealing with war and peace. The situation has dramatically changed.

The September 11, 2001, attacks of al Qaeda and the subsequent US-led War on Terror (officially the Global War on Terrorism, GWOT) generated an upsurge of interest in war among geographers. Notable collections were pub-lished in the 2000s and early 2010s on both sides of the Atlantic demonstrat-ing the plurality of approaches, both theoretically and methodologically. The US military interventions in Afghanistan and Iraq were the focus of much attention, but geographers also addressed in these volumes a wide range of conflicts in other parts of the world (including the securitization of urban spaces and domestic insecurities) and historical studies.

In 2004 Stephen Graham edited a collection that foregrounds urban sites of violence, both symbolically and materially. *Cities, War and Terrorism: Towards an Urban Geopolitics*. That same year *Sites of Violence: Gender and Conflict Zones*, edited by Wenona Giles and Jennifer Hyndman (2004), dealt with the highly gendered and racialized dimensions of nationalism, conflict, and asylum and analyzed violence against women in post-conflict societies.

Thereafter *The Geography of War and Peace* edited by Colin Flint (2005) offered a state of the art of the subfield. It explicitly targeted conventional ways of conceptualizing geography: "if there is one single purpose to this book, it is to debunk Nicholas Spykman's belief that 'Geography is the most important in foreign policy because it is the most permanent'" (Flint 2005, 4). Instead, the chapters explored each in its own way the possibilities that critical geographical knowledge can contribute to the study of war and peace.

Violent Geographies edited by Derek Gregory and Allan Pred (2007) disclosed more specifically the political imaginaries that generate the normality of war and political violence. The chapters situated in different conflicts demonstrated the pervasiveness of violence, terror, and terrorism, undermining the taken-for-granted distinction between us ("the peaceful ones") and them ("the violent ones"). *War, Citizenship, Territory*, edited by Deborah Cowen and Emily Gilbert (2007), brought together contributions that examine how conflicts over the control of people and places have reshaped the organization of collective human life.

In 2009 Audrey Kobayashi edited an annual special issue of the *Annals of the American Association of Geographers* (*AAG*) (the first ever) under the title *Geographies of Peace and Armed Conflict* (Kobayashi 2009). It featured twenty-three very diverse papers (although mostly from a "perspective critical of the political status quo – including everything from totalitarian to neoliberal regimes – that allows armed conflict to take place" (Kobayashi 2009, 824). They are loosely grouped into three clusters: articles that focused on issues of territory, nationality, and armed conflicts; articles addressing conflict resolution and peace processes; and articles dealing with post-conflict situations. The collection was also published in 2012 as an edited volume, except for the book review essay (Agnew 2009).

That same year *Spaces of Security and Insecurity: Geographies of the War on Terror*, edited by Alan Ingram and Klaus Dodds (2009), brought contributions drawing on critical geographical imaginations to interrogate the ongoing War on Terror across several domains (international law, migration, development, art, social movements). In 2011 Scott Kirsch and Colin Flint edited a collection foregrounding the transition from war to reconstruction: *Reconstructing Conflict: Integrating war and post-war geographies*.

In the same period notable special issues included:

- *Climate Change and Conflict* (Nordås & Gleditsch 2007) in *Political Geography* 26 (6),
- *The Geopolitical Economy of "Resource Wars"* (Le Billon 2004) in *Geopolitics* 9 (1),
- *Territorial Conflict and Resolution* (O'Lear et al. 2005) in *GeoJournal* 64 (4),
- *Military Natures: Militarism and the Environment* (Davis 2007) in *GeoJournal* 69 (3),
- *Geographies of Genocide* (O'Lear & Egbert 2009) in *Space & Polity* 13 (10),
- *War beyond the Battlefield* (Grondin 2011) in *Geopolitics* 16 (2).

These collections demonstrate the dynamism of the field and its diversity but were a small subset of an expanding literature on war, violence, and conflict, with a wealth of articles and monographs inspired by all geographical traditions. Different clusters and emerging subfields can be distinguished.

First are military geographies in which geographical knowledge is deployed for military purposes, i.e., the military active use of geography. An example of the latter has been *Military Geography from Peace to War* by Eugene Palka and Francis Galgano (2005) - a textbook for practitioners dealing with spatial dimensions of military operations, classic ones or so-called military operations other than war (MOOTW in military jargon) and peacetime operations carried out by the military. It is written from the perspective of the U.S. military. Military geography is mostly practiced in military academies and rarely interacts as such with academic geographies (with the noticeable exception of the Military Geography Specialty Group of AAG) and military geographers rarely publish in academic journals.

Among academic geographies of war (and peace), a second cluster of studies consists of spatial analytical approaches studying the distribution and the diffusion of conflict and violence in specific area or era (Raleigh et al. 2010 on the ACLED database; O'Loughlin et al. 2010; O'Loughlin and Witmer 2011; Radil et al. 2013; Radil and Flint 2013; Dulić, 2018 and also Mutschler and Bales 2023 on solid and liquid warfare in the Yemen war).

A third cluster consists of studies focusing on geographical imaginations and geopolitical representations and the analysis of discourses and practices justifying violence. This is probably the most prolific group. In his influential book *The Colonial Present* (2004) Derek Gregory discussed war and colonialism in Afghanistan, Iraq, and Palestine, and in the following years he addressed key aspects of contemporary war waging (Gregory 2006a, 2006b, 2006c, 2010a, 2010b, 2011a, 2011b, 2016). In *War and Peace* (Gregory 2010a), he analyzed the production of three spaces of advanced military violence "the abstract space of the target, the alien space of the enemy Other, and the legal-lethal space of the exception" and discussed counter-geographies that challenges their claims to legitimacy. Citing Edward Said's assessment of geography as "the art of war" and his call for it to become "the art of resistance" (see also Quiquivix 2014 for an emphasis on counter-cartography), Gregory concluded that

> The history of geography has long intersected with the history of war, but until recently these intersections have been seen as concrete rather than conceptual [...] The construction of a human geography that might also become one of the arts of peace requires us to think the two together – concrete and concept – and to act on their conjuncture.
>
> *(Gregory 2010a, 180–181)*

Many other studies investigated geopolitical representations in policies (Toal 2017) and foremost in media and popular culture (Hannah 2006; Kuus 2002, 2004, 2007; Dodds 2005, 2007, 2008; Ó Tuathail 2005; Dalby 2008; Benwell and Dodds 2011; Benwell 2021; Moore and Cartwright 2015; Sharp 2011; An 2020). Some have problematized the notion of "just war" (Flint and Falah 2004; Falah et al. 2006; Megoran 2008), including the attribution of the Nobel Peace Prize to U.S. President Obama in 2009 amid wars overseas (Holland 2011; Adams 2012); and the same Prize attributed to the European Union in 2012 amid wars in its so-called Neighbourhood (Mamadouh 2014). Other work explored the geographic imaginaries of violence and peacebuilding (Laliberté 2016) and of diaspora geopolitics (Hamdan 2021; Birka 2022). Nationalism is in this context the most powerful ideology justifying war and hampering postwar arrangements. Examples include Dahlman and Ó Tuathail (2005) and Toal and Dahlman (2011) on the return and resettlement after ethnic cleansing in Bosnia, Dempsey (2022a, 2022b) on nationalism and reconciliation in Northern Ireland, Klem (2014) on war's end and territorialization and circulation in postwar Eastern Sri Lanka, and Egbert et al. (2016) on territorial cleansing. Post-conflict activities also include dark tourism (Mahrouse 2011).

Another cluster consists of political economic approaches that highlight the relation between war and capitalism, between geopolitics and geoeconomics (Mercille and Jones 2009), including through outsourcing military logistics (Moore 2013, 2017, 2019) and the relations between war, capitalism, labor, and genocide in (South East) Asia (Tyner 2005, 2007, 2008, 2009a, 2009b).

Feminist approaches form another cluster with more attention to the impact on the lives of individuals: feminist geopolitics and violence (Hyndman 2010, 2019; Loyd 2012), domestic violence and intimate war (Pain and Staeheli 2014; Pain 2015), antiwar movements (Loyd 2014) but also beauty and militarism (Faria and Fluri 2022), violence toward migrants (Hyndman 2019; Dempsey 2020; Jacobsen 2022), militarization of gender (in a special section of *Gender, Place & Culture*, edited by Henry and Natanel 2016), gender in the British armed forces (Woodward and Winter 2004), gender, aid and the military (Fluri 2011), and a feminist take on the deployment of drones

(Jackman and Brickell 2021). Children's agency engaging with war and security (Hörschelmann 2008; Hopkins et al. 2019) and other peripheral groups (Horschelmann et al. 2019 and special section of *Geopolitics*) have been addressed too.

Building on feminist agendas, much attention has been devoted to the militarization of society, for example in local communities (Bernazzoli and Flint 2009a, 2009b, 2010), militarization and gender (see the feminist scholarship), the militarization of climate change (Gilbert 2012), or the geopolitics of militarism and humanitarianism (Bhungalia 2015; McCormack and Gilbert 2021).

A new subfield emerged under the label of *critical military studies* with a journal established in 2015 with a strong input from critical military geographies (Woodward 2004, 2005; Rech et al. 2015; Forsyth 2019; Adey et al. 2013 on the militarized aerial gaze). This interest in the military, its sites (in war and in peace time), and its impact (during war and long thereafter) also link to political ecologies of military landscapes (Woodward 2014), the critical study of the physical environment and the conduct of war (Harrison and Passmore 2021) as well as the environmental legacy of past military land use (Francis 2014; Havlick 2014; Coates 2013); see also political ecologies of war and forests (Peluso and Vandergeest 2011) and the enduring effect of war on carbon cost (Belcher et al. 2020) or health and ecosystems (Griffiths 2022). Resource wars have also been examined (Le Billon 2004, 2007, 2008).

Another new emerging subfield is *urban geopolitics*. Cities have been the focus of much attention, as sites and target of violence (Farish 2003 on American city planning under nuclear threat; Coaffee 2003 about British cities and terrorism; Graham 2004 for a general collection; Fregonese 2009, 2020 about Beirut; Hristova 2018 about Baghdad). Specific infrastructures have been studied extensively: the World War Two Mulberry Harbours (Flint 2016), bunkers in Albania (Lasserre et al. 2022), walls (Falah 2005; Cohen 2006; Alatout 2009 on Israeli wall construction and the Palestinian predicament; see also Joronen 2021). *Political Geography* published a set of interventions about walls (Till et al. 2013) and *L'Espace politique* published a special issue on *Borders of Wars and Borders of Peace* (Amilhat Szary and Cattaruzza 2017).

A third emerging subfield concerns digital geopolitics. Digitalization impacts war (and peace) as a technological game changer (think of the precision of weaponry and drones operating thousands of miles away from the operations). It also extends the realm of war (and peace) to a new online "territory." In *Cyberspace Is Used, First and Foremost, to Wage Wars* Frédérick Douzet and Aude Gery (2021) discuss the security dilemmas states face in cyberspace, paraphrasing the title of the landmark book *La géographie ca sert d'abord à faire la guerre* of the founder of their Institut de géopolitique (Lacoste 1976).

Finally, some publications were written for a broader public. They include atlases (Smith and Bræin 2003; Cattaruzza 2017 and many others focusing on a single conflict). In *Understanding Geography and War* International Relations scholar Steve Pickering (2017) addressed common misconceptions about geography, geopolitics, and war often circulated in bestsellers like *The Pentagon's Maps* (Barnett 2003), *The Revenge of Geography* (Kaplan 2012), or *Prisoners of Geography* and *The Power of Geography* (Marshall 2015, 2021). Another noteworthy effort to popularize insights from political geography, critical geopolitics, and critical border studies is *Border Wars* by Klaus Dodds (2021), inspired by a decade of writing short columns on geopolitical hotspots for a broad public in *The Geographical*, the monthly magazine of the Royal Geographical Society in London.

Geographers about peace (and conflict)

While from the 19th century onward some geographers have promoted peace, geographers had mostly studied peace as the overshadowed side of the war and peace binary (Mamadouh 2005). In the second decade of the 21st century, peace (finally!) gained the full attention of (political) geographers.

Three agenda-setting pieces were published independently in 2011, each calling for a more theoretical engagement with peace and for more robust research agendas for peace geographies (and peace geopolitics). While Gregory's plea for geography as an "art of peace" (see previous section) was still articulated in an analysis of late modern war (Gregory 2010a, 170–171), these articles discuss peace in its own right.

Reflecting on recent volumes on war on and peace discussed above (Flint 2005; Kobayashi 2009; Ingram and Dodds 2009), Philippa Williams and Fiona McConnell (2011) articulated their frustration with the focus on understanding war in geography and on the lack of a coherent research agenda "grounded in a nuanced and critical conception of peace" (p. 928). Building upon their research on everyday Hindu–Muslim interactions in North India (Williams 2007) and exile Tibetan politics (McConnell 2009), they argued that "geography has an important role to play in deconstructing normative assumptions about peace and exploring peace as situated knowledge within different cultural settings" (Williams and McConnell 2011, 929) and that peace should be understood as a process.

Building on his ethnographical work in the Ferghana Valley in Central Asia, Nick Megoran (2011) examined peace studies, biblical studies, and international relations theory before discussing the (lack of) conceptualization of peace in geography and advocating a commitment to peace, in his case through nonviolence inspired by Christian theology (see also Megoran 2008, 2010). This agenda-setting paper had been presented at the 2010 Annual Conference of the Royal Geographical Society as the *Political Geography* annual lecture and was published with a set of commentaries. The conversation was extended in *Space and Polity*, with an exchange with Simon Springer about nonviolence, religion, and anarchism (Springer 2014a, 2014b; Megoran 2014).

On the other side of the Atlantic, the call for a research agenda for peace was articulated by Joshua Inwood and James Tyner (2011), both working on violence (the first mostly within the United States, the second mostly overseas, in (South East Asia). They discussed how the military-industrial-academic complex perpetuates a war culture in American society and how the role of geography to promote peace could be enhanced with a more coherent research agenda with the goal of "destabilizing, contesting, and challenging a killing society, a 'war' culture that is dedicated to inequality, death, and the dehumanizing effects of violence" (Inwood and Tyner 2011, 444).

These three agenda-setting articles were followed by a wider turn to peace geographies, including two collections edited by the same geographers. Tyner and Inwood published the same year *Nonkilling Geography* with contributions by American geographers. In a later article, they also discussed violence as a fetish (Tyner and Inwood 2014; see also a similar point in Springer 2011). A few years later British geographers joined forces and coedited a more international volume, *The Geographies of Peace: New Approaches to Boundaries, Diplomacy and Conflict Resolution* (Megoran, McConnell and Williams 2014). All contributions see peace as a process grounded in particular places and geopolitical configurations. The stated agenda of the editors is "both to critically *conceptualize* peace, and also to be *committed* to some vision of peace" (Megoran, McConnell and Williams 2014, 18, original emphasis). In their conclusion, they highlight three crosscutting themes: the intertwining of violence and nonviolence, the role of agency, and power relations. They point at a twofold geographical contribution to the study of peace: revisiting traditional concerns (migration, borders, geopolitics, nature–human interactions) and reworking geographical concepts (place, space, scale). Moreover "the pursuit of 'peace' does not necessarily embody the realization of equality and justice" (Megoran, McConnell and Williams 2014, 206). Who benefits from a particular peace settlement, and who loses?

In April 2017 the International Geographical Union (IGU) convened a thematic conference on the *Geographies for Peace/Geografías por la paz* in La Paz, Bolivia. Apart from using the highly symbolic toponomy of the Bolivian city, the encounter was meant as a peacebuilding bridge between international geography marked by an anglophone hegemony and South American geographers resisting it (as they were convening the 16th *Encuentro de Geografías de América Latina* (EGAL) in La Paz a few days later). For this occasion, Nick Megoran and Simon Dalby reflected on almost one century of international encounters. They argued that geographers tend to produce small-scale studies based on detailed fieldwork that "are always in danger of overlooking the larger geopolitical contexts within which they are situated" (Megoran and Dalby 2018, 253). In his Spanish language lecture, Heriberto Cairo (2019) underlined a crucial distinction between different conceptualizations of peace. While *geografías de la paz* (geographies of peace) were based on the same logic of war and imperialist domination (according to the Latin adage *Si vis pacem, para bellum* "if you want peace, prepare for war"), *geografías pacifistas* (pacifist geographies) advocate more radical engagement with peace such as the philosophy of liberation of Argentine-Mexican Enrique Dussel stressing the political and ethical implication of his plea for solidarity (as a form of positive peace).

On the conceptual level it is remarkable that geographers have attempted to "rescue" geopolitics from its association with the militarized foreign policy agendas of powerful states. While earlier anti-geopolitics have been coined for grassroots resistance to geopolitical projects, new labels have been coined to enhance the potential of geopolitical approaches to peace: progressive geopolitics (Kearns 2008), pacific geopolitics (Megoran 2008), alter-geopolitics (Koopman 2011; Boyce et al. 2020), precarious geopolitics (Woon 2014a), or progressive and irenic geopolitics (Megoran and Dalby 2018). For example, Woon (2014a) builds on Judith Butler's notion of "precarious lives" to analyze the constitution of nonviolence through emotions, demonstrating how it can be deployed to interrogate critically his own research on nonviolent movements in the Philippines (such as Woon 2011). In other papers he explores the role of the military in peacebuilding (Woon 2015) again stressing the painstaking process of efforts regarding the harnessing of peace in a violence-plagued region, and analyzing how children deal with peace (Woon 2017).

Among the many geographies of peace we can recognize the plurality of approaches characterizing the geographies of war (save the military geographies proper): poststructuralist approaches dealing with geographical representations and geopolitical imaginations (for example media discourses (Korson 2015), discourses of liberal peace (Stokke 2009; Gonzalez-Vicente 2020) or popular geopolitics (Woon 2014b), as well as feminist approaches dealing with intimate geographies of peace (Brickell 2015), urban approaches to peacebuilding (Björkdahl 2013; Björkdahl and Buckley-Zistel 2016; Björkdahl and Kappler 2017; Danielsson 2023), road infrastructure (Ruwanpura et al. 2020), and political ecologies (Ramutsindela 2017; Duffy et al. 2019).

Peace can be said to have been acknowledged as an important topic. It has an entry in *Keywords in Radical Geography* in which Sara Koopman (2019) defines it as positive peace including justice, solidarity, care, well-being, dignity, and stressing once more that

Peace never clearly distinct from war. War is inside peace, shaping everyday political life, institutions and sociospatial order.

(Koopman 2019, 209)

This is demonstrated in Koopman's work on the uneven geographies of inclusive peace in Colombia (Koopman 2020; see also Cairo et al. 2018; and Georgi 2022). It is also evident in other studies. For example, Forde (2022) on peace as violence in post-apartheid Cape Town, Cante (2020) on anti-political peace in Abidjan, Penu and Esswa (2019) on the Alavanyo-Nkonya boundary dispute in

Ghana, and Feghali et al. (2021) on a Kenyan Somali community. Ideas about pluralism (peace meaning different things to different people) are also linked to a plea to decolonize geographies (Koopman 2019, 210). To acknowledge this pluralism Christopher Courtheyn (2018) proposed to expand modern-liberal peace to radical trans-relational peace, inspired by the "many peaces" framework, also known as the trans-rational school of peace research that emerged in the late 1990s in Innsbruck around Wolfgang Dietrich. Courtheyn (2018, 742) advances a concept of radical trans-relational peace, "ecological dignity created through social movement solidarity networks," and illustrates this with his empirical work with the San José Peace Community in Colombia (see also Courtheyn 2016). Elsewhere Macaspac (2019) speaks of an *insurgent peace* for community-led peacebuilding, based on a study of the Saga peace zone in the Philippines.

Moreover, critical geographies (both critical geopolitics and historical materialism) can be criticized for remaining reliant on social agonism – the idea that politics is inherently and unavoidably conflictual – and negating peace (Bregazzi and Jackson 2016). It is therefore key to "study and develop an understanding of the multiple kinds of sociospatial relations that produce nonviolence, justice, equality and compassion" (Bregazzi and Jackson 2016, 83). Developing their argument from Spinozist and decolonial positions, they conclude with a rejoinder with "geographies of peace" literature, with which they share a positive conceptualization of peace, an interest for the everyday (instead of the level of the nation state), for peace as a process, for the coexistence of peace and war in a spectrum of violence and nonviolence (with war within peace, and peace inside war) and for the enabling contextual understandings of peace referring to the unique socio-spatial contexts in which it develops (Bregazzi and Jackson 2016, 86–87).

In a similar vein, Jon Barnett (2019) uses a critique of the ontology of research on climate change and armed conflict (framing climate-induced violence as an uncontrollable consequence of our internal and external natures) to develop a positive and performative project of "the plausible geographies of climate resilient peace" (p. 931, but then again, contrast with Inwood and Tyner 2022). This rejection of agonism (see also Askins and Mason 2015) is in sharp contrast with critical geographies in other subfields that plea for the return of agonism to re-politicize issues in post-political societies.

Geographers *in* war and peace

During most of the 20th century geographers have been involved in war projects, advising policymakers and in military and other state institutions. However, peace has been a stated goal of many international exchanges between geographers (and geographical educationists; see Marsden 2000).

Geographical societies and their conferences have also been venues where geographers could circulate knowledge about war and peace and advance this research agenda. Sometimes these interventions have been highly visibly. For example, Derek Gregory's plenary lecture on war and peace at the 2008 RGS annual conference (see Gregory 2010a), the 2014 AAG's forum on geography and militarism (published in a special issue of the *Annals of the American Association of Geographers* (*AAG*) in 2016 (see Sheppard and Tyner 2016), or the 2017 IGU thematic conference on the *Geographies for peace/Geografías por la paz* in April 2017 La Paz (see Megoran and Dalby 2018, and Cairo 2019). Geography societies can also be arenas where more activist political agendas about (war and) peace materialize.

From the outset, peace was a stated aim of the international activities of geographers. However, war has interfered with the plans of the International Geography Congresses (beginning 1871) and the activities of the International Geographical Union (IGU), established in 1922 (Mamadouh 2022). The very first International Geographical Congress was postponed due to the French-Prussian war, but when it eventually took place in Brussels in 1871, it was described as "festival of peace and friendship" (Shimazu 2015). The most remarkable event in the past two decades has been the suspension of the Russian membership of the IGU with the exclusion of the Russian Geographical Society (RGS) in 2022, after the full-scale Russian invasion of Ukraine. This was due to the role of President Vladimir Putin as chair of the board of the RGS and of Minister of Defense Sergei Shoygu as Executive President (Mamadouh 2022).

In the past two decades activist geographers deployed collective efforts to mitigate war and to promote peace on several occasions. The Palestinian Campaign for the Academic and Cultural Boycott of Israel called on the IGU to cancel a regional conference scheduled in Tel Aviv in 2010. In 2011 concerned geographers petitioned against the venue chosen for the IGU 2011 regional conference in Santiago de Chile (Hirt and Palomino-Schalscha 2011; Till and Kuusisto-Arponen 2015; see also the influence of the *Escuela Militar* on Chilean geography Barton and Irarrázaval 2014).

The campaign to boycott Israel also led to a debate among political geographers in the early 2000s following the preliminary decision of one of the editors of *Political Geography* to refuse to assess a manuscript from a geographer affiliated to an Israeli university (O'Loughlin 2004, 2018; Slater 2004; Storey 2005; Waterman 2005). Another collective action emerged around the same journal to convince the publisher Reed Elsevier to disengage from the Defence and Security Equipment International (DSEI) and its investments in the weapon industry (Chatterton and Featherstone 2007; Hammett and Newsham 2007; Pringle 2007 and Kitchin 2007). This was part of a larger campaign successfully targeting larger Elsevier journals like *The Lancet*, and many geographers withdrew their submission to Elsevier journals and a major Elsevier publication, *The International Encyclopedia of Human Geography* (Megoran et al. 2016, 131).

Among American geographers, a sustained controversy pertained to the Bowman Expeditions Program of the American Geographical Society (AGS) in Central America funded by the U.S. military (Steinberg 2010; Bryan 2010; Herlihy 2010; Cruz 2010; Agnew 2010; Dobson 2012; Wainwright 2013; Bryan and Woods 2015). Another debate pertained to the Human Terrain System as ethnographic field research for the U.S. Army in Afghanistan and Iraq from 2007 onward that was also opposed by the Network of Concerned Anthropologists and the American Anthropological Association (Medina 2016). A controversy about the use of remote sensing as a counterterrorism tool, academic publishing, and ethical issues took place in *The Professional Geographer* (Beck 2003, 2005; O'Loughlin 2005; Shroder 2005).

The Forum edited by Sheppard and Tyner (2016) in the *Annals of the American Association of Geographers* dealt more specifically with the contemporary engagements of geographers with the U.S. military (and, to lesser extent, with Canada and the U.K.). Several contributors referred to the saying "war is God's way of teaching Americans geography." This aphorism is usually attributed to American author Ambrose Bierce around 1900 and often mentioned (see also Taylor 2004). Military interventions abroad and the associated casualties forced American citizens to learn about the rest of the world. But then again, the aphorism is a century old, a century in which "war has become America's way of teaching other people geography" (Gregory 2010a, 180). Moreover, "war structures [the discipline of] Geography, generating new knowledge of the world that, in turn, shapes new approaches to war" (Bryan 2016, 507), Sheppard and Tyner, following Wainwright 2016, called for the AAG to review the relationship between academic geography and militarism. An AAG special committee on Geography and the Military was eventually set up in 2017 in response to a petition by the Network of Concerned Geographers (Koopman 2019, 208).[4] Its reports and the reaction of the AAG Council were hotly debated (Wainwright and Weaver 2021, 2022; Rose-Redwood et al. 2022).

Peace activism was enacted too. A conference on *Peace in Geography and Politics: Critiques and Narratives of Peace* convened in Newcastle, England, in November 2011, featured a naming the dead ceremony, to mark the tenth anniversary of the start of the war, reading out the names and details of 50 U.K. servicemen and women killed in action in Afghanistan, and 50 Afghan civilians killed by NATO forces. A year later at the 2012 RGS Annual Conference held in Edinburgh, peace geographers contributed an on-site exhibition on civic geographies and organized an *Academic Seminar Blockade* at Faslane Peace Camp established since 1982 near Faslane Naval base in northwest Scotland, known as the home of British nuclear weapons (Askins and Mason 2015).

More generally, geographers' contribution to war has been widely studied (Mamadouh 2005; Barnes 2009, 2016; Klinke 2020), including Isaiah Bowman's role as President Theodore Roosevelt's geographer (Smith 2003), the Office of Strategic Services during World War Two, and the Cold War funding of Area Studies (Barnes and Farish 2006), the Ethnogeographic Board of the Smithsonian Institution during World War Two (Farish 2005), Project Revere (Pinkerton et al. 2011), all in the United States, and, in Britain, RAF recruitment (Rech 2014) and the British University Armed Service Units (Woodward et al. 2017).

Conclusion

More significantly, individual participation and agency to both war and peace-building operations are also discussed more openly than in the past. The precarious position of researchers is also addressed in the *Annals* special issue on the military academic complex already mentioned (Sheppard and Tyner 2016). Especially, see "Beware: Your Research May Be Weaponized!" (Koopman 2016). More recently Judith Verweijen (2022) has addressed the complexities of what she aptly calls the perilous engagement with public policy toward armed conflict that may unintentionally legitimize the institutions and status quo being criticized (Verweijen 2022, 129).

Navigating the dilemma of critique and legitimization requires asking uncomfortable answers, formulating unsatisfying answers, and underlining the importance of vocabulary and the need to reappropriate and reconceptualize key concepts like peace or genocide (Verweijen 2022, 129). Echoing Said's much quoted call for geography to become "the art of resistance" and Gregory's assessment discussed above, Verweijen concludes:

> The imperial and militaristic legacies that haunt the discipline of geography – like most other sciences – render both engagement and non-engagement with public policy daunting. In this context there is no "art of peace" – nor a "peace science '[...] – only a set of difficult, always imperfect choices regarding what constituencies to engage with and what vocabularies to use."
>
> *(2022, 133)*

Much of the literature discussed in this review demonstrates a strong collective effort among geographers to take on the challenge to develop geography as "an art of peace" – as well as much reflection about the perils of such engagements. A lot has been achieved since the previous review (Mamadouh 2005). Most recently, the commentaries on the Russian war in Ukraine collected in a virtual forum of *Political Geography* (Lizotte et al. 2022) demonstrates both the plurality of approaches and reactions – and the liabilities Verweijen underlines.

Notes

1 Instead, it is a personal report, partial and situated at the margin of anglophone geographies, (see Mamadouh 2020) from editorial positions in the community (such as Agnew et al. 2015 and academic journals in English, French, and Spanish) and service in the Commission on Political Geography of the International Geographical Union – IGU-CPG).
2 Finally literature on (critical) geopolitics, (critical) border studies, diplomacy, security and securitization, and literature addressing specific arrangements created to prevent conflicts (European integration, Antarctic Treaty, UNCLOS, etc.) without explicit reference to war and peace has not been included.
3 And Western states are reluctant participants. The United States did not sign the 1998 International Criminal Court treaty, and the border policies of the European Union and its member states patently contradict their commitments to the Refugee Conventions.
4 https://actionnetwork.org/petitions/network-of-concerned-geographers (last accessed 28 March 2023). The letter includes a list of useful references of published works on the involvement of the US military in geography.

Bibliography

Adams, P.C. 2012. Trajectories of the Nobel Peace Prize. *Geopolitics* 17:553–577.

Adey, P., M. Whitehead, and A. Williams. eds. 2013. *From above: War, violence, and verticality*. Oxford University Press.

Agnew, J. 2009. Killing for Cause? Geographies of War and Peace. *Annals of the Association of American Geographers* 99:1054–1059.

Agnew, J. 2010. Ethics or Militarism? The Role of the AAG in What Was Originally a Dispute Over Informed Consent. *Political Geography* 29:422–423.

Agnew, J., V. Mamadouh, A. Secor, and J. Sharp. eds. 2015. *The Wiley Blackwell Companion to Political Geography*. Wiley Blackwell.

Alatout, S. 2009. Walls as Technologies of Government: The Double Construction of Geographies of Peace and Conflict in Israeli Politics, 2002–Present. *Annals of the Association of American Geographers* 99:956–968.

Amilhat Szary, A.L. and A. Cattaruzza. 2018. Frontières de guerre, frontières de paix: nouvelles explorations des espaces et temporalités des conflits, *L'Espace Politique* 33, 4403.

An, N. 2020. *Confucian Geopolitics: Chinese Geopolitical Imaginations of the US War on Terror*. Springer.

Askins, K., and K. Mason. 2015. Us and Us: Agonism, Non-Violence and the Relational Spaces of Civic Activism. *ACME: An International Journal for Critical Geographies* 14:422–430.

Barnes, T. J. 2009. Obituaries, War, 'Corporeal Remains', and Life: History and Philosophy of Geography. *Progress in Human Geography* 33:693–701.

Barnes, T.J. 2016. American Geographers and World War II: Spies, Teachers, and Occupiers. *Annals of the American Association of Geographers* 106:543–550.

Barnes, T. J., and M. Farish. 2006. Between Regions: Science, Militarism, and American Geography from World War to Cold War. *Annals of the Association of American Geographers* 96:807–826.

Barnett, J. 2019. Global Environmental Change I: Climate Resilient Peace? *Progress in Human Geography* 43:927–936.

Barnett, T.P. M. 2003. *The Pentagon's New Map. War and Peace in the Twenty-First Century*. Penguin Books.

Barton, J.R., and F. Irarrázaval. 2014. Geographical Representations: The Role of the Military in the Development of Contemporary Chilean Geography. *Area* 46: 129–136.

Beck, R. 2005. Reply to Commentaries by O'Loughlin and Shroder. *The Professional Geographer* 57:598–608.

Beck, R.A. 2003. Remote Sensing and GIS as Counterterrorism Tools in the Afghanistan War: A Case Study of the Zhawar Kili Region. *The Professional Geographer* 55 (2):170–179.

Belcher, O., P. Bigger, B. Neimark, and C. Kennelly. 2020. Hidden Carbon Costs of the "everywhere war": Logistics, Geopolitical Ecology, and the Carbon Boot-Print of the US Military. *Transactions of the Institute of British Geographers* 45:65–80.

Benwell, M.C. 2021. Going Back to School: Engaging Veterans' Memories of the Malvinas War in Secondary Schools in Santa Fe, Argentina. *Political Geography* 86:102351.

Benwell, M. C., and K. Dodds. 2011. Argentine Territorial Nationalism Revisited: The Malvinas/Falklands Dispute and Geographies of Everyday Nationalism. *Political Geography* 30:441–449.

Bernazzoli, R. M., and C. Flint. 2009a. Power, Place, and Militarism: Toward a Comparative Geographic Analysis of Militarization. *Geography Compass* 3:393–411.

Bernazzoli, R. M., and C. Flint. 2009b. From Militarization to Securitization: Finding a Concept that Works. *Political Geography* 28:449–450.

Bernazzoli, R. M., and C. Flint. 2010. Embodying the Garrison State? Everyday Geographies of Militarization in American Society. *Political Geography* 29:157–166.

Bhungalia, L. 2015. Managing Violence: Aid, Counterinsurgency, and the Humanitarian Present in Palestine. *Environment and Planning A: Economy and Space* 47: 2308–2323.

Birka, I. 2022. Thinking Diaspora Diplomacy after Russia's War in Ukraine. *Space and Polity* 26:53–61.

Björkdahl, A. 2013. Urban peacebuilding. *Peacebuilding* 1:207–221.

Björkdahl, A., and S. Buckley-Zistel. eds. 2016. *Spatialising Peace and Conflict: Mapping the Production of Places, Sites and Scales of Violence*: Palgrave Macmillan.

Björkdahl, A., and S. Kappler. 2017. *Peacebuilding and Spatial Transformation. Peace, Space and Place*. Routledge.

Boyce, G. A., S. Launius, J. Williams, and T. Miller. 2020. Alter-Geopolitics and the Feminist Challenge to the Securitization of Climate Policy. *Gender, Place & Culture* 27:394–411.

Bregazzi, H., and M. Jackson. 2016. Agonism, Critical Political Geography, and the New Geographies of Peace. *Progress in Human Geography* 42:72–91.

Brickell, K. 2015. Towards Intimate Geographies of Peace? Local Reconciliation of Domestic Violence in Cambodia. *Transactions of the Institute of British Geographers* 40:321–333.

Bryan, J. 2010. Force Multipliers: Geography, Militarism, and the Bowman Expeditions. *Political Geography* 29:414–416.

Bryan, J. 2016. Geography and the Military: Notes for a Debate. *Annals of the American Association of Geographers* 106:506–512.

Bryan, J., and D. Woods. 2015. *Weaponizing Maps: Indigenous Peoples and Counterinsurgency in the Americas*. Guilford Press.

Cairo, H. 2019. Geografías de la paz y geografías pacifistas en la Guerra Fría: una diferenciación conceptual y ético-política. *ACME: An International Journal for Critical Geographies* 18:1167–1183.

Cairo, H., U. Oslender, C. E. Piazzini Suárez, J. Ríos, S. Koopman, V. Montoya Arango, F. B. Rodríguez Muñoz, and L. Zambrano Quintero. 2018. "Territorial Peace": The Emergence of a Concept in Colombia's Peace Negotiations. *Geopolitics* 23: 464–488.

Cante, F. 2020. Mediating Anti-Political Peace in Abidjan: Radio, Place and Power. *Political Geography* 83:102282.

Cattaruzza, A. 2017. *Atlas des guerres et des conflits. Un tour du monde géopolitique.* Nouvelle edition ed. Autrement.

Chatterton, P., and D. Featherstone. 2007. Intervention: Elsevier, Critical Geography and the Arms Trade. *Political Geography* 26:3–7.

Coaffee, J. 2003. *Terrorism, Risk and the City: The Making of a Contemporary Urban Landscape.* Ashgate.

Coates, P. 2013. From Hazard to Habitat (or Hazardous Habitat): The Lively and Lethal Afterlife of Rocky Flats, Colorado. *Progress in Physical Geography: Earth and Environment* 38:286–300.

Cohen, S. E. 2006. Israel's West Bank Barrier: An Impediment to Peace? *Geographical Review* 96:682–695.

Courtheyn, C. 2016. 'Memory Is the Strength of Our Resistance': An 'Other Politics' Through Embodied and Material Commemoration in the San José Peace Community, Colombia. *Social & Cultural Geography* 17:933–958.

Courtheyn, C. 2018. Peace Geographies: Expanding from Modern-Liberal Peace to Radical Trans-Relational Peace. *Progress in Human Geography* 42:741–758.

Cowen, D., and E. Gilbert. eds. 2007. *War, Citizenship, Territory.* Routledge.

Cruz, M. 2010. A Living Space: The Relationship between Land and Property in the Community. *Political Geography* 29:420–421.

Dahlman, C., and G. Ó. Tuathail. 2005. The Legacy of Ethnic Cleansing: The International Community and the Returns Process in post-Dayton Bosnia-Herzegovina. *Political Geography* 24:569–600.

Dalby, S. 2008. Warrior Geopolitics: Gladiator, Black Hawk Down and THE Kingdom of Heaven. *Political Geography* 27:439–455.

Danielsson, A. 2023. Minecraft as a Technology of Postwar Urban Ordering: the Situated-Portable Epistemic Nexus of Urban Peacebuilding in Pristina. *Territory, Politics, Governance*: DOI: 10.1080/21622671.2023.2189610

Davis, J. S. 2007. Introduction: Military natures: Militarism and the Environment. *GeoJournal* 69:131–134.

Dempsey, K.E. 2020. Spaces of Violence: A Typology of the Political Geography of Violence against Migrants Seeking Asylum in the EU. *Political Geography* 79:102157.

Dempsey, K.E. 2022a. Fostering Grassroots Civic Nationalism in an Ethno-Nationally Divided Community in Northern Ireland. *Geopolitics* 27:292–308.

Dempsey, K.E. 2022b. *An Introduction to the Geopolitics of Conflict, Nationalism, and Reconciliation in Ireland.* Routledge.

Dobson, J. E. 2012. The Why, What, and Where of Bowman Expeditions. *Focus on Geography* 55:117–118.

Dodds, K. 2005. Screening Geopolitics: James Bond and the Early Cold War films (1962–1967). *Geopolitics* 10:266–289.

Dodds, K. 2007. Steve Bell's Eye: Cartoons, Geopolitics and the Visualization of the 'War on Terror'. *Security Dialogue* 38:157–177.

Dodds, K. 2008. Hollywood and the Popular Geopolitics of the War on Terror. *Third World Quarterly* 29:1621–1637.

Dodds, K. 2021. *Border Wars: The Conflicts that Will Define Our Future.* Ebury Press.

Douzet, F., and A. Gery. 2021. Cyberspace Is Used, First and Foremost, to Wage Wars: Proliferation, Security and Stability in Cyberspace. *Journal of Cyber Policy* 6: 96–113.

Duffy, R., F. Massé, E. Smidt, E. Marijnen, B. Büscher, J. Verweijen, M. Ramutsindela, T. Simlai, L. Joanny, and E. Lunstrum. 2019. Why We Must Question the Militarisation of Conservation. *Biological Conservation* 232:66–73.

Dulić, T. 2018. The Patterns of violence in Bosnia and Herzegovina: Security, Geography and the Killing of Civilians During the War of the 1990s. *Political Geography* 63:148–158.

Egbert, S. L., N. R. Pickett, N. Reiz, W. Price, A. Thelen, and V. Artman. 2016. Territorial Cleansing: A Geopolitical Approach to Understanding Mass Violence. *Territory, Politics, Governance* 4:297–318.

Falah, G.W. 2005. The Geopolitics of 'Enclavisation' and the Demise of a Two-State Solution to the Israeli-Palestinian Conflict. *Third World Quarterly* 26:1341–1372.

Falah, G.W., C. Flint, and V. Mamadouh. 2006. Just War and Extraterritoriality: The Popular Geopolitics of the United States' War on Iraq as Reflected in Newspapers of the Arab World. *Annals of the Association of American Geographers* 96:142–164.

Faria, C. V., and J. L. Fluri. 2022. Allure and the Spatialities of Nationalism, War and Development: Towards a Geography of Beauty. *Geography Compass* 16: e12652.

Farish, M. 2003. Disaster and Decentralization: American Cities and the Cold War. *Cultural geographies* 10:125–148.

Farish, M. 2005. Archiving Areas: The Ethnogeographic Board and the Second World War. *Annals of the Association of American Geographers* 95:663–679.

Feghali, S., C. Faria, and F. Jama. 2021. "Let Us Create Space": Reclaiming Peace and Security in a Kenyan Somali Community. *Political Geography* 90:102453.

Flint, C. ed. 2005. *The Geography of War and Peace: From Death Camps to Diplomats.* Oxford University Press.

Flint, C. 2016. *Geopolitical Constructs: the Mulberry Harbours, World War Two, and the Making of a Militarized Transatlantic.* Rowman and Littlefield.

Flint, C., and G.W. Falah. 2004. How the United States Justified Its War on Terrorism: Prime Morality and the Construction of a 'Just War'. *Third World Quarterly* 25:1379–1399.

Fluri, J. 2011. Armored Peacocks and Proxy Bodies: Gender Geopolitics in Aid/Development Spaces of Afghanistan. *Gender, Place & Culture* 18:519–536.

Forde, S. 2022. The Violence of Space and Spaces of Violence: Peace as Violence in Unequal and Divided Spaces. *Political Geography* 93:102529.

Forsyth, I. 2019. A Genealogy of Military Geographies: Complicities, Entanglements, and Legacies. *Geography Compass* 13:e12422.

Francis, R.A. 2014. On War (and Geography): Engaging with an Environmental Frontier. *Progress in Physical Geography: Earth and Environment* 38:265–270.

Fregonese, S. 2009. The Urbicide of Beirut? Geopolitics and the Built Environment in the Lebanese Civil War (1975–1976). *Political Geography* 28:309–318.

Fregonese, S. 2020. *War and the City: Urban Geopolitics in Lebanon.* I.B. Tauris.

Georgi, F.R. 2022. Peace Through the Lens of Human Rights: Mapping Spaces of Peace in the Advocacy of Colombian Human Rights Defenders. *Political Geography* 99:102780.

Gilbert, E. 2012. The Militarization of Climate Change. *ACME: An International Journal for Critical Geographies* 11:1–14.

Giles, W., and J. Hyndman. eds. 2004. *Sites of Violence: Gender and Conflict Zones.* University of California Press.

Gonzalez-Vicente, R. 2020. The Liberal Peace Fallacy: Violent Neoliberalism and the Temporal and Spatial Traps of State-Based Approaches to Peace. *Territory, Politics, Governance* 8:100–116.

Graham, S. ed. 2004. *Cities, War and Terrorism; Towards an Urban Geopolitics.* Blackwell.

Gregory, D. 2004. *The Colonial Present: Afghanistan, Palestine, Iraq.* Blackwell.

Gregory, D. 2006a. The Death of the Civilian? *Environment & Planning D: Society & Space* 24:633–638.

Gregory, D. 2006b. "In Another Time-Zone, the Bombs Fall Unsafely ….": Targets, Civilians, and Late Modern War. *The Arab World Geographer* 9:88–111.

Gregory, D. 2006c. The Black Flag: Guantánamo Bay and the Space of Exception. *Geografiska Annaler: Series B, Human Geography* 88:405–427.

Gregory, D. 2010a. War and Peace. *Transactions of the Institute of British Geographers* 35:154–186.

Gregory, D. 2010b. Seeing Red: Baghdad and the Event-Ful City. *Political Geography* 29:266–279.

Gregory, D. 2011a. The Everywhere War. *The Geographical Journal* 177:238–250.

Gregory, D. 2011b. From a View to a Kill: Drones and Late Modern War. *Theory, Culture & Society* 28:188–215.

Gregory, D. 2016. The Natures of War. *Antipode* 48:3–56.

Gregory, D., and A. Pred. (eds.). 2007. *Violent Geographies: Fear, Terror, and Political Violence.* Routledge.

Griffiths, M. 2022. The Geontological Time-Spaces of Late Modern War. *Progress in Human Geography* 46:282–298.

Grondin, D. 2011. The Other Spaces of War: War beyond the Battlefield in the War on Terror. *Geopolitics* 16:253–279.

Hamdan, A. 2021. Ephemeral Geopolitics: Tracing the Role of Refugees in Syria's Transnational Opposition. *Political Geography* 84:102299.

Hammett, D., and A. Newsham. 2007. Intervention: Widening the Ethical Debate – Academia, activism, and the Arms Trade. *Political Geography* 26:10–12.

Hannah, M. 2006. Torture and the Ticking Bomb: The War on Terrorism as a Geographical Imagination of Power/Knowledge. *Annals of the Association of American Geographers* 96:622–640.

Harrison, S., and D. G. Passmore. 2021. On Geography and War: New Perspectives on the Ardennes Campaigns of 1940 and 1944. *Annals of the American Association of Geographers* 111:1079–1093.

Havlick, D.G. 2014. Opportunistic Conservation at Former Military Sites in the United States. *Progress in Physical Geography: Earth and Environment* 38:271–285.

Henry, M., and K. Natanel. 2016. Militarisation as Diffusion: The Politics of Gender, Space and the Everyday. *Gender, Place & Culture* 23:850–856.

Herlihy, P. H. 2010. Self-Appointed gatekeepers Attack the American Geographical Society's First Bowman Expedition. *Political Geography* 29:417–419.

Hirt, I., and M. Palomino-Schalscha. 2011. Geography, the Military and Critique on the Occasion of the 2011 IGU Regional Meeting in Santiago de Chile. *Political Geography* 30:355–357.

Holland, E. C. 2011. Barack Obama's Foreign Policy, Just War, and the Irony of Political Geography. *Political Geography* 30:59–60.

Hopkins, P., K. Hörschelmann, M. C. Benwell, and C. Studemeyer. 2019. Young People's Everyday Landscapes of Security and Insecurity. *Social & Cultural Geography* 20:435–444.

Hörschelmann, K. 2008. Populating the Landscapes of Critical Geopolitics – Young People's Responses to the War in Iraq (2003) *Political Geography* 27:587–610.

Hörschelmann, K., C. Cottrell Studemeyer, P. Hopkins, and M. Benwell. 2019. Special Section Introduction: "Peripheral Visions: Security By, and For, Whom?". *Geopolitics* 24:777–786.

Hristova, S. 2018. Charting the Territory: Space and Power in the Iraq War. *ACME: An International Journal for Critical Geographies* 17:939–957.

Hyndman, J. 2010. The Question of 'the Political' in critical geopolitics: Querying the 'child Soldier' in the 'War on Terror.' *Political Geography* 29:247–256.

Hyndman, J. 2019. Unsettling Feminist Geopolitics: Forging Feminist Political Geographies of Violence and Displacement. *Gender, Place & Culture* 26:3–29.

Ingram, A., and K. Dodds. eds. 2009. *Spaces of Security and Insecurity: Geographies of the War on Terror*. Ashgate.

Inwood, J., and J. Tyner. 2011. Geography's Pro-Peace Agenda: An Unfinished Project. *ACME: An International Journal for Critical Geographies* 10:442–457.

Inwood, J. and J. A. Tyner. 2022. Militarism and the Mutually Assured Destruction of Climate Change. *Space and Polity* 26:62–66.

Jackman, A., and K. Brickell. 2021. 'Everyday Droning': Towards a Feminist Geopolitics of the Drone-Home. *Progress in Human Geography* 46:156–178.

Jacobsen, M.H. 2022. Wars in Refuge: Locating Syrians' Intimate Knowledges of Violence Across Time and Space. *Political Geography* 92:102488.

Joronen, M. 2021. Unspectacular Spaces of Slow Wounding in Palestine. *Transactions of the Institute of British Geographers* 46:995–1007.

Kaplan, R.D. 2012. *The Revenge of Geography. What the Map Tells Us About the Coming Conflict and the Battle Against Fate*. Random House.

Kearns, G. 2008. Progressive Geopolitics. *Geography Compass* 2:1599–1620.

Kitchin, R. 2007. Intervention: Elsevier, the Arms Trade, and Forms of Protest – A Response to Chatterton and Featherstone. *Political Geography* 26:499–503.

Klem, B. 2014. The Political Geography of War's End: Territorialisation, Circulation, and Moral Anxiety in Trincomalee, Sri Lanka. *Political Geography* 38:33–45.

Klinke, I. 2020. Geography at War. In *The SAGE Handbook of Historical Geography* eds. M. Domosh, M. Heffernan and C. W. J. Withers: 449–465. SAGE.

Kobayashi, A. 2009. Geographies of Peace and Armed Conflict: Introduction. *Annals of the Association of American Geographers* 99:819–826.

Kobayashi, A. ed. 2012. *Geographies of Peace and Armed Conflict*. Routledge.

Koopman, S. 2011. Alter-Geopolitics: Other Securities Are Happening. *Geoforum* 42:274–284.

Koopman, S. 2016. Beware: Your Research May Be Weaponized. *Annals of the American Association of Geographers* 106:530–535.

Koopman, S. 2019. Peace. In *Keywords in Radical Geography: Antipode at 50*: 207–211.

Koopman, S. 2020. Building an Inclusive Peace Is an Uneven Socio-Spatial Process: Colombia's Differential Approach. *Political Geography* 83:102252.

Korson, C. 2015. Framing Peace: The Role of Media, Perceptions, and United Nations Peacekeeping Operations in Haiti and Côte d'Ivoire. *Geopolitics* 20:354–380.

Kuus, M. 2002. Toward Cooperative Security? International Integration and the Construction of Security in Estonia. *Millennium* 31:297–317.

Kuus, M. 2004. "Those Goody-Goody Estonians": Toward Rethinking Security in the European Union Candidate States. *Environment and Planning D: Society and Space* 22:191–208.

Kuus, M. 2007. "Love, peace and Nato": Imperial Subject-Making in Central Europe. *Antipode* 39:269–290.

Lacoste, Y. 1976. *La géographie, ça sert, d'abord, à faire la guerre*. François Maspéro.

Laliberté, N. 2016. 'Peace Begins at Home': Geographic Imaginaries of Violence and Peacebuilding in Northern Uganda. *Political Geography*:24–33.

Lasserre, F., E. Arapi, and M. Bennett. 2022. Bunker Mentalities: The Shifting Imaginaries of Albania's Fortified Landscape. *Borders in Globalization Review* 3:66–76.

Le Billon, P. 2004. The Geopolitical Economy of 'resource wars'. *Geopolitics* 9:1–28.

Le Billon, P. 2007. Geographies of War: Perspectives on 'Resource Wars'. *Geography Compass* 1.

Le Billon, P. 2008. Diamond Wars? Conflict Diamonds and Geographies of Resource Wars. *Annals of the Association of American Geographers* 98:345–372.

Lizotte, C., M. M. Bennett, and K. Grove. 2022. Introduction: Virtual Forum on the Russian Invasion of Ukraine. *Political Geography* 97:102673.

Loyd, J. M. 2012. Geographies of Peace and Antiviolence. *Geography Compass* 6:477–489.

Loyd, J. M. 2014. *Health Rights Are Civil Rights: Peace and justice activism in Los Angeles, 1963–1978*: University of Minnesota Press.

Macaspac, N. V. 2019. Insurgent Peace: Community-Led Peacebuilding of Indigenous Peoples in Sagada, Philippines. *Geopolitics* 24:839–877.

Macaspac, N. V., and A. Moore. 2022. Peace Geographies and the Spatial Turn in Peace and Conflict Studies: Integrating Parallel Conversations through Spatial Practices. *Geography Compass* 16:e12614.

Mahrouse, G. 2011. War-Zone Tourism: Thinking Beyond Voyeurism and Danger. *ACME: An International Journal for Critical Geographies* 15:330–345.

Mamadouh, V. 2005. Geography and War, Geographers and Peace. In *The Geography of War and Peace: From Death Camps to diplomats* ed. C. Flint: 26–60. Oxford University Press.

Mamadouh, V. 2014. One Union, Two Speakers, Three Presidents, an… 500 Million Eu Citizens: The European Union and the 2012 Nobel Peace Prize. *Political Geography* 42 (A1–a3).

Mamadouh, V. 2020. A View from the Borderlands of Anglophone (political) Geography. *GeoJournal*.

Mamadouh, V. 2022. A hundred years in the shadow of war and peace: The Internal Geographical Union. *Newsletter of the IGU Commission on Political Geography –* June 2022.

Marsden, W.E. 2000. Geography and Two Centuries of Education for Peace and International Understanding. *Geography* 85:289–302.

Marshall, T. 2015. *Prisoners of Geography: Ten Maps That Explain Everything About the World*. Scribner.

Marshall, T. 2021. *The Power of Geography: Ten Maps That Reveal the Future of Our World*. Elliot & Thompson.

McConnell, F. 2009. De Facto, Displaced, Tacit: The Sovereign Articulations of the Tibetan Government-in-Exile. *Political Geography* 28:343–352.

McConnell, F., N. Megoran, and P. Williams. eds. 2014. *Geography of Peace*. I.B. Tauris.

McCormack, K., and E. Gilbert. 2021. The Geopolitics of Militarism and Humanitarianism. *Progress in Human Geography* 46:179–197.

Medina, R. M. 2016. From Anthropology to Human Geography: Human Terrain and the Evolution of Operational Sociocultural Understanding. *Intelligence and National Security* 31:137–153.

Megoran, N. 2008. Militarism, Realism, Just War, or Nonviolence? Critical Geopolitics and the Problem of Normativity. *Geopolitics* 13:473–497.

Megoran, N. 2010. Towards a Geography of Peace: Pacific Geopolitics and Evangelical Christian Crusade Apologies. *Transactions of the Institute of British Geographers* 35:382–398.

Megoran, N. 2011. War and Peace? An Agenda for Peace Research and Practice in Geography. *Political Geography* 30:178–189.

Megoran, N. 2014. On (Christian) Anarchism and (Non)Violence: A Response to Simon Springer. *Space and Polity* 18 (1):97–105.

Megoran, N., and S. Dalby. 2018. Geopolitics and Peace: A Century of Change in the Discipline of Geography. *Geopolitics* 23:251–276.

Megoran, N., F. McConnell, and P. Williams. eds. 2014. *Geography and Peace*. Palgrave Macmillan.

Megoran, N., F. McConnell, and P. Williams. 2016. Geography and Peace. In *The Palgrave Handbook of Disciplinary and Regional Approaches to Peace*. eds. O. P. Richmond, S. Pogodda and J. Ramović: 123–138. Palgrave Macmillan.

Mercille, J., and A. Jones. 2009. Practicing Radical Geopolitics: Logics of Power and the Iranian Nuclear "Crisis". *Annals of the Association of American Geographers* 99:856–862.

Moore, A. 2013. *Peacebuilding in Practice: Local Experience in Two Bosnian Towns*. Cornell University Press.

Moore, A. 2017. US Military Logistics Outsourcing and the Everywhere of War. *Territory, Politics, Governance* 5:5–27.

Moore, A. 2019. *Empire's Labor: The Global Army That Supports US Wars*. Cornell University Press.

Moore, A., and W. Cartwright. 2015. Special Feature: Extracting Geography from Cartoons in a War Context. *Journal of Spatial Science* 60:19–36.

Mutschler, M., and M. Bales. 2023. Liquid or Solid Warfare? Autocratic States, Non-State Armed Groups and the Socio-Spatial Dimension of Warfare in Yemen. *Geopolitics*:1–29.

Nordås, R., and N. P. Gleditsch. 2007. Climate Change and Conflict. *Political Geography* 26:627–638.

Ó Tuathail, G. 2005. The Frustrations of Geopolitics and the Pleasures of War: *Behind Enemy Lines* and American Geopolitical Culture. *Geopolitics*:356–377.

O'Lear, S., and S. L. Egbert. 2009. Introduction: Geographies of Genocide. *Space and Polity* 13:1–8.

O'Loughlin, J. 2004. Academic Openness, Boycotts and Journal Policy. *Political Geography* 23:641–643.

O'Loughlin, J. 2005. The War on Terrorism, Academic Publication Norms, and Replication. *The Professional Geographer* 57:588–591.

O'Loughlin, J. 2018. Thirty-Five Years of Political Geography and Political Geography: The Good, the Bad and the Ugly. *Political Geography* 65:143–151.

O'Loughlin, J., and F. D. W. Witmer. 2011. The Localized Geographies of Violence in the North Caucasus of Russia, 1999–2007. *Annals of the Association of American Geographers* 101:178–201.

O'Loughlin, J., W. Witmer, A. Linke, and N. Thorwardson. 2010. Peering into the Fog of War: The Geography of the WikiLeaks Afghanistan War Logs, 2004–2009. *Eurasian Geography and Economics* 51:472–496.

O'Lear, S., P. F. Diehl, D. V. Frazier, and T. L. Allee. 2005. Dimensions of Territorial Conflict and Resolution: Tangible and Intangible Values of Territory. *GeoJournal* 64:259–261.

Paasi, A. 2015. Academic Capitalism and the Geopolitics of Knowledge. In *The Wiley-Blackwell Companion to Political Geography*, eds. J. Agnew, V. Mamadouh, A. Secor and J. Sharp, 507–523. Wiley Blackwell.

Pain, R. 2015. Intimate War. *Political Geography* 44:64–73.

Pain, R., and L. Staeheli. 2014. Introduction: Intimacy-Geopolitics and Violence. *Area* 46:344–347.

Palka, E. J., and F.A. Galgano Jr. eds. 2005. *Military Geography: From Peace to War*. McGraw-Hill.

Peluso, N.L., and P. Vandergeest. 2011. Political Ecologies of War and Forests: Counterinsurgencies and the Making of National Natures. *Annals of the Association of American Geographers* 101:587–608.

Penu, D. A. K., and D. W. Essaw. 2019. Geographies of Peace and Violence During Conflict: The Case of the Alavanyo-Nkonya Boundary Dispute in Ghana. *Political Geography*: 91–102.

Pickering, S. 2017. *Understanding Geography and War: Misperceptions, Foundations, Prospects*: Palgrave Macmillan.

Pinkerton, A., S. Young, and K. Dodds. 2011. Postcards from Heaven: Critical Geographies of the Cold War Military–Industrial–Academic Complex. *Antipode* 43:820–844.

Pringle, C. 2007. Response to Chatterton and Featherstone, "Intervention: Elsevier, Critical Geography and the Arms Trade". *Political Geography* 26:8–9.

Quiquivix, L. 2014. Art of War, Art of Resistance: Palestinian Counter-Cartography on Google Earth. *Annals of the Association of American Geographers* 104:444–459.

Radil, S.M., and C. Flint. 2013. Exiles and Arms: The Territorial Practices of State Making and War Diffusion in Post–Cold War Africa. *Territory, Politics, Governance* 1:183–202.

Radil, S.M., C. Flint, and S.H. Chi. 2013. A Relational Geography of War: Actor–Context Interaction and the Spread of World War I. *Annals of the Association of American Geographers* 103:1468–1484.

Raleigh, C., A. Linke, H. Hegre, and J. Karlsen. 2010. Introducing ACLED: An Armed Conflict Location and Event Dataset. *Journal of Peace Research* 47:651–660.

Ramutsindela, M. 2017. Greening Africa's Borderlands: The Symbiotic Politics of Land and Borders in Peace Parks. *Political Geography*:106–113.

Rech, M., D. Bos, K. N. Jenkings, A. Williams, and R. Woodward. 2015. Geography, Military Geography, and Critical Military Studies. *Critical Military Studies* 1:47–60.

Rech, M.F. 2014. Be Part of the Story: A Popular Geopolitics of War Comics Aesthetics and Royal Air Force Recruitment. *Political Geography* 39:36–47.

Rose-Redwood, R., E. Sheppard, G. Pratt, S. M. Roberts, M. R. Read, C. Fuhriman, and E. T. Yeh. 2022. Ethics and the Geography–Military Nexus: Responses to Wainwright and Weaver. *Annals of the American Association of Geographers* 112:e-i–e-vi.

Ruwanpura, K.N., L. Chan, B. Brown, and V. Kajotha. 2020. Unsettled Peace? The Territorial Politics of Roadbuilding in Post-War Sri Lanka. *Political Geography* 76:102092.

Sharp, J. 2011. A Subaltern Critical Geopolitics of the War on Terror. *Geoforum* 42:297–306.

Sheppard, E., and J. Tyner. 2016. Forum on Geography and Militarism: An Introduction. *Annals of the American Association of Geographers* 106:503–505.

Shimazu, T. J. G. 2015. War, Peace, and a Geographical Internationalism: The 1871 Antwerp International Geographical Congress. *Geographical reports of Tokyo Metropolitan University* 50:97–105.

Shroder, J. 2005. Remote Sensing and GIS as Counterterrorism Tools in the Afghanistan War: Reality, Plus the Results of Media Hyperbole. *The Professional Geographer* 57 (4):592–597.

Slater, D. 2004. Editorial Comment: Academic Politics and Israel/Palestine. *Political Geography* 23 (6):645–646.

Smith, D., and A. Bræin. 2003. *The Atlas of War and Peace.* Fourth Edition Earthscan.

Smith, N. 2003. *American Empire: Roosevelt's Geographer and the Prelude to Globalization.* University of California Press.

Springer, S. 2011. Violence Sits in Places? Cultural Practice, Neoliberal Rationalism, and Virulent Imaginative Geographies. *Political Geography* 30:90–98.

Springer, S. 2014a. War and Pieces. *Space and Polity* 18:85–96.

Springer, S. 2014b. God Dethroned: A Reply to Nick Megoran. *Space and Polity* 18:106–109.

Steinberg, P. E. 2010. Professional Ethics and the Politics of Geographic Knowledge: The Bowman Expeditions. *Political Geography* 29:413.

Stokke, K. 2009. Crafting Liberal Peace? International Peace Promotion and the Contextual Politics of Peace in Sri Lanka. *Annals of the Association of American Geographers* 99:932–939.

Storey, D. 2005. Academic Boycotts, Activism and the Academy. *Political Geography* 24:992–997.

Taylor, P. J. 2004. God Invented War to Teach Americans Geography. *Political Geography* 23:487–492.

Till, K.E., and A.-K. Kuusisto-Arponen. 2015. Towards Responsible Geographies of Memory: Complexities of Place and the Ethics of Remembering. *Erdkunde* 69:291–306.

Till, K.E., J. Sundberg, W. Pullan, C. Psaltis, C. Makriyianni, R. Zincir Celal, M. O. Samani, and L. Dowler. 2013. Interventions in the Political Geographies of Walls. *Political Geography* 33:52–62.

Toal, G., and C. T. Dahlman. 2011. *Bosnia Remade: Ethnic Cleansing and Its Reversal.* Oxford University Press.

Tyner, J. 2005. *Iraq, Terror, and the Philippines' Will to War.* Rowman & Littlefield.

Tyner, J. 2007. *America's Strategy in Southeast Asia: From the Cold War to the Terror War.* Rowman & Littlefield.

Tyner, J. 2008. *The Killing of Cambodia: Geography, Genocide, and the Un-Making of Space*. Ashgate.

Tyner, J. 2009a. *Military Legacies: A World Made by War*. Routledge.

Tyner, J. 2009b. *War, Violence and Population: Making the Body Count*. Guilford Press.

Tyner, J., and J. Inwood eds. 2011. *Nonkilling Geography*. Honolulu: Centre for Global Nonkilling.

Tyner, J., and J. Inwood. 2014. Violence as Fetish: Geography, Marxism, and Dialectics. *Progress in Human Geography* 38:771–784.

Verweijen, J. 2022. War, Peace and Geography: The Perilous Engagement with Public Policy Toward Armed Conflict. *Space and Polity* 26:128–134.

Wainwright, J. 2013. *Geopiracy: Oaxaca, Militant Empiricism, and Geographical Thought*. Palgrave Macmillan.

Wainwright, J., and B. R. Weaver. 2021. A Critical Commentary on the AAG Geography and Military Study Committee Report. *Annals of the American Association of Geographers* 111:1137–1146.

Wainwright, J., and B. R. Weaver. 2022. The Ethics of Geography–Military Relations: A Reply to Our Interlocutors. *Annals of the American Association of Geographers*:1–5.

Wainwright, J. D. 2016. The U.S. Military and Human Geography: Reflections on Our Conjuncture. *Annals of the American Association of Geographers* 106:513–520.

Waterman, S. 2005. True Exchanges or Thought Crime? *Political Geography* 24:998–1001.

Williams, P. 2007. Hindu–Muslim Brotherhood: Exploring the Dynamics of Communal Relations in Varanasi, North India. *Journal of South Asian Development* 2:153–176.

Williams, P., and F. McConnell. 2011. Critical Geographies of Peace. *Antipode* 43:927–931.

Woodward, R. 2004. *Military Geographies*. Blackwell.

Woodward, R 2005. From Military Geography to Militarism's Geographies: Disciplinary Engagements with the Geographies of Militarism and Military Activities. *Progress in Human Geography* 29:718–740.

Woodward, R. 2013. Military Landscapes: Agendas and Approaches for Future Research. *Progress in Human Geography* 38:40–61.

Woodward, R., K. N. Jenkings, and A. J. Williams. 2017. Militarisation, Universities and the University Armed Service Units. *Political Geography* 60:203–212.

Woodward, R., and P. Winter. 2004. Discourses of Gender in the Contemporary British Army. *Armed Forces & Society* 30:279–301.

Woon, C.Y. 2011. Undoing Violence, Unbounding Precarity: Beyond the Frames of Terror in the Philippines. *Geoforum* 42:285–296.

Woon, C.Y. 2014a. Precarious Geopolitics and the Possibilities of Nonviolence. *Progress in Human Geography* 38:654–670.

Woon, C.Y. 2014b. Popular Geopolitics, Audiences and Identities: Reading the 'War on Terror' in the Philippines. *Geopolitics* 19:656–683.

Woon, C.Y. 2015. 'Peopling' Geographies of Peace: The Role of the Military in Peacebuilding in the Philippines. *Transactions of the Institute of British Geographers* 40:14–27.

Woon, C.Y. 2017. Children, Critical Geopolitics, and Peace: Mapping and Mobilizing Children's Hopes for Peace in the Philippines. *Annals of the American Association of Geographers* 107:200–217.

3

GEOGRAPHIES OF PEACE

Nerve V. Macaspac and Adam Moore

Introduction

Geography as a discipline has a long history of serving the cause of war and empire, a fact recognized by both proponents and critics of this state of affairs (Lacoste 1976; Mackinder 1911). This has changed in recent decades with the emergence of radical and heterodox journals such as *Antipode* and *Herodote*, the development of the field of critical geopolitics, and the publication of a number of edited collections that have explored the geographies of both war and peace (i.e., Flint 2005; Kliot and Waterman 1991; Pepper and Jenkins 1985), indicating a shift from a "war geography" toward becoming "a science for peace" (Mamadouh 2005, 41).

Peace geographies—which we trace to a series of articles that appeared in the early 2010s (i.e., Inwood and Tyner 2011; Loyd 2012; Megoran 2010, 2011; Williams and McConnell 2011)—represent the most recent manifestation of this shift in orientation. These articles advanced two critiques. First, they rehearsed criticisms of the discipline's long history of entanglement with empire and the pursuit of war (Megoran 2010). Second, and more significantly, they argued that geography still focused inordinately on war, militarism, and violence at the expense of researching and theorizing peace. Consequently "geography is better at studying war than peace" (Megoran 2011, 178). Thus, they argued, there is a need to further rebalance geographic research agendas toward the study of peace. In addition to these supposed analytical shortcomings, geographers were called to pursue a much more explicitly normative "pro-peace agenda" that challenges the culture of war and "focuses on the interlinkages between violence, militarism, and inequality" (Inwood and Tyner 2011, 443).

DOI: 10.4324/9781003345794-3

Beyond critique, these articles also argued that geography needed to craft a disciplinary definition of peace distinct from the absence or endpoint of war. This challenges the implication in the discipline that "peace exists only as a point of reference, an empty 'other' defined by an absence of violence" (Williams and McConnell 2011, 928). It also foregrounds the reality that peace, like war, must be understood as a process (Loyd 2012).

These early interventions opened up new avenues of geographical theorization of and research on peace. We highlight four key strands. First, an understanding of peace and conflict as relational, place-specific, processes. Second, an examination of diverse political practices and ideologies that animate the project of peace. Third, the development of an integrative definition of peace as a set of holistic social and ecological (human and nonhuman) agendas for a better world. And fourth, an empirical focus on everyday peace. We turn now to these elements of the peace geographies literature.

Peace and conflict as relational, place-specific processes

One of the main contributions of recent work on peace in geography is the emphasis on peace and conflict as relational, place-specific processes. As Sara Koopman (2011a, 193) argues, "the two are intertwined, and we cannot understand one without the other." As place-specific processes, peace geographies are multiple, uneven, and dynamic. They embody multiple meanings, and these meanings differ across time, space, and place (Koopman 2011a, 2020). Thus peace, as with violence, can be better understood as situated knowledges within specific historical and cultural settings (Williams and McConnell 2011, 929). As such, peace geographies are permeated with power dynamics. Situated in place, peace geographies also embody local and global encounters and the relationships of power and power struggles that come along with these encounters.

Research on postwar reconstruction and international peacebuilding reveals the blurred boundaries between the geographies of war and peace, and between wartime and peacetime. Kristian Stokke's (2009) study of the peace dividend as a component of the peace process between the Sri Lankan government and the Liberation Tigers of Tamil Eelam (LTTE) demonstrated how postwar reconstruction efforts are deeply connected with wartime political relationships and structures wherein the peace dividend was transformed into a vehicle for the government and rebels to gather political support. Gerard Toal and Carl Dahlman's (2011) study of the geopolitics of refugee returns in Bosnia and Herzegovina revealed the power of wartime legacies, as shown by the fraught political dynamics that shaped the return of displaced persons and refugees to their former communities as part of peace agreements and postwar recovery. Scott Kirsch and Colin Flint (2011) also highlighted wartime legacies during post-conflict reconstruction wherein a new set of relationships and geopolitical landscapes shaped by militarized power relations take place even during a time of peace. Adam Moore's (2013) comparative study of peacebuilding in two

Bosnian towns, Mostar and Brčko, revealed the ways in which place-specific wartime social processes actively shape the success or failure of postwar peace-building projects.

Geographers have also extensively examined challenges and frictions between local and international peacebuilding practices. Phillipe Le Billon (2008) focused upon the issue of corruption to reveal the challenges and contentious politics of international peacebuilding projects within post-conflict states. Paul Higate and Marsha Henry (2009) investigated the ways in which peacekeeping missions of the United Nations (UN) produce and reproduce hegemonic and unequal global relations of power in many countries in the global South. Nicole Laliberté (2016) revealed the differential politics and visions of peace-building work among international, national, and regional actors in Uganda as shaped by differences in gender, race, and class.

Finally, research on peace geographies has attended to the unique and meaningful material constructions and co-productions of peace and place. From peace parks to peace communities and peace zones, geographers have examined the material and discursive struggles over the upkeep or control of a place, over its meaning, or over concerns around belonging and who gets excluded and included. Brian King (2010) traced the colonial and settler-colonial legacies of peace parks in sub-Saharan Africa that led to the displacement of indigenous peoples. Chris Courtheyn (2016, 2022) highlighted the important role of memory and commemoration as spatial and place-based practices in the production of a peace community in Colombia. Nerve Macaspac (2018, 2022) underscored the ways in which rules of place and the set of internal social and cultural norms are integral sources of the upkeep of a peace zone by an indigenous community amid insurgency and counterinsurgency in the Philippines.

The politics of peace

Geographers have reframed peace as politics that confronts hegemonic power. Inspired by transdisciplinary literature on agency and social movements, they have increasingly examined peace beyond the context of international, technical, or state-led projects of conflict resolution or peacebuilding. Many have argued that the making of peace relies upon collective efforts in fostering solidarity and addressing shared experiences of violence or marginalization. Sara Koopman's (2011b) framework of alter-geopolitics underscored the important role of transnational solidarity and political alliance between privileged and marginalized communities between the global North and South in creating alternative spaces of peace amid violent conflict. Chih Yuan Woon (2011, 286) highlighted emotional and interpersonal bonds as sources of collective action and expressions of a "radical praxis of peace" in response to state-induced fear, terror, and violence.

Other geographers have examined the political ontology of peace to unpack the analytical trap of binary thinking that positions peace as an opposite to violence. Harry Bregazzi and Mark Jackson (2018) underscored the limits of studying violence alone as a proxy for revealing what peace might look like in contrast to violence. Rather, they argued, scholars must attend to the human and nonhuman interactions and relationships that are "always-already producing peace" (Bregazzi and Jackson 2018, 74). Relatedly, Chris Courtheyn (2018a, 743) suggested a framework of "radical trans-relational peace" to theorize positive peace relationally and beyond the binary thinking of negative and positive peace. Through this approach, he examined issues concerning race, the biosphere, and solidarity movements to reveal the interconnectedness of diverse theories and practices of peace, which he calls "ecological dignity" to refer to the flourishing of human and nonhuman life amid different forms of violence (Courtheyn 2018a, 743).

Finally, some scholars have suggested that peace geographies can be better understood through a variety of already existing ideological and world-making perspectives, from the religious to anarchism. Nick Megoran (2011, 186) re-conceptualized peace as "nonviolence" drawing from a Christian proverb that "violence is to be confronted through love." He also proposed the framework of "pacific geopolitics" to reorient global geopolitics toward the promotion of "peaceful and mutually enriching human coexistence" (2010, 382). Ethan Yorgason (2011) and Carl Grundy-Warr (2011) responded by foregrounding Mormon and Buddhist perspectives on peace activism. Simon Springer (2016, 145) examined negative and positive peace within the project of anarchism suggesting that equality is "a *sine qua non* of peace" and that visions of a peaceful world are synonymous to anarchism's promise of "a world without war, without domination, without bondage, and without violence." Other geographers have engaged with the concept and practical techniques of "nonviolence" as a lens to understand peace geographies. Chih Yuan Woon (2011) investigated the emotional sources that facilitate the deployment of nonviolent approaches in enacting political change. He proposed "precarious geopolitics" to signal that emotional and interpersonal bonds over shared precarity and vulnerability drive collective enactments of nonviolence. Jenna Loyd (2012, 486) made a distinction between nonviolence and what she referred to as "antiviolence" to suture simultaneous processes of confronting multiple forms of societal violence and different social movements against colonial, racialized, and gendered violence. For Harry Bregazzi and Mark Jackson (2018), the making of peace geographies can be better understood through already existing lived experiences of human collectivity.

Peace as a set of holistic social and ecological (human and nonhuman) agendas

In response to the calls to develop a disciplinary perspective on peace, geographers have developed two lines of research that re-think peace as a set of holistic social and ecological, and human and nonhuman, agendas for a better world. First, they have challenged the concepts of "negative peace" and "positive peace" as the traditional definition of peace. Joshua Inwood and James Tyner (2011) attended to the interconnectedness of structural and institutional violence that legitimizes militarism, direct violence, and human deaths as a requirement of geopolitical, geoeconomic, or geosocial power. Amy Ross (2011, 197) reframed peace as a critique of power toward equity. More generally, research has taken a more capacious perspective on peace and violence, drawing the attention of peace geographers to domestic violence (Brickell 2015), to health rights (Loyd 2014), and to education (Ferretti 2016), reinforcing Sara Koopman's (2019, 209) argument that peace is the "root for social and political change."

Second, efforts to better integrate multiple dimensions of peace through critical, intersectional, or relational perspectives extend to urgent concerns toward ecological and planetary life and human–nature interactions. While geographers have long been invested in concerns related to political economy, political ecology, oil-led development and violence, and extractive resources and armed conflicts (Le Billon 2001, 2008; Peet and Watts 2004; Peluso and Watts 2001; Watts 2001, 2013), recent research is increasingly oriented toward climate-conflict linkages (Barnett 2019). Geographers have explored a range of consequences of climate change toward human security, particularly in driving existing or new violent conflicts related to resource scarcity, and the risks that climate change poses to peace and societal development (Barnett and Adger 2007). Clionadh Raleigh and Henrik Urdal (2007) argued that climate-related environmental change dramatically reduces freshwater availability and the productive capacity of soils that affect human lives, settlement patterns, and political systems. Jon Barnett and W. Neil Adger (2007) suggested that conflicts are likely to emerge due to the consequences of efforts to reduce fossil fuel emissions in the global political economy. Alongside climate-related risks to human security, attention has also been given to the uneven geographies of vulnerability. Geographers examined health risks and outbreaks of certain diseases induced by changes in the climate (Comrie 2007), extreme weather events (Pandey and Bardsley 2015), and extreme hydroclimatic hazards indicated by flood, heat wave, and drought events (Binita, Shepherd and Gaither 2015). Increasingly, geographers also are foregrounding the consequences of the changing climate toward indigenous peoples (Bennett 2018; Cameron 2012; Smith and Rhiney 2016). This research underscores the social inequalities wherein populations who are least responsible for climate change are the ones most at risk from its deadly impacts and the ways in which states and powerful institutions simultaneously mobilize and depoliticize vulnerability (Barnett 2020).

More recently, geographers have shifted attention toward environmental peacebuilding. Formerly referred to as "environmental peacemaking" (see Conca and Dabelko 2002), the literature on environmental peacebuilding operates within an assumption that global environmental change may provide incentives for global cooperation and peace (see Alleson and Schoenfeld 2007; Dresse et al. 2019). While the premise of environmental management or governance as an effective pathway for conflict resolution or prevention has gained traction among scholars and institutions, including the UN (see Conca and Wallace 2009), many geographers have approached environmental peacebuilding critically. Tobias Ide (2019) argued that environmental peacebuilding has its "dark side," which includes depoliticization, displacement, discrimination, deterioration into conflict, delegitimization of the state, and environment degradation. Tobias Ide, Lisa Palmer, and Jon Barnett (2021) have suggested that top-down approaches of environmental peacebuilding tend to reproduce the pitfalls and counterproductive consequences of liberal peacebuilding, calling for localized studies of environmental peacebuilding practices.

Finally, geographers have explored the intersections of environmental catastrophe, violence, and peace, drawing from feminist and Science and Technology Studies (STS) perspectives. Shannon O'Lear (2016) examined the politics of technoscientific understanding of climate change and its implications for political geography, in particular how selective interpretations of climate change potentially result in "slow violence" or indirect and prolonged human suffering, further foreclosing other emancipatory perspectives toward understanding human–nature relationships.

Chris Courtheyn (2018a, 743) traced the intersections of "race, the biosphere, and autonomous-solidarity politics," suggesting the concept of "radical trans-relational peace" as a normative framework in studying peace.

Everyday peace

The fourth prominent strand in the peace geographies literature focuses on everyday peace. Everyday peace refers to the kinds of quotidian work that ordinary people do in making peace amid ongoing physical, structural, or symbolic violence. The analytical work focused on everyday peace and our understanding of the making of peace geographies is two-pronged. First, it foregrounds the agentive capacities, techniques, skills, and social and spatial practices of ordinary people or marginalized communities. Everyday peace research decenters the making of peace away from formal, official, technical, or expert-driven peace processes and peacebuilding projects designed and implemented by the state, military, humanitarian or civil society organizations, UN experts and their agencies, or other transnational institutions and stakeholders. Rather, spaces of peace emerge from the daily initiatives of ordinary people in navigating violent conflicts or other forms of societal violence, particularly when states, non-state actors, or institutions fail to do so or are the sources of violence.

Research in this vein tends to focus on local scale, "bottom-up" initiatives rather than "top-down" projects initiated and controlled by international and state actors. Philippa Williams and Fiona McConnell (2011, 927) focused upon "peace at the margins" while Carl Grundy-Warr (2011) examined the "bottom-up" political geographies of peace. And inspired in part by postcolonial and subaltern studies, geographers have plumbed the "hidden transcripts" of the production of peace geographies and examined everyday peace dynamics akin to James Scott's (1985) notion of "weapons of the weak."

Philippa Williams (2015, 181), for instance, traced the ways in which members of the Muslim community navigate Muslim–Hindu relations and political economies in the city of Varanasi, India, and showed the making of everyday peace relies upon "local capacity to create real and imagined spaces of connection, tolerance and civility." Everyday peace, for Williams, aims to capture how subaltern communities make peace from positions of marginality. It reveals the situated politics of peace, pragmatism, and resilience as tactics of everyday peacemaking, how ordinary people simultaneously imagine and enact positive social change in the present, and how processes of everyday peacemaking are intimately related with the articulation and realization of citizenship practices (Williams 2015).

Second, everyday peace signals the distinct sociality, spatiality, and temporality of what can be viewed as ordinary, pervasive, yet taken-for-granted, relative to what might be considered as spectacular, a rupture, or historic moment in the context of peace and conflict. As an epistemological approach, everyday peace builds upon social theories that examine everyday life and production of space under capitalist modernity (i.e., Bourdieu, De Certeau, Engels, Foucault, Lefebvre, and Marx) and intersectional feminist epistemologies that unsettle normative or conventional conceptualizations of scale as a set of hierarchical categories, and illustrates the relationality of social phenomena across different places and within both public and private spaces (Katz 2001; Low and Maguire 2019; Nagar, Lawson, McDowell and Hanson 2002; Pratt 2004; Staeheli, Kofman, and Peake 2004).

Empirically, these theoretical interventions have oriented research toward understanding the daily maintenance or upkeep of spaces of peace, especially the perspectives and practices of local communities, such as local peace communities in Colombia (Courtheyn 2016, 2018b), indigenous-led peace zones in the Philippines (Macaspac 2018, 2022), and Somali cultural centers in Kenya (Feghali, Faria, and Jama 2021). One of the many contributions of this literature is to investigate peace through the lens of social reproduction (Katz 2004) to foreground the messy, complex, constant, and contentious kinds of work required of ordinary people in producing and sustaining everyday peace as sites of struggle, resistance, and positive social change amid ongoing violence. Finally, a focus on everyday peace has also led to a greater focus on emotions (Woon 2011, 2017) and embodied perspectives (Courtheyn 2016; Hayes-Conroy and Montoya 2017; Koopman 2014; Woon 2017), including how peace and conflict dynamics are differentially experienced (Koopman 2020; Woon 2017).

Geographers' research on everyday peace mirrors a similar move in the broader field of peace and conflict studies that has been dubbed the "local turn" (e.g., Mac Ginty and Richmond 2013; Paffenholz 2015). This research shares an interest in bottom-up or grassroots peacebuilding initiatives (Autesserre 2014a, 2021; Firchow 2018; Mac Ginty 2011, 2013, 2014, 2021; Mac Ginty and Richmond 2013; Mitchell and Hancock 2012; see also Ware and Ware 2022), but it also examines topics such as foreign peacebuilders' everyday practices and lives (Autesserre 2014b; Smirl 2015), interactions between local populations and international actors (Pouligny 2006), subnational variations in peace outcomes (Moore 2013), and why peacebuilding interventions frequently fail to address local sources of conflict (Autesserre 2014b, 2021). In recent years peace and conflict scholars outside of geography—especially those associated with the local turn—have begun to engage more substantially with geographic ideas and thinkers, leading to the announcement of a "spatial turn" within the field, which presents new opportunities for transdisciplinary conversations on the geographies of peace.

Engaging with the "spatial turn" in peace and conflict studies

Concurrent with the development of peace geographies over the past decade, there has also been increased interest in spatial dynamics and concepts within the broader, transdisciplinary field of peace and conflict studies. This "spatial turn" can be traced to the publication of the edited volume, *Spatializing Peace and Conflict*, by Annika Björkdahl and Susanne Buckley-Zistel (2016a), which was announced as "a *cri de coeur* for research to investigate the interconnectedness between space, peace and conflict" (Björkdahl and Buckley-Zistel 2016b, 1). Invoking Ed Soja's (1989) declaration of a spatial turn in the social sciences and humanities they called for a similar spatial turn in the field of peace and conflict studies that investigates "how space, scale and sites can be conceptualized and analytically employed." In addition to Soja, spatial turn scholars frequently cite Lefebvre's (1991) ideas about the production of space and Doreen Massey's (2005) feminist and relational approach to space and place (Björkdahl and Buckley-Zistel 2016b, 2022; Brigg and George 2020; Carabelli 2018; Engel 2020; Forde 2019a; Gusic 2019, 2022).

Somewhat surprisingly, to date there has been relatively little substantive engagement in theoretical and conceptual ideas animating the respective conversations of peace geographers and spatial turn scholars (Macaspac and Moore 2022). One reason for this is that to non-geographers the peace geographies literature is perceived as "mostly an internal debate within political geography that provides few theoretical and analytical ideas" (Gusic 2019, 48) that speak to their concerns. There is some truth to this observation given that peace geographies emerged as a critique of the field of political geography. Thus, as successful as this critique has been within geography its broader impact has been limited due to the lack of resonance that these concerns have beyond the field. Additionally, peace geographers' emphasis on (re)theorizing peace does not speak to the conversations that animate spatial turn scholars at the moment. Consequently, the latter often wonder why geographers fail to sufficiently theorize the spatiality of peace and conflict (Björkdahl and Kappler 2017; Gusic 2019).

We believe that it is time for peace geographers to move beyond these narrow disciplinary concerns and engage more substantially with broader conversations among peace and conflict scholars. When one surveys empirical research, there is much common ground to be found. A central point of overlapping interest, for instance, is developing a better understanding of the spatial conditions, behaviors, and perspectives that shape everyday dynamics of peace and conflict. Methodologically, this has translated into a shared emphasis on ethnographic and participatory research, such as community mapping (Forde 2016, 2019a, 2019b; Koopman 2014), working with local communities to construct everyday peace indicators (Firchow 2018; Firchow and Mac Ginty 2017), and accompaniment (Courtheyn 2016, 2018b; Koopman 2011b, 2014), as well as an emphasis on embodiment and reflexivity (Arias López et al. 2023; Brigg and Bleiker 2010; Macaspac 2017; Millar 2021). One conceptual framework that usefully ties together these common empirical and methodological orientations is spatial practices (more on this below). More broadly, as we argue in the following section, future research would benefit from attention to a broader range of spatialities and temporalities that shape peace and conflict processes.

The multiple spatialities and temporalities of peace and conflict

An empirical focus on everyday peace has greatly enriched geographers' understanding of the dynamics of peace and conflict. At the same time, it has narrowed the spatial and temporal registers that tend to be foregrounded, with an emphasis on local actors, places, and processes and relatively recent events.[1] We suggest that a more capacious orientation toward the multiple spatialities and temporalities of peace and conflict—both methodologically and theoretically— offers opportunities for future research.

One can discern a turn in this direction with recent work. Nick Megoran and Simon Dalby (2018) have called for connecting local peace processes to wider geopolitical agendas and actors. Jake Hodder (2015, 2017, 2019) has drawn on archival research to trace broader histories and geographies of pacifist movements and struggles around the meaning of peace. Sara Fregonese (2019, 6) has combined archival research with oral histories to examine the linkages between "large scale geopolitical discourses" and "micro-geographies of violence" in Lebanon from the 1975–6 war to the present. Patricia Ehrkamp, Jenna Loyd, and Anna Secor (2022) have argued that trauma and displacement from war should be studied as a series of displacements and emplacements across multiple sites rather than a onetime event. Geographers have also mobilized a range of spatial concepts and practices beyond space, place, and scale to analyze the dynamics of peace and conflict, including territory and territoriality (Anderson 2008; Moore 2016; Cairo et al. 2018; Le Billon, Roa-García and López-Granada 2020), borders, barriers, and boundary-making (Pullan 2013; Jones 2016; Megoran 2017), and networks and assemblages (Williams 2015; Courtheyn 2016; Hamdan 2021).

Elsewhere we have presented spatial practices as a conceptual framework to examine the different spatialities and temporalities of peace and conflict (Macaspac and Moore 2022). This aligns with a broader shift in human geography over the past fifteen years toward theorizing sociospatial relations along multiple spatialities rather than privileging a single dimension like space or scale (Jessop, Brenner, and Jones 2008; Leitner, Sheppard, and Sziarto 2008). Spatial practices influence the dynamics of peace and conflict in profound ways, from daily, localized, and embodied actions to spatial framings and narratives that inform organizational behaviors and peace agreements. One illustrative example is the creation of zones, a frequently deployed spatial practice used by actors in a range of contexts, from peace zones developed by local communities in the midst of armed conflict (Hancock and Mitchell 2007; Kaplan 2017; Macaspac 2018) to internationally designated safe areas or safe zones for civilian protection during war (Hyndman 2003; Yamashita 2004; Orchard 2014). Zones are sustained by an ensemble of actors and practices, and are tenuous achievements that require constant maintenance work, negotiation, and collaboration. They also enroll multiple spatialities, as Nerve Macaspac (2022) has illustrated with his research on the Sagada Peace Zone in the Philippines. At the same time a range of zoning practices and projects— such as sanctuary, safe areas, peace zones and accompaniment—share a common spatial logic: the promotion of spaces that are demarcated as outside certain aspects of the arena of violence as a means of facilitating peace and protecting civilians. Here we see how a spatial practice(s) framework allows us to draw fruitful connections across diverse literatures and facilitates the development of more generalizable theoretical insights into the spatial dynamics of peace and conflict.

Conclusion

Geographic research on the making of peace is on the rise. And for a very clear reason: wars, violent conflicts, disproportionate deaths, and social crises amid a global COVID-19 pandemic, rapid and irreversible destruction of the planet, and multiple, related forms of societal violence, including racialized, gendered, settler-colonial, and colonial violence, continue to structure, shape, and create uneven geographies of human suffering, injustice, and premature deaths. Notions of both peace and geography share important characteristics, as we have illustrated above. They are both dynamic processes. And they are also integrative concepts that weave together social, spatial, and temporal processes.

As scholars of peace and conflict, it is a compelling time to study and promote peace geographies that co-exist within, and in contention with, existing geographies of violence. Indeed, as we have shown above, geographers have broadened and deepened our understanding of peace across different scales, and mobilized a variety of theoretical, analytical, and critical frameworks to help reveal, visualize, and strengthen the making of peace geographies. However, there is still much work to be done. We have identified two directions for future research: engaging more productively with parallel conversations happening in our neighboring disciplines, specifically the spatial turn in peace and conflict studies (Macaspac and Moore 2022); and foregrounding the multiple, integrative aspects of the social, spatial, and temporal processes required in making a peaceful world.

Note

1 Others have noted the foreshortening of time frames and "presentism" of geographical research more broadly in recent decades (see Jones 2004, 2011).

References

Alleson, I. and S. Schoenfeld. 2007. Environmental justice and peacebuilding in the Middle East. *Peace Review* 19: 371–379.

Anderson, J. 2008. Partition, consociation, border-crossing: Some lessons from the national conflict in Ireland/Northern Ireland. *Nations and Nationalism* 14: 85–104.

Arias López, B., C. Andrä and B. Bliesemann de Guevara. 2023. Reflexivity in research teams through narrative practice and textile-making. *Qualitative Research* 23: 306–312.

Autesserre, S. 2014a. Going micro: Emerging and future peacekeeping research. *International Peacekeeping* 21: 492–500.

Autesserre, S. 2014b. *Peaceland: Conflict resolution and the everyday politics of international intervention.* Cambridge University Press.

Autesserre, S. 2021. *Frontlines of peace: An insider's guide to changing the world.* Oxford University Press.

Barnett, J. 2019. Global environmental change I: Climate resilient peace? *Progress in Human Geography* 43: 927–936.

Barnett, J. 2020. Global environmental change II: Political economies of vulnerability to climate change. *Progress in Human Geography* 44: 1172–1184.

Barnett, J. and W.N. Adger. 2007. Climate change, human security and violent conflict. *Political Geography* 26: 639–655.

Bennett, M. 2018. From state-initiated to Indigenous-driven infrastructure: The Inuvialuit and Canada's first highway to the Arctic Ocean. *World Development* 109: 134–148.

Binita, K.C., J. Marshall Shepherd and C. Johnson Gaither. 2015. Climate change vulnerability assessment in Georgia. *Applied Geography* 62: 62–74.

Björkdahl, A. and S. Buckley-Zistel. 2016a. *Spatializing Peace and Conflict: Mapping the Production of Places, Sites and Scales of Violence*. Palgrave Macmillan.

Björkdahl, A. and S. Buckley-Zistel. 2016b. Spatializing peace and conflict: An introduction. In *Spatializing Peace and Conflict: Mapping the Production of Place, Sites and Scales of Violence*, eds. A. Björkdahl and S. Buckley-Zistel, 1–22. Palgrave Macmillan.

Björkdahl, A. and S. Buckley-Zistel. 2022. Space for peace: A research agenda. *Journal of Intervention and Statebuilding*.

Björkdahl, A. and S. Kappler. 2017. *Peacebuilding and Spatial Transformation: Peace, Space and Place*. Routledge.

Bregazzi, H. and M. Jackson. 2018. Agonism, critical political geography, and the new geographies of peace. *Progress in Human Geography* 42: 72–91.

Brickell, K. 2015. Towards intimate geographies of peace? Local reconciliation of domestic violence in Cambodia. *Transactions of the Institute of British Geographers* 40: 321–333.

Brigg, M. and R. Bleiker. 2010. Autoethnographic international relations: Exploring the self as a source of knowledge. *Review of International Studies* 36: 779–798.

Brigg, M. and N. George. 2020. Emplacing the spatial turn in peace and conflict studies. *Cooperation and Conflict* 55: 409–420.

Cairo, H., U. Oslender, C. Suárez, J. Ríos, S. Koopman, V. Arango, F. Rodríguez Muñoz, and L. Zambrano Quintero. 2018. 'Territorial Peace': The emergence of a concept in Colombia's peace negotiations. *Geopolitics* 23: 464–488.

Cameron, E. 2012. Securing Indigenous politics: A critique of the vulnerability and adaptation approach to the human dimensions of climate change in the Canadian Arctic. *Global Environmental Change* 22: 103–114.

Carabelli, G. 2018. *The Divided City and the Grassroots: The (Un)making of Ethnic Divisions in Mostar*. Palgrave Macmillan.

Comrie, A. 2007. Climate change and human health. *Geography Compass* 1: 325–339.

Conca, K. and G. Dabelko. 2002. *Environmental Peacemaking*. Woodrow Wilson Center Press.

Conca, K. and J. Wallace. 2009. Environment and peacebuilding in war-torn societies: Lessons from the UN environment programme's experience with postconflict assessment. *Global Governance* 15: 485–504.

Courtheyn, C. 2016. 'Memory is the strength of our resistance': An 'other politics' through embodied and material commemoration in the San José peace community, Colombia. *Social and Cultural Geography* 17: 933–958.

Courtheyn, C. 2018a. Peace geographies: Expanding from modern-liberal peace to radical trans-relational peace. *Progress in Human Geography* 42: 741–758.

Courtheyn, C. 2018b. Territories of peace: alter-territorialities in Colombia's San José de Apartadó Peace Community. *The Journal of Peasant Studies* 45: 1432–1459.

Courtheyn, C. 2022. *Community of Peace: Performing Geographies of Ecological Dignity in Colombia*. University of Pittsburgh Press.

Dresse, A., I. Fischhendler, J. Ø. Nielsen and D. Zikos. 2019. Environmental peacebuilding: Towards a theoretical framework. *Cooperation and Conflict* 54: 99–119.

Ehrkamp, P., J. Loyd and A. Secor. 2022. Trauma as displacement: Observations from refugee settlement. *Annals of the American Association of Geographers* 112: 715–722.

Engel, U. 2020. Peace-building through space-making: The spatializing effects of the African union's peace and security policies. *Journal of Intervention and Statebuilding* 14: 221–236.

Feghali, S., C. Faria and F. Jama. 2021. "Let us create space": Reclaiming peace and security in a Kenyan Somali community. *Political Geography* 90.

Ferretti, F. 2016. Geographies of peace and the teaching of internationalism: Marie-Thérèse Maurette and Paul Dupuy in the Geneva school (1924–1948). *Transactions of the Institute of British Geographers* 41: 570–584.

Firchow, P. 2018. *Reclaiming Everyday Peace: Local Voices in Measurement and Evaluation After War*. Cambridge University Press.

Firchow, P. and R. Mac Ginty. 2017. Measuring peace: Comparability, commensurability, and complementarity using bottom-up indicators. *International Studies Review* 19: 6–27.

Flint, C., ed. 2005. *The Geography of War and Peace: From Death Camps to Diplomats*. Oxford University Press.

Forde, S. 2016. The bridge on the Neretva: Stari Most as a stage of memory in post-conflict Mostar, Bosnia-Herzegovina. *Cooperation and Conflict* 51: 467–483.

Forde, S. 2019a. *Movement as Conflict Transformation: Rescripting Mostar, Bosnia-Herzegovina*. Palgrave MacMillan.

Forde, S. 2019b. Cartographies of transformation in Mostar and Cape Town: Mapping as methodology in divided cities. *Journal of Intervention and Statebuilding*, 13: 139–157.

Fregonese, S. 2019. *War and the City: Urban Geopolitics in Lebanon*. Bloomsbury.

Grundy-Warr, C. 2011. Pacific geographies and the politics of Buddhist peace activism. *Political Geography* 30: 190–192.

Gusic, I. 2019. The relational spatiality of the postwar condition: A study of the city of Mitrovica. *Political Geography* 71: 47–55.

Gusic, I. 2022. Peace between peace(s)? Urban peace and coexistence of antagonists in city spaces. *Journal of Intervention and Statebuilding* 16: 619–640.

Hamdan, A. 2021. Ephemeral geopolitics: Tracing the role of refugees in Syria's transnational opposition. *Political Geography* 84: 102299.

Hancock, L. and C. Mitchell. 2007. *Zones of Peace*. Kumarian Press.

Hayes-Conroy, A. and A. Saenz Montoya. 2017. Peace building with the body: Resonance and reflexivity in Colombia's Legion del Afecto. *Space and Polity* 21: 144–157.

Higate, P. and M. Henry. 2009. *Insecure Spaces: Peacekeeping, Power, and Performance in Haiti, Kosovo, and Liberia*. Zed Books.

Hodder, J. 2015. Conferencing the international at the world pacifist meeting, 1949. *Political Geography* 49: 40–50.

Hodder, J. 2017. Waging peace: Militarising pacifism in Central Africa and the problem of geography, 1962. *Transactions of the British Institute of Geographers* 42: 29–43.

Hodder, J. 2019. Casting a black Gandhi: Martin Luther King Jr., American Pacifists and the global dynamics of race. *Journal of American Studies* 55: 48–74.

Hyndman, J. 2003. Preventative, palliative, or punitive? Safe spaces in Bosnia-Herzegovina, Somalia, and Sri Lanka. *Journal of Refugee Studies* 16: 167–185.

Ide, T. 2019. The impact of environmental cooperation on peacemaking: Definitions, mechanisms, and empirical evidence. *International Studies Review* 21: 327–346.

Ide, T., L. Palmer and J. Barnett. 2021. Environmental peacebuilding from below: Customary approaches in Timor-Leste. *International Affairs* 97: 103–117.

Inwood, J. and J. Tyner. 2011. Geography's pro-peace agenda: An unfinished project. *ACME: An International Journal for Critical Geographies* 10: 442–457.

Jessop, B., N. Brenner and M. Jones. 2008. Theorizing sociospatial relations. *Environment and Planning. D, Society and Space* 26: 389–401.

Jones, O. 2011. Geography, memory, and non-representational geographies. *Geography Compass* 5: 875–885.

Jones, R. 2004. What time human geography? *Progress in Human Geography* 28: 287–304.

Jones, R. 2016. *Violent Borders: Refugees and the Right to Move.* Verso.

Kaplan, O. 2017. *Resisting War: How Communities Protect Themselves.* Cambridge University Press.

Katz, C. 2001. On the grounds of globalization: A topography for feminist political engagement. *Signs: Journal of Women in Culture and Society* 26: 1213–1234.

Katz, C. 2004. *Growing Up Global: Economic Restructuring and Children's Everyday Lives.* University of Minnesota Press.

King, B. 2010. Conservation geographies in Sub-Saharan Africa: The politics of national parks, community conservation and peace parks. *Geography Compass* 4: 14–27.

Kirsch, S. and C. Flint, eds. 2011. *Reconstructing Conflict: Integrating War and Post-War Geographies.* Ashgate.

Kliot, N. and S. Waterman, eds. 1991. *The political geography of conflict and peace.* Belhaven.

Koopman, S. 2011a. Let's take peace to pieces. *Political Geography* 30: 193–194.

Koopman, S. 2011b. Alter-geopolitics: Other securities are happening. *Geoforum* 42: 274–284.

Koopman, S. 2014. Making space for peace: International protective accompaniment in Colombia. In *Geographies of Peace*, eds. F. McConnell, N. Megoran and P. Williams, 109–130. I.B. Tauris.

Koopman, S. 2019. Peace. In *Keywords in Radical Geography: Antipode at 50*, eds. Antipode Editorial Collective et al., 207–211.

Koopman, S. 2020. Building an inclusive peace is an uneven spatial process: Colombia's differential approach. *Political Geography* 83.

Lacoste, Y. 1976. *La géographie, ça sert, d'abord, à faire la guerre.* Maspero.

Laliberté, N. 2016. 'Peace begins at home': Geographic imaginaries of violence and peacebuilding in Northern Uganda. *Political Geography* 52: 24–33.

Le Billon, P. 2001. The political ecology of war: Natural resources and armed conflicts. *Political Geography* 20: 561–584.

Le Billon, P. 2008. Diamond wars? Conflict diamonds and geographies of resource wars. *Annals of the Association of American Geographers* 98: 345–372.

Le Billon, P., M. Roa-García and A. López-Granada. 2020. Territorial peace and gold mining in Colombia: Local peacebuilding, bottom-up development and the defense of territories. *Conflict, Security and Development* 20: 303–333.

Lefebvre, H. 1991. *The Production of Space.* Blackwell Publishing.

Leitner, H., E. Sheppard and K. Sziarto. 2008. The spatialities of contentious politics. *Transactions of the Institute of British Geographers* 33: 157–172.

Low, S. and M. Maguire, eds. 2019. *Spaces of security: Ethnographies of securityscapes, surveillance, and control.* New York University Press.

Loyd, J. 2012. Geographies of peace and antiviolence. *Geography Compass* 6: 477–489.

Loyd, J. 2014. *Health Rights are Civil Rights: Peace and Justice Activism in Los Angeles, 1963–1978.* University of Minnesota Press.

Mac Ginty, R. 2011. *International Peacebuilding and Local Resistance: Hybrid Forms of Peace.* Palgrave Macmillan.

Mac Ginty, R. 2013. Indicators+: A proposal for everyday peace indicators. *Evaluation and Program Planning* 36: 56–63.

Mac Ginty, R. 2014. Everyday peace: Bottom-up and local agency in conflict-affected societies. *Security Dialogue* 45: 548–564.

Mac Ginty, R. 2021. *Everyday Peace: How So-Called Ordinary People Can Disrupt Violent Conflict*. Oxford University Press.

Mac Ginty, R. and O. Richmond. 2013. The local turn in peace building: A critical agenda for peace. *Third World Quarterly* 34: 763–783.

Macaspac, N. 2017. Suspicion and ethnographic peace research (notes from a local researcher). *International Peacekeeping* 25: 677–694.

Macaspac, N. 2018. Insurgent peace: Community-led peacebuilding of indigenous peoples in Sagada, Philippines. *Geopolitics* 24: 839–877.

Macaspac, N. 2022. Spatialities of peace zones. *Cooperation and Conflict*.

Macaspac, N. and A. Moore. 2022. Peace geographies and the spatial turn in peace and conflict studies: integrating parallel conversations through spatial practices. *Geography Compass* 16: e12614.

Mackinder, H. 1911. The teaching of geography from an imperial point of view, and the use which could and should be made of visual instruction. *The Geographical Teacher* 6: 79–86.

Mamadouh, V. 2005. Geography and war, geographers and peace. In *The Geography of War and Peace: From Death Camps to Diplomats*, ed. C. Flint, 26–60. Oxford University Press.

Massey, D. 2005. *For Space*. Sage.

Megoran, N. 2010. Towards a geography of peace: Pacific geopolitics and evangelical Christian Crusade apologies. *Transactions of the Institute of British Geographer* 35: 382–398.

Megoran, N. 2011. War and peace? An agenda for peace research and practice in geography. *Political Geography* 30: 178–189.

Megoran, N. 2017. *Nationalism in Central Asia: A Biography of the Uzbekistan-Kyrgyzstan Boundary*. University of Pittsburgh Press.

Megoran, N. and S. Dalby. 2018. Geopolitics and peace: A century of change in the discipline. *Geopolitics* 23: 251–276.

Millar, G. 2021. Coordinated ethnographic peace research: Assessing complex peace interventions writ large and over time. *Peacebuilding* 9: 145–159.

Mitchell, C. and L. Hancock, eds. 2012. *Local Peacebuilding and National Peace: Interaction between Grassroots and Elite Processes*. Continuum.

Moore, A. 2013. *Peacebuilding in Practice: Local Experience in Two Bosnian Towns*. Cornell University Press.

Moore, A. 2016. Ethno-territoriality and ethnic conflict. *Geographical Review* 106: 92–108.

Nagar, R., V. Lawson, L. McDowell and S. Hanson. 2002: Locating globalization: Feminist (re)readings of the subjects and spaces of globalization. *Economic Geography* 78: 257–284.

O'Lear, S. 2016. Climate science and slow violence: A view from political geography and STS on mobilizing technoscientific ontologies of climate change. *Political Geography* 52: 4–13.

Orchard, P. 2014. Revisiting humanitarian safe areas for civilian protection. *Global Governance* 20: 55–75.

Paffenholz, T. 2015. Unpacking the local turn in peacebuilding: A critical assessment towards an agenda for future research. *Third World Quarterly* 36: 857–874.

Pandey, R. and D. Bardsley. 2015. Social-ecological vulnerability to climate change in the Nepali Himalaya. *Applied Geography* 64: 74–86.

Peet, R. and M. Watts. 2004. *Liberation Ecologies: Environment, Development, Social Movements*. 2nd ed. Routledge.

Peluso, N. and M. Watts, eds. 2001. *Violent Environments*. Cornell University Press.

Pepper, D. and A. Jenkins, eds. 1985. *The Geography of Peace and War*. Blackwell.

Pouligny, B. 2006. *Peace Operations Seen From Below: UN Missions and Local People*. Kumarian Press.

Pratt, G. 2004. *Working Feminism*. Temple University Press.

Pullan, W. 2013. Conflict's tools. Borders, boundaries, and mobility in Jerusalem's spatial structures. *Mobilities* 8: 125–147.

Raleigh, C. and H. Urdal. 2007. Climate change, environmental degradation and armed conflict. *Political Geography* 26: 674–694.

Ross, A. 2011. Geographies of war and the putative peace. *Political Geography* 30: 197–199.

Scott, J. C. 1985. *Weapons of the Weak: Everyday Forms of Peasant Resistance*. Yale University Press.

Smirl, L. 2015. *Spaces of Aid: How Cars, Compounds and Hotels Shape Humanitarianism*. Zed Books.

Smith, R and K. Rhiney. 2016. Climate (in)justice, vulnerability and livelihoods in the Caribbean: The case of the indigenous Caribs in northeastern St. Vincent. *Geoforum* 73: 22–31.

Soja, E. 1989. *Postmodern Geographies: The Reassertion of Space in Critical Social Theory*. Verso.

Springer, S. 2016. *The Anarchist Roots of Geography: Toward Spatial Emancipation*. University of Minnesota Press.

Staeheli, L.A., E. Kofman and L. Peake, eds. 2004. *Mapping Women, Making Politics: Feminist Perspectives on Political Geography*. Routledge.

Stokke, K. 2009. Crafting liberal peace? International peace promotion and the contextual politics of peace in Sri Lanka. *Annals of the Association of American Geographers* 99: 932–939.

Toal, G. and C. Dahlman. 2011. *Bosnia Remade: Ethnic Cleansing and Its Reversal*. Oxford University Press.

Ware, A. and V. Ware. 2022. Everyday peace: Rethinking typologies of social practice and local agency. *Peacebuilding* 10: 222–241.

Watts, M. 2001. Violent geographies. Speaking the unspeakable and the politics of Space. *City and Society* 13: 85–117.

Watts, M. 2013. *Silent Violence: Food, Famine and Peasantry in Northern Nigeria*. 2nd ed. University of Georgia Press.

Williams, P. 2015. *Everyday Peace? Politics, Citizenship, and Muslim Lives in India*. John Wiley and Sons, Inc.

Williams, P. and F. McConnell. 2011. Critical geographies of peace. *Antipode* 43: 927–931.

Woon, C. 2011. Undoing violence, unbounding precarity: Beyond the frames of terror in the Philippines. *Geoforum* 42: 285–296.

Woon, C. 2017. Children, critical geopolitics, and peace: Mapping and mobilizing children's hopes for peace in the Philippines. *Annals of the Association of American Geographers* 107: 1–18.

Yamashita, H. 2004. *Humanitarian Space and International Politics: The Creation of Safe Areas.* Ashgate.

Yorgason, E. 2011. The normative commitments behind peacemaking. *Political Geography* 30: 200–201.

4

SPATIALIZING PEACE AND PEACEBUILDING

Where is knowledge about peace and peacebuilding produced?

Annika Björkdahl

Introduction: Taking the spatial turn to research peace and peacebuilding agency

War and peace always take place somewhere, and peace and conflict scholars have long assumed that they can localize peace and situate it in space (and time). Often it is also assumed that spaces where peace is present, war is absent. However, close analyses of conflict dynamics and peace processes reveal that in the midst of conflict there are islands of peace and in times of peace there are outbreaks of violent conflict (Macaspac 2018; Saulich and Werthes 2020). This means that peace and war often co-exist in time and space. The ambiguousness of these post-conflict spaces has been described by Richards (2005) as no peace, no war, and by Mac Ginty (2006) as no war, no peace. Flint (2011, 31) discloses that peace and war are intrinsically intertwined manifestations of dynamic social processes and cannot be treated separately. Thus, the space-time connections need to be considered to understand societies transitioning from war. However, to capture where peace is being built, peace scholars need to upgrade the importance of spatiality to offer a perspective in which space is recognized as intimately involved in the production of peace.

Because space as an analytical tool provides new and important insights into the relationship between processes of peace and spatiality, such analysis of peace is steadily gaining traction in the field of peace and conflict studies. As an innovative approach, it focuses on how peace is spatially constructed, and the mutual construction of spaces and agency in a field that has thus far merely considered space as a backdrop against which war and peace unfold. Theoretically and conceptually borrowing from peace geography and anthropology the spatial turn in peace and conflict research brings to the fore the purchase of

DOI: 10.4324/9781003345794-4

employing space and place as analytical vehicles to theorize peace and peace-building. It thus contributes to the growing transdisciplinary research which explores the making geographies of peace.

To demonstrate the impact of spatial approaches on peace research, this chapter explores some of the novel understandings of peace informed by spatial approaches, such as corporeal peace, everyday peace, urban peace, mobile peace, and transscalar peace. Moreover, it makes visible the spatial practices of peacebuilding agency in space- and place-making processes that transform spaces and places of violent conflict and war. It concludes by mapping the contributions of transdisciplinary peace research in several subfields, such as urban studies, migration studies, memory studies, gender studies, and novel critical questions raised about where knowledge about peace and conflict is produced.

Overview of the spatial turn in peace studies

Since the end of the Cold War, the overarching goal of international peace-building efforts in societies transitioning from war has been to implement the idea of the liberal peace. The idea of the liberal peace builds on the democratic peace thesis, which suggests that democracies do not go to war with each other because democracy ensures that domestic politics within states are peaceful, and that together with free trade, it also ensures that states do not go to war with each other (Richmond and Franks 2009). This has been the dominant intellectual framework of the post–Cold War era informing policies and practices of peacebuilding intervention in post-conflict societies. However, a key conclusion of critical peace analysis, which has gained significant traction in last decades, is that the abstract notion of "liberal peace" often fails to take root in societies transition from war (Campbell, Chandler and Sabaratnam 2011). One reason is that this liberal notion of peace is not localized, does rarely reflect local understandings of peace, and rarely are the voices of the citizens heard in the politics shaping the peace and the post-conflict space. This tends to create "peace gaps," or shortfalls between internationally brokered peace accords and local expectations of the peace dividend (Mac Ginty 2008).

Consequently, peacebuilding scholarship shifted focus from investigating international, top-down peacebuilding to studying local, bottom-up peace-building (Richmond 2009; Mac Ginty 2011; Moore 2008). The "local" became regarded as a savior for peacebuilding, by providing local legitimacy and own-ership of international peacebuilding, and by localizing peace, thus providing a fast track to success, and an easy exit from the intervention space (Donais 2009; Richmond 2011; Mac Ginty 2015). The trend to enhance the importance of local spaces and local agency was referred to as the "local turn" in peace research (Mac Ginty and Richmond 2013; Leonardsson and Rudd 2015). Yet, recently there has been a backlash against what is seen as research and practices romanticizing the local assuming that local peace initiatives are in-herently inclusive and harmonious (Wiegink 2018). Based on fieldwork and analysis of peacebuilding Solomon Islands and Timor Leste, Richmond (2011) concludes that the international peacebuilding needs to be "de-mystified" and the local "de-romanticized" to rethink the cooperation between the local and the international. Mac Ginty (2015) for example, separated the concept of the local from territory and viewed it in terms of activity, networks, and relation-ships and deterritorialized.

Yet, the dichotomy as well as a homogenized understanding of "the local" and "the international" is often maintained (Paffenholz 2015; Buckley-Zistel 2016). As Heathershaw argues, the local turn seems unable to capture the co-constitution between the local and the international and the ways that "the local is partially produced by what internationals find, initiate or are willing to fund" (Heathershaw 2013, 280 cited in Brigg 2020). This critique has sparked interest in exploring frictions between this "global" notion of peace and locally constructed ideas of peace(s) (Björkdahl and Höglund 2013; Björkdahl et al. 2016), and the hybrid outcomes of the interplay between international and lo-cal visions of peace (Mac Ginty 2010; Richmond 2014). Since the critical, local peacebuilding agenda was set out by Mac Ginty and Richmond (2013), more peace scholars have come to engage with peace in relation to space, place, and scale. Consequently, spatial categories are now referred to in peace and conflict research, but most studies do not fully utilize the potential of spatial concepts for the analysis, and they do not analyze space or place itself in a theoretically informed way (cf. Chojnacki and Engels 2016, 12).

So, to spatialize conflict and peace research and to fully make use of the analytical tools available, peace scholars have turned to geography for insights and inspiration (Flint 2004; Thrift 2003; Cresswell 2004). The three key con-cepts – space, place, and scale – that are most frequently borrowed from geog-raphy, assist in much needed analysis of spaces for peace and peacebuilding. Studying peace through the lens of space thus requires attending to the ways multiple spatialities such as scale, space, and place are co-implicated in build-ing sustainable peace (Macaspac 2023). A central contention of the spatial turn is that the organization of space is significant for the structure and func-tion of peace as well as war (Flint 2004). Warf and Arias (2008) point to an understanding of space that exceeds its materiality by stressing its significance for social relations, thus also for peace and conflict.

Yet, the relationship between peace and space is not a linear, causal one (Björkdahl and Kappler 2017). Instead, the spatial dimensions of peace re-quire a relational and contextual reading to appreciate the nonlinear ways in which peace impacts spatiality. Thus, space is both the product of and the context through which peace is built and negotiated across scale, articulated through different forms of peace narratives, and informed by uneven geogra-phies of power. Understanding space as a sphere of multiplicity and co-existing heterogeneity (Massey 2005) makes visible peace(s) and conflicts in the same space and time, because "peace(s) are always shaped in and through spaces and times through which they are made" (Koopman 2014, 197). Thus, there is a growing consensus among the spatial turn scholars that space and peace are co-constitutive. The spatial turn also makes use of place as an analytical category arguing that without being based in concrete places peace is simply an abstraction. Thus, peace conceived as a set of place-specific processes (Macaspac and Moore 2022) on a small scale seems more contextualized, tangible, and achievable. Place is thus seen a material, physical and bounded entity, a location, representing where peace is emplaced (cf. Cresswell 2004).

Recently, spatial practices, spatial dynamics, and space formations have been explored as key entry roads for understanding the constitution of spaces for peace (Björkdahl and Buckley-Zistel 2022). Spatial practices capture the agents and activities through which spaces are created, transformed, or dissolved in relation to peace and conflict. Spatial practices make social space happen; they produce social realities. As expressed by Macaspac and Moore (2022, 6): "[s]patial practices influence the dynamics of peace and conflict in profound ways, from daily, localized, and embodied actions to spatial framings and narratives that inform organizational behaviors and peace agreements." Spatial practices then reveal the everyday activities through which spaces of peace can be created, transformed, or dissolved. Hence, researching spatial practices brings fore notions such as identity, exclusion, segregation, belonging, territoriality, and cognitive space to understand peace and conflict as an emplaced practice. Spatial practices (peace acts) are thus constitutive of spatial dynamics. Spatial dynamics relate to the social construction of spaces. Understanding spatial dynamics helps challenge a static understanding of spaces of peace, and demonstrate that such spaces are socially constructed, and relational regarding other spaces. Space formations ask what form of spatiality is involved, i.e., how space matters and manifests itself. By linking these spatial practices, spatial dynamics, and space formation into a loosely fit framework we find that spatial formations reveal underlying assumptions of spaces of peace.

Thinking of peace in spatial terms makes visible peace as a contested spatial process composed of practices and actors at multiple scales and opens up the possibility that peace can move between different scales, such as the local and the global. However, rarely are such processes exclusively local or global, but rather glocal, as the local and the global are mutually constitutive. Thus globalization suggests the possibility that processes of peacebuilding simultaneously construct new political circumstances at different scales, such as the local and global (Swyngedouw 1997). Inspired by the concepts of scale, scalar politics (Megoran and Dalby 2018) research on peacebuilding is paying attention to the social construction of spatial scales as a way of governing peace.

A scalar approach to peace offers a means to establish where peace occurs and to uncover how various agencies strive to define and possibly change the politics of peace at different scales (Moore 2008). Spatial scales are the outcome of social practices. Thus, it is possible to envision multiple spaces of peace and ways in which both conflict dynamics and peace practices interact at various scales (Bank and Van Heur 2007, 596). Consequently, peacebuilding unfolds in different spaces and places, and on different scales, which are not ontologically given. Where peacebuilding takes place reflects assumptions about which scale is best suited for the implementation of particular policies of peace (cf. McCann 2003, 163) as well as about the types of agents who have the right to reshape and transform a place (McCann 2003, 172). The establishment of scales is in and of itself a political process impregnated with power (McCann 2003, 160). Attempts to rescale peacebuilding from one level to another, and the resistance against this process, can all be considered spatial practices of power (Buckley-Zistel 2016). Inspired by the concepts of scale, scalar politics (Moore 2008; Megoran and Dalby 2018) and the notion of transscalarity (Scholte 2014), the term "transscalar peace" has emerged as useful tool to investigate peacebuilding across various spaces and places. Peace practices that take place at multiple scales at once and peacebuilding process that intersect with simultaneous processes at other scale can be seen as transscalar peace (cf. Millar 2020, 2021a, 2021b; Hellmüller 2022). Thus, this approach allows us to view spaces of peace as situated in transscalar networks of agency, and we can move beyond the binary opposition of the local as a space needing peacebuilding and the global as a space providing peacebuilding.

Spatial concepts of peace

Peace plays out in the multiple spaces and places in which people live and move. To understand this better critical peace research has engaged in conversations with peace geography, anthropology, international relations, urban studies, migration studies, and memory studies. Efforts to unpack peace by "taking peace into pieces, not just by place, but by space" have helped to think a new about the geographies of peace (Koopman 2011).

Traditionally, International Relations scholars viewed the traditional Westphalian state as the main, and often only, point of reference for peace. Recent theorizing, however, has described the state withering as an effective political space. There is now a belated realization that new peace theories are necessary, which calls for a rethinking of the state in peacebuilding and to search for peace dynamics beyond the state. Critical peace research has long challenged the state-centrism of peacebuilding and suggests a rescaling of peace research to map and investigate peace where it takes place. Thus, novel terms and understandings of peace have emerged; emplaced peace (George 2020), mundane peace, the peaceable body, corporeal peace (Väyrynen 2019), mobile peace

(Mannergren Selimovic 2022; Richmond and Mac Ginty 2019), shared space for peace (Dempsey 2022), everyday peace (Williams 2015; Mac Ginty 2021, O'Driscoll 2021), urban peace (Bollens 1999; Björkdahl 2013; Gusic 2022), as well as notion of transscalar peace (Hellmüller 2022; Millar 2020).

Taking the urban as a point of reference makes it possible to study urban peace and peacebuilding. Urban peacebuilding is not peacebuilding only adapted to scale, but it focuses on material places and structures, urban politics and institutions, and urban dynamics. It involves place-based and people-related processes such as, mediating intergroup competition over territory, construct safe shared public spaces and limiting spatial expressions of conflictual discourses. All in all, it aims to build an urban peace and to realize the peace dividend to the urban dwellers (Bollens 1999; Björkdahl 2020). Urban peacebuilding research reveals that the formation of space is central to the structure and function of urban peace (Björkdahl 2013; Forde 2019). The relational combined with the spatial is at the core of an understanding of urban peace, and it means that the political becomes connected to the spatial. The notion of urban peace is about a sociospatial relation that is always made and re-made. It is closely aligned with plural forms of peace emplaced in the urban space. Gusic (2022, 2) understands urban peace to be attuned to city spaces. These spaces are multifaceted and flexible, and are envisioned to enable coexistence between starkly diverging peace(s), including polemic peace(s).

To rethink peace in terms of the mundane and the ordinary has turned the everyday into a site of peace analysis (Mannergren Selimovic 2019; Mac Ginty 2021; Richmond 2009, 2010). Richmond (2010, 670) defines the everyday as "a space in which local individuals and communities live and develop political strategies in their local environment, towards the state and towards international models of order…it is representative of the deeper local-local … often transversal and transnational." Thus, the notion of everyday peace is centered on practices that are not performed by top-level politicians in peace negotiations or international mediators, but emplaced actions of individuals and communities (cf. Williams 2015; Mac Ginty 2021). Such peace is socially produced and reproduced through everyday practices. A series of social practices involving sociality, reciprocity, and solidarity influence "the (re)production of peace on the ground" (Williams 2015, 32). However, for "everyday peace to be significant it relies on the scaling up of entanglements between individuals and small groups so that they will have a broader effect across the community and society," Mac Ginty concludes (2021, 77).

Taking a different look at the everyday, feminist geographers demonstrate how everyday spaces are gendered, and bound into the power structures that confine women (Rose 1993). Fear, insecurity, and threats of violence are forces that structure women's spatial practices and everyday use of space. Thus, women's interpretation and experience of both public and private space are significantly different from those of men (Mackenzie and Rose 1983). This is key to understanding why peace often fails to materialize in women's everyday experiences. However, when women can take possession of space, their everyday spatial practices may help avoid conflict and build peace (Koskela 1997). Such peace building is the carrying out of daily, embodied activities which perpetuates the relationships and practices of everyday life.

To deepen the analysis of peace we zoom in on the body, because "war and peace are socio-political institutions that begin and end with the body" (Mannergren Selimovic 2022, 585). Thus, rescaling peace to the body produces an understanding of peace as corporeal, i.e., the embodied peace and of the peaceable body. Attention to bodies and embodied practice shows that peace is a lived experience embedded in a notion of embodiment. Approaching peace through the lens of feminist peace research Tarja Väyrynen (2019) places the body front and center in theorizing peacebuilding. She advances a corporeal peace and argues that the body is vital to peace research as it can be seen to be a nexus through which the potentialities for emancipatory politics, e.g. sustainable peace, actualize. Corporeality provides an understanding of the body as both a subject and an object of peacebuilding policies and practice. Peace is an event that emerges in mundane and corporeal encounters between strange and familiar bodies (Wilson 2017). Corporeal peace brings together affect, emotions, and the somatic (Väyrynen 2019; Mannergren Selimovic 2019). Moreover, this lens reveals how peace and war touch our bodies in multiple ways and how the body is constitutive of war and peace and help to unpack transgenerational memory, silence, post-conflict governance of subaltern bodies as well as to think through issues such as agency, temporality and spatiality. So, in a way, people and their embodied experience repopulate the discourse of peacebuilding (cf. Berents 2015).

Let us now turn to the notion of mobile peace, which challenges the idea that peace take place somewhere and can be emplaced, as it captures the implications of mobility and immobility on patterns of peace (Mac Ginty and Richmond 2013). Mobile peace as a concept brings to the fore a mobile form of peace forged and embodied through everyday actions. Thus, the mobile political agency is connected to mobilities and may come in the form of the movement of migrants from conflict-affected zones to Europe. Mannergren Selimovic (2022) provides a spatiotemporal rethinking of peace research that challenges the Global North's spatial imaginations of war "there" and peace as "here" and by employing the concepts of streams of violence and streams of peace she theorizes and reconfigures where and when peace takes place.

As much as rescaling to the body, the everyday, the urban, have been completely necessary, scholars also recognize that the field cannot only focus on the place-specific, localized, bottom-up peace practices and processes and exclude the regional and global spaces, and international actors and practices. A series of international crises and conflicts in the first two decades of the 21st century, and a worrying number of new and old challenges to peace, evidence the continuing need for peacebuilding to overcome both latent and overt conflict. Indeed, as Millar (2021a) argues, complex dynamics produce global systems of conflict that incentivize violence across scales. Contemporary understandings of conflicts as intra- or interstate, or as internationalized fail to capture the truly complex global dynamics that drive, inspire, and exacerbate specific cases of direct violence at different scales or that transcends scale. Such global conflict dynamics will demand peace activities not only within, but across scales. A few important and ambitious theoretical approaches have been launched, appropriate for addressing today's challenges (Richmond 2022; Millar 2021a). Richmond (2022) traces the evolution of the contemporary, layered peace architecture and provides insights to the complex relationship between power and peace at different scales, while exposing counter-peace processes, agents, and frameworks that act as blockages of peace. To address contemporary conflicts the international peace architecture needs to be strengthened, not dismantled, and to connect to peace(s) produced on several different scales and to connect with transscalar peace dynamics. In contrast, Millar (2020, 2021a, 2021b) argues that the contemporary peace system at best produces a negative peace, at worst contributes to global conflict dynamics. To move beyond the negative peace, Millar develops a transscalar peace system to address structural violence and inequalities across scale and to build a positive peace.

Peacebuilding agency and the spaces and places of peace

Agency can be expressed in various ways, and it is situated in diverse enabling or disabling spaces and places (Giddens 1984; Flint 2003). By studying where peace takes place it is possible to show how spaces and places of peace are made meaningful through agency. Thus, peacebuilding agency is constituted by and constitutes spaces and places of peace, and spatial practices are expressions of peacebuilding agency (Björkdahl and Kappler 2017, 2018). Peacebuilders, peacekeepers, and peace activists, and people going about their everyday lives, as well as armies, rebel groups and militias can be regarded as "spatial entrepreneurs" (Engel 2020, 24). The spatial entrepreneurs of interest here are the peacebuilding agents that construct, maintain, challenge, and dissolve various spaces for peace, i.e., buffer zones, security corridors, safe areas, peace zones, peace spaces, and everyday spaces of peace, through a broad spectrum of spatial practices (Higate and Henry 2010; Vogel 2018; Lemay-Hébert 2018; Mannergren Selimovic 2019).

Recent research in peace and conflict studies explores the agentive practices of peacekeepers and peacebuilders in post-conflict spaces. Contained within the international community's approach to post-conflict spaces is the assumption that places can be reconstructed, states made, and peace built, and that peacekeeping practices produce territoriality and reconfigure post-conflict spaces in ways that also shape perceptions of security. Engel (2020) for example, investigates the spatializing effects of the African Union's (AU) attempts to address the scourge of terrorism and violent extremism and to (re-establish) the territorial sovereignty of its member states. The intended and unintended consequences of the AUs spatial practices affect the way social processes relating to peace and security on the continent are spatialized (Engel 2020). Higate and Henry's (2010) work is another case in point, and it demonstrates how UN peacekeepers demarcation of space into zones and enclaves transform spaces and reveals the implications for local populations. In a similar vein, Lemay-Hébert (2018) uncovers how the security mapping practices of the UN peace mission to Haiti were part of the securitization of the everyday that structured the peacekeepers' relations with the local population. Such analysis of agency in conflict zones reveals the contradictory mix of insecurity and security that peacekeepers produce in spaces of intervention, and the political nature and consequences of their spatial security-seeking practices, including their production of the material and ideational structures of intervention spaces and places (Brigg and George 2020; Wallis 2020).

In the margins hidden agents also seek to shape space to resist violence, manage conflict, and build peace (Heathershaw and Lambach 2008). For example, peace activists appropriate space for peace activism (Vogel 2018). In such peace spaces marginalized voice of peace are enabled and can "challenge hegemonic discourses of violence prevalent in (post-) conflict spaces" (Vogel 2018, 432). A case in point is the peace activists that appropriate parts of the physical locale of the Buffer Zone to contribute to peacebuilding. Particularly, the Home for Cooperation located in the Buffer zone is such enabling peace space built by various agents that collectively make up the bi-communal peace movement of Cyprus. However, the bi-communal movement seems locked into their peace space, and as many Cypriots show a reclining interest to interact with the other, the movement fails to expand this space (Hadjipavlou 2012; Demetriou and Hadjipavlou 2016).

Agency is not just the preserve of peace activists, in particular, but is seen in the actions of people and their everyday activities that may support peace. Clearly, "agents that make up the micropolitics of the everyday," and as research shows, such agents are able to overcome obstacles to peace (Mannergren Selimovic 2019, 132). This agency is grounded in the lived space, embodied and contingent on relationship with others. It is a subtle form of agency that is not necessarily purpose-driven, but may "shift and slide from unintentional and *ad hoc* to intentional and organized" (Mannergren Selimovic 2019, 134). Palestinians of East Jerusalem use their agency to find employment or to have a family life, and in doing so defy the security barrier and the politics of division. Although their agency may not change the dynamics of the Israel–Palestinian conflict, it may contribute to building everyday peace in the midst of conflict. Another illustration of people becoming agentive subjects is evident through participation in peace marches. For example, the Peace March commemorating the Srebrenica genocide starts on the first Sunday in July in the village of Nezuk to arrive, three days later, at Potočari Memorial Center. The walking procession follows the reverse journey of the 100 kilometers, so-called, Death March that around 15,000 Bosniaks embarked on after fleeing the Srebrenica safe area in July 1995 to try to reach the liberated territories under the control of the Army of the Republic of Bosnia and Herzegovina near Tuzla (Ljubojević 2022). To commemorate is a means of manifesting and emplacing peace, which also reinterprets and reworks the notion of mobile peace. While studying movement and embodiment of memory through engagement with space and place, the peace march can be interpreted as a way of spatializing both peace and memory through walking.

Building peace: Space-making and place-making

We know that peacebuilding agency can be expressed in spatial terms. In exploring peacebuilding agency, we have in previous research demonstrated the spatial component of peacebuilding and situate peacebuilding agency between space and place (Björkdahl and Kappler 2017, 2018). Peacebuilders' spatial practices such as space-making and place-making create, transform, or dissolve space and places in relation to peace and conflict.

Place is where peace and conflict affect people in their everyday. Place can be understood as a material, physical and bounded entity, a location. It defines who is "in place" and who is "out of place" (Cresswell 1996, 2004). Place may stabilize and provide durability to social structures such as peace and rearrange post-war power relations. Place can sustain both imagined peace(s) and the materialized everyday peace by daily practices. Space, instead, refers to the ideational, symbolic counterpart of place, its relational qualities and the meanings associated with it. It denotes the meanings ascribed to any spatial practice. In that it is a product of social interactions and their associated understandings of a particular place (Björkdahl and Kappler 2021).

In human geography, place-making is seen as a "set of social, political and material processes by which people iteratively create and recreate the experienced geographies in which they live" (Pierce et al. 2011, 54). Building on this, research peace and conflict studies employ the concept of place-making to the process through which a material presence is given to an ideational space (Björkdahl and Kappler 2017). In the context of peacebuilding, this is the materialization of the idea of peace(s). Clearly, a place is not neutral, but reflects power. A conflict-affected society may become an arena where a dominant idea of peace may be enforced, but also a ground for the contestation of that idea (Gusic 2022; Olivius and Hedström 2021). Place-making is an inherently networked process, constituted by the socio-spatial relationships that link individuals together through a common place-frame. Thus, peace is understood as emplaced and constituted and populated by the way people experience the place. This place-making is an expression of agency and reflects the power relations of the postwar society.

Space-making, on the other hand, refers to the creation of a symbolic, ideational counterpart of material place-making (Björkdahl and Kappler 2017). It is the ability to turn a physical place into an ideational, discursive political space giving it new meaning. With respect to peacebuilding, it extends to the creation of peace. Thus, space-making processes make physical places relevant and useful for social and political processes (Björkdahl and Kappler 2017). Both place-making and space-making are conceptualized as expressions of agency. Place-making and space-making are processes that are not limited to the scale of the nation-state, but reach across different scales of analysis. Place-making processes often comprise an interplay of local, national, and international actors (Björkdahl and Kappler 2018). In a similar vein, Ulf Engel (2020, 223) suggests that "the making of spatial order out of competing spatial formats is a dialectic process of imagination and materialization." Also Heathershaw and Lambach (2008) make use of the concept space-making in their analysis of transformation of post-conflict spaces and to demonstrate that state-making is about space-making. The processes of place-making and space-making are thus not the same, but may have to take place jointly if peace is to emerge (Olivius and Hedström 2021).

Advancing this research on place-making and space-making further, Olivius and Hedström (2021) employ place-making and space-making to provide a detailed empirical account of how competing visions of peace have shaped, and been mobilized in, the conflict around the Aung San statue in Myanmar's Kayah state. They demonstrate that the meanings ascribed to such places are bound to be challenged by other narratives and ideas that are subsequently projected onto them. Physical places, and the ideational spaces associated with them, can be appropriated and inhabited in multiple ways. Moreover, Olivius and Hedström (2021) highlight the importance of access to different forms of power and resources. Their research demonstrates that place-making requires a position of institutional political power as well as funding. In such processes actors such as state governments and international peacebuilders often become key actors. Whilst the making of place expresses dominant power relations and political visions, these very places may also become sites for mobilization where their meaning can be contested and changed. Space-making, Olivius and Hedström (2021) argue, is a form of agency deployed "from below."

The place-making and spatial agency challenging dominant power relations is also visible in specific places that represent the legacy of the war, as, for example, in Višegrad, Bosnia-Herzegovina. In the everyday use of the bridge across the river Drina the absence of the extraordinary may be the most striking. There is no memorial plaque on the bridge, nor any other physical reminder of the atrocities committed on the bridge during the war that disrupted the relationship between place and belonging in eastern Bosnia-Herzegovina.

Once a year, displaced Bosniaks return to Višegrad and appropriate the bridge to remember their dead relatives and family members and their presence on the bridge breaks the silence upheld by the Bosniak-Serb majority now inhabiting Višegrad. At the commemoration 3,000 red roses are thrown into the Drina; one rose for each person believed to have been killed in Višegrad and its surroundings. This temporary transformation of the bridge reveals competing claims of belonging and competing notions of peace(s) in the post-conflict town. The returning Bosniaks invest the bridge with meaning, belonging, and identity, and their spatial practices emplace an agonistic peace in Višegrad (Björkdahl and Mannergren Selimovic 2016).

A novel addition to space-making in the everyday is couchsurfing as it may turn a home into a space for peace. Peer-to-peer accommodation, such as couchsurfing, is an intimate spatial activity where one opens one's home to strangers. In societies where prolonged ethnic conflict affects most aspects of everyday life, such as Cyprus inviting couchsurfers representing "the other" to spend a night on one's couch is in a sense turning one's home into a space for peace. Couchsurfing can be read as a space-making in the everyday as it brings together members of the intercommunal Cyprus-based network from across the island's divide into each other's homes. Such spatial practice transcends material and sociopolitical barriers, and in Cyprus couch-surfing seems to foster positive intercultural relations (Antoniou 2022).

In sum, space-making and place-making adds to the critical debate about materialization of peace and the emplacement of peacebuilding agents. To situate peacebuilding agency in space and place, and to link spatial practices to expressions of agency, the spatial turn contributes to re-theorizing agency.

Where does knowledge production of peace take place?

Peace needs to be understood as "situated knowledges" embedded within specific temporal, spatial, and political settings (Williams and McConnell 2011, 929). Critical research addresses the important question where knowledge about peace and conflict is produced. Here transdisciplinary research is essential to advance the knowledge production about spaces of peace. The spatial turn in peace research connects to calls for situated knowledge of peace where more research on different space, places, and scales is needed to demonstrate how peace is differently constructed, materialized, and interpreted through space and time.

Peace research often focuses on the Global South because peacebuilding is often seen as taking place in the Global South. There are numerous case studies of international peacebuilding in Namibia, East Timor, and DR Congo (Paris 2004; Autesserre 2014), and of everyday peacebuilding in locales in sub-Saharan countries; for example Sri Lanka (Firchow and Mac Ginty 2017, 2020), India (Williams 2015), Colombia (Berents 2015), Iraq (O'Driscoll 2021).

Local or emplaced knowers have so far limited influence on the academic knowledge production. Studies of peacebuilding in, for example, Northern Ireland, Cyprus, Kosovo, Bosnia-Herzegovina, and Finland, on the other hand, demonstrate that knowledge production is emplaced, and this knowledge is very present in the canon of peace research cited (Mac Ginty 2009; Visoka 2016, 2017; Hadjpavlou 2012; Demetriou and Hadjipavlou 2016; Simic 2012; Väyrynen 2019).

Clearly, specific and emplaced knowledge and expertise are important to advance our understanding of where peace takes place. Mannergren Selimovic (2019, 136) suggests that corporeal experiences are expressions of specific knowledge and expertise. Her study of East Jerusalem shows that Palestinians hold useful everyday knowledge about for example, the temporal patterns of "flying checkpoints," of who mans the checkpoint, and how to dress to avoid attention. Such mundane knowledge holds a powerful agential subjectivity. "In places of shifting violence and insecurity these embodied practices closely reveal that in their very inconsistency and unpredictability they produce place-specific and, for example, gender-specific knowledge necessary to cope with and possibly transgress the obstacles of the everyday" (Mannergren Selimovic 2019, 136). Also, the researcher "being in the place" doing fieldwork, Mannergren Selimovic (2019, 136) notes "needs to gain an embodied experience herself of the material place in a reflexive process." Thus, to spatially analyze a place means being there, reflecting upon one's own embodied experience.

Recently, the "spatial turn" in peace and conflict studies has been criticized as the knowledge is mainly produced by "peace scholars and practitioners of the Global North" and that they 'take up space'," which "means that Indigenous approaches to spatiality tend to be overlooked as part of a broader problem in the politics of knowledge and peacebuilding practice" according to Brigg et al. (2022, 410). To address this challenge and to advance the spatial understandings of peace, we need to engage knowledge producers from geography and peace studies as well as from the Global North and the Global South in our efforts to probe where peace takes place (Macaspac and Moore 2022; Macaspac 2023). Brigg and George (2020, 410) suggest that by considering "peoples and knowers as 'emplaced', the spatial turn can assist the peace and conflict studies field to develop not only more technically valuable analyses, but also more expansive – and perhaps more ethical – understandings of … peace."

Conclusion: Advancing the spatial a research agenda across disciplines

The merits of studying peace through the lens of space are many and they vary greatly. All the diverse (sub-)disciplines, perspectives, and approaches that engage with peace through the lens of space constitute a vibrant and insightful research field.

The spatial turn in peace and conflict studies has opened the way for understanding spaces of peace as constituted socially, politically, and discursively, as well as considerations of how space and place shape, and are shaped by, peacebuilding agency. It has provided peace and conflict scholars with a spatial perspective as well as a set of analytical tools such as place- and space-making that has been able to bring clarity and purchase on where peace, peacebuilding, and peacebuilding are situated and how spaces and places can be transformed in the transition from war to peace.

Peace and conflict scholars in conversation with urban scholars have managed to urbanize peace and rescale peacebuilding to the urban, making it possible to think of a spatially divided peace, i.e., two peace(s) co-existing in a divided city like Nicosia. Moreover, it has also helped map peace activities and agents that frequently are under the radar of peace and conflict scholars. Notions of space, locality, and territory are confronted to delineate socio-spatial practices juxtaposed in the urban peace.

Research on memory politics and cultural heritage uses spatial approaches to investigate the legacy of war. Memorial sites, for example, pin memory politics to place and help scholars investigate where memory politics take place (cf. McDowell and Braniff 2014). Through a spatial perspective we can explore how places and spaces are negotiated and how they shape peoples' experiences, memories, feelings, and interpretations of conflicts and peace(s). Such places may include memorials, museums, and parks which commemorate past violence or encourage peace. The process of spatializing memory connects space and time with social and mnemonic practices. It is a process of localizing memory, of drawing boundaries and putting memory on the map.

Feminist geographers informs gender studies and feminist peace research by reminding us that spaces of peace are produced by and serves existing gender hierarchies and thereby marginalizing women's experiences of war and peace (Valentine 1989; Massey 1994; Bondi 1990; Davidson et al. 2007. Consequently, this could mean that spaces for peace may exclude women and disadvantaged groups if these spaces are produced by those that signed the peace deal, as these are often the same men who conducted the warfare.

A constructive dialogue between migration and peace studies analyses the implications of mobility and immobility and how this patterns peace (Mac Ginty and Richmond 2013). Movement and mobility as spatial practices have thus challenged our understanding of where peace and conflict take place. Recent research has used mobility to make us rethink the notion of peace and to demonstrate how movement challenges understandings of peace, conflict, and security.

Another important contribution to the spatial analysis of peace and conflict is derived from the field of critical security studies and the research on "emplacement," which pays attention to "emplaced security" and efforts and processes of conflict transition, security, governance, and political ordering (Brigg and George 2020). This novel research demonstrates how spatial analysis opens the way for closer consideration of the ways diverse peoples conceive of spaces and places as sites of collective knowledge, practice, and heritage, as well as the accompanying politics of knowledge production.

Thus, to advance the field more researchers from disciplines such as geography, migration studies, memory studies, urban studies, anthropology, and peace and conflict studies and others are invited to engage in a conversation to explore the making geographies of peace. By taking space seriously they may contribute to advance new theoretical explorations of and empirical insights into peace, where peace takes place, and its situated knowledges processes.

References

Antoniou, K. 2022. Peer-to-peer accommodation as a peacebuilding tool: Community resilience and group membership. In *Peer-to-Peer Accommodation and Community Resilience*, eds. A. Farmaki, D. Ioannides and S. Kladou. 111–122. CABI.

Autesserre, S. 2014. *Peaceland: Conflict Resolution and the Everyday Politics of International Intervention*. Cambridge University Press.

Bank, A. and B. Van Heur. 2007. Transnational conflict and the politics of scalar networks: Evidence from Northern Africa. *Third World Quarterly* 28: 593–612.

Berents, H. 2015. An embodied everyday peace in the midst of violence. *Peacebuilding* 3: 1–14.

Björkdahl, A. 2013. Urban peacebuilding. *Peacebuilding* 1: 207–221.

Björkdahl, A. 2020. Urban peacebuilding. In *The Palgrave Encyclopedia of Peace and Conflict Studies*, eds. O. Richmond and G. Visoka. 1629–1634. Palgrave Macmillan.

Björkdahl, A. and S. Buckley-Zistel. 2022. Space for peace: A research agenda. *Journal of Intervention and Statebuilding* 16: 659–676.

Björkdahl, A. and K. Höglund. 2013. Precarious peacebuilding: Friction in global–local encounters. *Peacebuilding* 1: 289–299.

Björkdahl, A. and S. Kappler. 2017. *Peacebuilding and Spatial Transformation. Peace, Space and Place*. Routledge.

Björkdahl, A. and S. Kappler 2018. Peacebuilding and spatial transformation. In *The Palgrave Encyclopedia of Peace and Conflict Studies* eds. O. Richmond and G. Visoka. Palgrave Macmillan. https://doi.org/10.1007/978-3-030-11795-5_23-1

Björkdahl, A. and S. Kappler. 2021. The spatial dimensions of state-building. In *Handbook of Intervention and Statebuilding* ed. N. Lemay-Hebert, 3–219. Edward Elgar.

Björkdahl, A and J. Mannergren Selimovic. 2016. A tale of three bridges: agency and agonism in peace building, *Third World Quarterly* 37: 321–335.

Bollens, S. 1999. *Urban Peacebuilding in Divided Societies*. Routledge.

Bondi, L. 1990. Progress in geography and gender: feminism and difference. *Progress in Human Geography* 14: 438–445.

Brigg, M. 2020. The spatial-relational challenge: Emplacing the spatial turn in peace and conflict studies. *Cooperation and Conflict* 55: 535–552.

Brigg, M. and N. George 2020. Emplacing the spatial turn in peace and conflict studies. *Cooperation and Conflict* 55: 409–420.

Brigg, M., N. George and K. Higgins. 2022. Making space for indigenous approaches in the Southwest Pacific? The spatial politics of peace scholarship and practice. *Journal of Intervention and Statebuilding* 16: 545–562.

Buckley-Zistel, S. 2016. Frictional spaces: Transitional justice between the global and the local. In *Peacebuilding and Friction. Global and Local Encounters in Post-conflict Societies*, eds. A. Björkdahl, K. Höglund, G. Millar, J. van der Lijn and W. Verkoren, 17–31. Routledge.

Campbell, S., D. Chandler and M. Sabaratnam eds. 2011. *A Liberal Peace. The Problems and Practices of Peacebuilding*. Zed Press.

Chojnacki, S. and B. Engels. 2016. Overcoming the material/social divide: Conflict studies from the perspective of spatial theory. In *Spatializing Peace and Conflict: Mapping the Production of Places, Sites and Scales of Violence*, eds. A. Björkdahl and S. Buckley-Zistel, 25–40. Palgrave Macmillan.

Cresswell, T. 1996. *In Place/Out of Place: Geography, Ideology and Transgression*. University of Minnesota Press.

Cresswell, T. 2004. *Place: A Short Introduction*. Blackwell Publishing Ltd.

Davidson, J., L. Bondi, and M. Smith eds. 2007. *Emotional Geographies*. Ashgate.

Demetriou, O. and M. Hadjipavlou. 2016. Engendering the post-liberal peace in Cyprus: UNSC Resolution 1325 as a tool. In *Post-Liberal Peace Transitions: Between Peace Formation and State Formation*, eds. O.P. Richmond and S. Pogodda, 83–104. Edinburgh University Press.

Dempsey, K.E. 2022. Fostering grassroots civic nationalism in an ethno-nationally divided community in Northern Ireland. *Geopolitics* 27: 292–308.

Donais, T. 2009. Empowerment or imposition? Dilemmas of local ownership in post-conflict peacebuilding processes. *Peace and Change* 34: 3–23.

Engel, U. 2020. Peace-building through space-making: The spatializing effects of the African Union's peace and security policies. *Journal of Intervention and Statebuilding* 14 221–236.

Firchow, P. and R. Mac Ginty. 2017. Measuring peace: Comparability, commensurability and complementarity using bottom-up indicators. *International Studies Review* 19: 6–27.

Firchow, P. and Mac Ginty, R. 2020. Including hard-to-access populations using mobile phone surveys and participatory indicators. *Sociological Methods & Research* 49: 133–160.

Flint, C. 2003. Political geography: Context and agency in a multiscalar framework. *Progress in Human Geography* 27: 627–636.

Flint, C. 2004. *The Geography of War and Peace*. Oxford University Press.

Flint, C. 2011. Intertwined spaces of peace and war: The perpetual dynamism of geopolitical landscapes. In *Reconstructing Conflict: Integrating War and Post-War Geographies*, eds. S. Kirsch and C. Flint, 31–48. Ashgate.

Forde, S. 2019. Socio-spatial agency and positive peace in Mostar, Bosnia and Herzegovina. *Space and Polity* 23: 154–167.

George, N. 2020. Conflict transition, emplaced identity, and the gendered politics of scale in Solomon Islands. *Cooperation and Conflict*, 55: 518–534.

Giddens, A. 1984. *The Constitution of Society: Outline of the Theory of Structuration*. University of California Press.

Gusic, I. 2022. Peace between peace(s)? Urban peace and the coexistence of antagonists in city spaces. *Journal of Intervention and Statebuilding* 16: 619–640.

Hadjopavlou, M. 2012. The third alternative space: Bi-communal work in divided Cyprus. *Israel Journal of Politics, Economics and Culture* 18: 102–112.

Heathershaw, J. and D. Lambach. 2008. Introduction: Post-conflict spaces and approaches to statebuilding. *Journal of Intervention and Statebuilding* 2: 269–289.

Hellmüller, S. 2022. A trans-scalar approach to peacebuilding and transitional justice: Insights from the Democratic Republic of Congo. *Cooperation and Conflict* 57: 415–434.

Higate, P. and M. Henry. 2010. Space, performance and everyday security in the peacekeeping context. *International Peacekeeping* 17: 32–48.

Koopman, S. 2011. Let's take peace to pieces. *Political Geography*, 30: 193–194.

Koopman, S. 2014. Making space for peace: International protective accompaniment in Colombia. In *Geographies of Peace*, eds. N. Megoran and P. Williams, 109–130. I.B. Tauris.

Koskela, H. 1997. 'Bold walk and breakings': Women's spatial confidence versus fear of violence. *Gender, Place and Culture: A Journal of Feminist Geography* 4: 301–320.

Lemay-Hébert, N. 2018. Living in the Yellow Zone: The political geography of intervention in Haiti. *Political Geography* 67: 88–99.

Leonardsson, H. and G. Rudd. 2015. The 'local turn' in peacebuilding: A literature review of effective and emancipatory local peacebuilding. *Third World Quarterly* 36: 825–839.

Ljubojević, A. 2022. Walking the past, acting the past? Peace march to Srebrenica commemoration. *Nationalities Papers* 50: 1125–1142.

Mac Ginty, R. 2006. *No War, No Peace: The Rejuvenation of Stalled Peace Processes and Peace Accords*. Palgrave Macmillan.

Mac Ginty, R. 2008. Indigenous peace-making versus the liberal peace. *Cooperation and Conflict* 43: 139–163.

Mac Ginty, R. 2009. The liberal peace at home and abroad: Northern Ireland and liberal internationalism. *The British Journal of Politics and International Relations* 11: 690–708.

Mac Ginty, R. 2010. Hybrid peace: The interaction between top-down and bottom-up peace. *Security Dialogue* 41: 391–412.

Mac Ginty, R. 2011. *International Peacebuilding and Local Resistance: Hybrid Forms of Peace*. Palgrave Macmillan.

Mac Ginty, R. 2015. Where is the local? Critical localism and peacebuilding. *Third World Quarterly* 36: 840–856.

Mac Ginty, R. 2021. *Everyday Peace: How So-called Ordinary People Can Disrupt Violent Conflict*. Oxford University Press.

Mac Ginty, R. and O. P. Richmond. 2013. The local turn in peace building: A critical agenda for peace. *Third World Quarterly* 34: 763–783.

Macaspac, N.V. 2018. Insurgent peace: Community-led peacebuilding of indigenous peoples in Sagada, Philippines. *Geopolitics* 24: 839–877.

Macaspac, N. V. 2023. Spatialities of peace zones. *Cooperation and Conflict*. Advance online publication.

Macaspac, N. V. and A. Moore. 2022. Peace geographies and the spatial turn in peace and conflict studies: Integrating parallel conversations through spatial practices. *Geography Compass* 16: e12614.

Mackenzie, S. and D. Rose. 1983. Industrial change, the domestic economy and home life. In *Redundant Spaces in Cities and regions: Studies in Industrial Decline and Social Change*, eds. J. Anderson, S. Duncan, and R. Hudson, Academic Press, 155–200.

Mannergren Selimovic, J. 2019. Everyday agency and transformation: Place, body and story in the divided city. *Cooperation and Conflict* 54: 131–148.

Mannergren Selimovic, J. 2022. Challenging the 'here' and 'there' of peace and conflict research: Migrants' encounters with streams of violence and streams of peace. *Journal of Intervention and Statebuilding* 16: 584–599.

Massey, D. 1994. *Space, Place and Gender*. University of Minnesota Press.

Massey, D. 2005. *For Space*. Sage.

McCann, E. J. 2003. Framing space and time in the city: Urban policy and the politics of spatial and temporal scale. *Journal of Urban Affairs* 25:159–178.

McDowell, S. and M. Braniff. 2014. *Commemoration as Conflict: Space, Memory and Identity in Peace Processes*. Palgrave Macmillan.

Megoran, N. and S. Dalby. 2018. Geopolitics and peace: A century of change in the discipline of geography. *Geopolitics* 23: 251–276.

Millar, G. 2020. Toward a trans-scalar peace system: Challenging complex global conflict systems. *Peacebuilding* 8: 261–278.

Millar, G. 2021a. Ambition and ambivalence: Reconsidering positive peace as a trans-scalar peace system. *Journal of Peace Research* 58: 640–654.

Millar, G. 2021b. Trans-scalar ethnographic peace research: Understanding the invisible drivers of complex conflict and complex peace. *Journal of Intervention and Statebuilding* 15: 289–308.

Moore, A. 2008. Rethinking scale as a geographical category: From analysis to practice. *Progress in Human Geography* 32: 203–225.

O'Driscoll, D. 2021. Everyday peace and conflict: (un)privileged interactions in Kirkuk, Iraq. *Third World Quarterly* 42: 2227–2246.

Olivius, E. and J. Hedström 2021. Spatial struggles and the politics of peace: The Aung San statue as a site for post-war conflict in Myanmar's Kayah state. *Journal of Peacebuilding and Development* 16: 275–288.

Paffenholz, T. 2015. Unpacking the local turn in peacebuilding: A critical assessment towards an agenda for future research. *Third World Quarterly* 36: 857–874.

Paris, R. 2004. *At Wars End*. Cambridge University Press.

Pierce, J., D.G. Martin, and J.T. Murphy. 2011. Relational place-making: the networked politics of place. *Transactions of the Institute of British Geographers* 36: 54–70.

Richards, P. 2005. *No peace, no war: An anthropology of contemporary armed conflicts*. Ohio University Press/James Currey.

Richmond, O. P. 2009. A post-liberal peace: Eirenism and the everyday. *Review of International Studies* 35: 557–580.

Richmond, O. P. 2010. Resistance and the post-liberal peace. *Millennium – Journal of International Studies* 38: 665–692.

Richmond, O. P. 2011. De-romanticising the local, de-mystifying the international: Hybridity in Timor Leste and the Solomon Islands. *The Pacific Review* 24:115–136.

Richmond, O. P. 2014. The dilemmas of hybrid peace: Negative or positive? *Cooperation and Conflict* 50: 50–68.

Richmond, O. P. 2022. *The grand design. The evolution of the international peace architecture*. Oxford University Press.

Richmond, O. P. and J. Franks. 2009. *Liberal Peace Transitions. Between Peacebuilding and Statebuilding*. Edinburg University Press.

Richmond, O. P. and R. Mac Ginty. 2019. Mobilities and peace. *Globalizations* 16: 606–624.

Rose, G. 1993. *Feminism and Geography: The Limits of Geographical Knowledge*. University of Minnesota Press.

Saulich, C. and S. Werthes. 2020. Exploring local potentials for peace: Strategies to sustain peace in times of war. *Peacebuilding* 8: 32–53.

Scholte, J.A. 2014. Reinventing global democracy. *European Journal of International Relations* 20: 3–28.

Simic, O. 2012. Challenging Bosnian women's identity as rape victims, as unending victims: The 'other' sex in times of war. *Journal of International Women's Studies* 13: 129–142.

Swyngedouw, E. 1997. Neither global nor local: Glocalization and the politics of scale. In *Spaces of Globalization: Reasserting the Power of the Local*, ed. K. Cox, 137–166. Guilford Press.

Thrift, N. 2003. Space: The fundamental stuff of geography. In *Key Concepts in Geography*, eds. S. L. Holloway, S. P. Rice, and G. Valentine, 95–107. Sage.

Valentine, G. 1989. The geography of women's fear. *Area* 21: 385–390.

Väyrynen, T. 2019. *Corporeal Peacebuilding: Mundane Bodies and Temporal Transitions.* Palgrave.

Visoka, G. 2016. *Peace Figuration after International Intervention: Intentions, Events and Consequences of Liberal Peacebuilding.* Routledge.

Visoka, G. 2017. *Shaping Peace in Kosovo: The Politics of Peacebuilding and Statehood.* Palgrave Macmillan.

Vogel, B. 2018. Understanding the impact of geographies and space on the possibilities of peace activism. *Cooperation and Conflict* 53: 431–448.

Wallis, J. 2020. Displaced security? The Relationships, routines and rhythms of peacebuilding interveners. *Cooperation and Conflict* 55(4): 479–496.

Warf, B. and S. Arias, eds. 2008. *The Spatial Turn: Interdisciplinary Perspectives.* Routledge.

Wiegink, N. 2018. Violent spirits and a messy peace: Against romanticizing local understandings and practices of peace in Mozambique. In *Ethnographic Peace Research* ed. G. Millar, 137–157. Palgrave Macmillan.

Williams, P. 2015. *Everyday Peace? Politics, Citizenship and Muslim Lives in India.* Wiley-Blackwell.

Williams, P. and McConnell, F. 2011. Critical geographies of peace. *Antipode* 43(4): 927–931.

Wilson, H. 2017. On geography and encounter: Bodies, borders and difference. *Progress in Human Geography* 41: 451–471.

5

NAVIGATING THE AMBIGUOUS GEOGRAPHIES OF WAR AND PEACE

James A. Tyner

The objective I undertake in this brief chapter is straightforward: to provide an understanding of the myriad manifestations of war and peace at this present moment. The path toward this objective, however, is fraught with many conceptual thickets and brambles; the danger is that I stray too far from my destination and leave readers to wander in the dark. There is, in fact, no firm footing on which to begin, but, as necessary, begin we must.

The pathway is difficult because the meanings of war and peace are seemingly self-evident but have been fluid over time. In one definition, war consists of two or more organized armed groups engaged in protracted and intense armed hostilities (O'Connell 2005–2006, 537). Barring such activities, peace is at hand. However, as Christopher Greenwood (1987, 284) explains, "war was not simply a state of affairs or fact, it was a legal condition." Indeed, the concept of war prior to the creation of the United Nations in 1945 was an important feature of international law (Greenwood 1987, 284). In the nineteenth century, for example, the creation of a condition of war, that is, following a formal declaration of war, had three main consequences in international law: (1) the laws of war became applicable to govern the conduct of hostilities between the parties; (2) the non-hostile relations of the parties, such as the application of treaties between them, were affected; and (3) relations between the belligerents and other states became subject to the laws of neutrality (Greenwood 1987, 284). In turn, myriad and related treaties and conventions were promulgated to establish rules and regulations governing the conduct of belligerent states, of military commanders, and of individual soldiers with reference to the conduct of hostilities (Walzer 2000, 127; see also Byers 2005; Slim 2008).

DOI: 10.4324/9781003345794-5

In the latter half of the twentieth century the emphasis shifted from legal definition. The result was that many wars have been fought but not declared. It was at the United Nations that human rights were placed at the core of the quest for world security (Heyns and Gravett 2017, 577). Indeed, the preamble of the United Nations charter declared the lofty goal "to save succeeding generations from the scourge of war."[1] Over the subsequent decades, wars were not averted—although our definitions and conceptions of war have been radically—perhaps irrevocably—transformed. Notably, while there have been many instances of hostilities since 1945, there have been no formal declarations of war, and only a handful of these conflicts have been characterized as 'war' (Greenwood 1987, 283). This is not to imply the paucity of international law and lawyers. As Michael Byers (2005, 3) clarifies, "since 1945, governments that use force have almost always sought to justify their actions in legal terms, however tenuously." The difference is that formal declarations of war have for all intents and purposes been avoided. Consider, for example, the long history of military force enacted by the United States. Since 1798, the U.S. Congress has authorized the president to wage war only eleven times. Indeed, the last formal declaration of war enacted by Congress was in 1941–1942, marking the entry of the United States into World War Two. In this regard, the major conflagrations in which the United States participated since World War Two—Korea, Vietnam, Iraq, and Afghanistan—were not, legally, wars.

If, however, we broaden our understanding of war to include any participation in armed conflict against a foreign adversary, we find that the United States has been continuously at war. Barbara Salazar Torreon and Sofie Plagakis (2022), for example, document over 200 instances in which the United States has used military forces abroad in situations of military conflict or potential conflict to protect U.S. citizens or to promote U.S. interests. Remarkably, this list does *not* include covert operations, or the many times U.S. forces have engaged in joint operations with foreign militaries. Indeed, in the 2010s the United States Special Operations Command (SOCOM) acknowledged having forces on the ground in over 120 countries. In other words, the United States conducted military or paramilitary operations—including assassinations, counterterrorist raids, special reconnaissance, unconventional warfare, psychological operations, foreign troop training, and weapons of mass destruction counterproliferation operations—in approximately two-thirds of the world's countries (Turse 2014). My point is not to focus excessively or exclusively on the pervasiveness of U.S. military operations, although it is simply not possible to provide a comprehensive overview of contemporary geographies of war and peace without reference to America's global military presence. Rather, my purpose is to underscore the futility of clearly delineating between the conditions of war and peace. As demonstrated by the actions of the United States—and certainly the United States is not exceptional in this regard—the absence of formal declarations of war should not beguile us to assume peace is at hand.

As long as states and non-state actors continue to use means other than diplomacy to achieve their objectives, we cannot limit our understanding of war to legal definitions. As such, efforts to document and discuss the many and varied incidences where violence exists must extend beyond law. The same holds for our understanding of peace. Far from being war's opposite, peace as a process is also an elusive concept that remains overburdened, at risk of collapsing under the weight of academic largess. As Elisabeth Olivius and Malin Åkebo (2021, 7) caution, simplified and poorly conceptualized notions of peace can seriously impede, paradoxically, efforts at promoting peace. So forewarned, to reaffirm, in this chapter I attempt to navigate the ambiguous geographies of war and peace for our present, moment. My purpose, I underscore, is not to catalogue each and every instance of armed conflict—or lack therefore—for all countries and territories. Rather, my goal is to provide a walking stick by which we can more effectively negotiate the tangled manifestations of war and peace.

Understanding war and peace

In Lewis Carroll's (1946, 94), *Through the Looking Glass*, Humpty Dumpty declares in a rather scornful tone, "When I use a word it means just what I choose it to mean—neither more nor less." The young Alice, ever-perceptive, responds, "The question is whether you *can* make words mean many different things." Unruffled, Humpty Dumpty retorts, "the question is … which is to be master—that's all."[2] Geopolitics, we surmise from Humpty Dumpty, is a matter of mastery over language; and for understanding war and peace, this makes all the difference. To realize this, however, it is necessary to go beyond the *Looking Glass* and return to our real-world and the realities of how war and peace are constructed.

Standing before a crowd numbering in the tens of thousands, on December 25, 2022, Pope Francis delivered the traditional *Urbi et Orbi* message from the balcony of St. Peter's Basilica in the Vatican. The Pontiff warned that "our time is experiencing a grave *famine of peace*" as "the icy winds of war continue to buffet humanity." The immediate focus of his concern was the ongoing war in Ukraine and the needless suffering of those threatened by the "thunder of weapons" in a "senseless war." However, he implored his audience to reflect also upon those who suffered elsewhere: of the people of Syria, "scarred by a conflict that has receded into the background but has not ended"; of Israel and Palestine, "where … violence and confrontations have increased, bringing death and injury in their wake"; of the Sahel, "where peaceful coexistence between peoples and traditions is disrupted by conflict and acts of violence"; and to Yemen, Myanmar, and Iran, all marked by "bloodshed." And to "a world severely sick with indifference," Pope Francis appealed to his listeners to "not forget the many displaced persons and refugees who knock at our door in search of some comfort, warmth and food" or the "marginalized, those living alone, the orphans, the elderly … who risk being set aside, and prisoners, whom we regard solely for the mistakes they have made and not as our fellow men and women."[3] The solemn remarks of Pope Francis speak strongly with our contemporary moment—a moment marked by suffering, misery, injury, and death. War, famines, and genocide, as the Pope recites, are familiar specters that haunt the present-day; but so too the scourges of want and neglect. War makes the headlines, but harm and suffering result also from prejudice, neglect, and indifference.

For many scholars, pundits, and politicians, peace is portrayed as the absence of war (Diehl 2016). Both Pinker (2011) and Muchembled (2012), for example, argue that the world is a more peaceful place, in part because large-scale *conventional* wars have declined. What these scholars and others fail to consider are the myriad ways in which people remain subject to violence in the supposed absence of war. But this is to get ahead of ourselves. If we are to master the words of war (and peace), it is necessary to proceed more cautiously, for Alice is correct: It is possible to make words, including war and peace, mean anything we want.

Jessica Wolfendale (2017, 19–20) provides some clarity. Historically, she finds, war has been defined along three registers: morality, law, and politics. In practice, these are not necessarily exclusive; however, for heuristic purposes, it is helpful to consider these individually. Just war definitions identify the conditions under which waging war is morally permitted rather than the conditions that define war as such. Both Augustine and Thomas Aquinas are notably in this regard. Juridical definitions, conversely, emerged in the seventeenth century in the context of international law; as introduced earlier in this chapter, these definitions attempt to identify the conditions under which the laws of war apply. The seventeenth-century Dutch jurist Hugo Grotius was influential in the establishment of the legal definition of war and the lawful conduct of warfare. A third register, and the one I carry forward in this chapter, centers on the politics of war and traces its lineage to the nineteenth-century Prussian general, Carl von Clausewitz. In a well-known passage, Clausewitz (1956, 2) defines war simply as "an act of violence intended to compel our opponents to fulfill our will." Decidedly more capacious than most legal definitions, the approach undertaken by Clausewitz calls attention to the multifaceted reality of governance. Notably, Clausewitz defines war not by asking, "what it is" but instead by asking, "what does it do" (Smith 2020, 10). To that end, Clausewitz conceived of war as having two elements, fused in the German word *politik*, referring both to "policy"—the aims and ambitions of individual states—and to "politics"—the workings of human interaction on a large scale (Smith 2020, 10).

War is not, from a Clausewitzian perspective, a condition but instead both a relation and a process. As Quincy Wright (1924, 759) premised a century ago, the existence of war is not dependent upon the type of operations undertaken by the belligerents. There may be no actual hostilities at all for considerable periods, or there may be barbarities in violation of the law of war, but war, if begun in fact and not stopped, will nevertheless exist.

Clausewitz captures this quality when he famously argued that war is the continuation of *politiks* by other means.

If war is the continuation of politics by other means, what do we make of peace? Notably, the Chinese Communist leader Mao Zedong in 1938 expounded upon Clausewitz's maxim, stating that "politics is war without bloodshed, while war is politics with bloodshed" (Blattman 2022, 11). Following Mao, therefore, politics is always and already marked and marred by war; the difference supposedly lies in the presence or absence of *direct* violence that causes grievous bodily injury and death—the spilling of blood. Peace seemingly has no place in politics and we are condemned, as do the characters in Shakespeare's *Coriolanus*, to declare simply: "Let us have war, say I." Otherwise, it is helpful to refine both Clausewitz and Mao to consider other forms of organized violence to better discriminate between war and peace.

In 1969, Johan Galtung forwarded the position that direct, or "physical," violence occurs when there is an identifiable actor who commits (or commands) an act of violence; indirect, or "structural," violence, conversely, is premised to occur when no such actor is identifiable. In addition, Susan Opotow (2001, 151) explains, structural violence is manifest "as inequalities structured into a society so that some have access to social resources that foster individual and community well-being—high quality education and health care, social status, wealth, comfortable and adequate housing, and efficient civic services—while others do not." In other words, even in the absence of bloody wars, during times of so-called peace, people are still subject to violence. And it is for this reason Galtung provided the conceptual couplet of *negative peace* and *positive peace*. Here, negative peace refers to the common formulation that peace equates with the absence of violence. However, Galtung explains that the absence of (often) physical violence does not necessarily equate with well-being. Simply put, negative peace may exist in the absence of war; positive peace, conversely, must be actively pursued. As Galtung (1969, 183) writes, "the absence of structural violence is what we [refer] to as social justice, which is a positively defined condition." So conceived, positive peace comprises efforts "to improve unjust social, political or economic structures" and is "pro-active in that it aims to not just eliminate direct (physical) violence, but also to address situations in which structural violence prevails" (Kappler 2017, 640). As such, any effort to catalogue the contemporary geographies of war and peace is to confront an infinitely more complex world than a simple listing of countries at war or not at war.

Modalities of warfare and welfare

Although often used interchangeable, war and warfare are not synonymous. Following Speller and Tuck (2008, 2–3), the study of war implies an investigation that is not limited to the battlefield but that caters for the intrusion of political factors. The study of warfare, conversely, is a subset within the study of war and directs attention toward the conduct of war. Warfare is thus primarily about the employment of organized violence—the operations, strategies, tactics, and the weapons and armaments. Simply stated, as Speller and Tuck (2008, 3) posit, "The study of warfare implies a particular focus on armies, campaigns, battles and engagements, the basic hard currency of war." This does not mean that warfare is conducted in the absence of, for example, political or economic contexts. Indeed, exogenous factors significantly condition the modalities of warfare. Too often, however, scholars of war and peace blur the conceptual boundaries between war and warfare, mistaking the former for the latter.

Conventional wars, civil wars, limited wars, stateless wars, imperial wars, asymmetric wars, irregular wars, hybrid wars, postmodern wars: recent decades have witnessed an effloresce of writings on wars. Strictly speaking, however, our conceptual cornucopia describes different modalities of warfare rather than war itself. In other words, "war" can still be understood as the continuation of *politiks*—that is, policy and politics—by other means. To be sure, both war and warfare are chameleon-like, able to adapt to changing conditions. Belligerents are not limited to states but include a panoply of non-state actors; likewise, wars are not restricted to bombs and bullets but instead may be waged with bytes and bandwidth. The *politiks* of war, however, remain the same: to compel, to not compromise, in the pursuit of one's objectives. Here, we can turn to Thomas Schelling's (2008 [1966], 72–76) forwarding of *compellence* as the diplomacy of violence. Unlike deterrence, the neologism "compellence" is active, aggressive; compellence is an action engineered to affect compliance.

Even if he were to be so inclined, Clausewitz died in 1831 and thus never had a chance to read Carroll's *Through the Looking Glass*. No doubt, though, he would have found much to admire in Humpty Dumpty's admonition to master words and discourse. When Clausewitz penned *On War*, he did so at a particular historical moment—a period seemingly far removed from the present-day. Several historians of *modern* war have largely relegated Clausewitz to the dustbin of history: Martin van Creveld (2009), Mary Kaldor (2013), and John Keegan (2011), among others. Criticisms vary. Clausewitz promotes an antiquated state-centric concept of geopolitics and international relations. Clausewitz predates tanks and airplanes and weapons of mass destruction and so is irrelevant for understanding contemporary forms of combat. These are certainly accurate assessments. Yet, as David Lonsdale (2008) writes, Clausewitz's writings remain salient in that he identified war's rational element as emanating from its relationship with policy and that, as a tool of policy, war has to be controlled to serve its political masters. Building on this edifice, we can now turn to the transformation of warfare in the twenty-first century and the consequences of these changes on our conception of peace.

In 1989, William Lind coauthored a highly influential article in the *Marine Corps Gazette* on the "changing face of war." Composed at the denouement of the Cold War—the Soviet Union was formally dissolved in 1991—and twelve years *before* the so-called "Global War on Terror," Lind and his colleagues premised that modern warfare was undergoing a profound and irrevocable transformation. In brief, they argued that previous modalities of modern warfare comprised three distinct forms or "generations." First-generation warfare, symbolized by the smoothbore musket, entailed the tactics of line and column, and was embodied by commanders such as Napoleon Bonaparte. This form of warfare was largely rendered obsolete with the replacement of the smoothbore by the rifled musket and subsequently ushered in a second generation of modern warfare defined by fire and movement. World War One, with its attendant trenches, barbed wire, machine guns, chemical weapons, tanks, and airplanes, demonstrated the brutality of second-generation warfare. Tactics remained essentially linear; the major innovation was the greater reliance on indirect fire and the replacement of massed manpower by massed firepower, summed up by the French maxim, "the artillery conquers, the infantry occupies" (Lind et al. 1989).

Third-generation warfare, based on maneuver rather than attrition, came to fruition during World War Two. In this modality, military operations and tactics became predominantly nonlinear and, notably, emphasis shifted from spatial to temporal concerns. This is seen most clearly in the German army's strategy of *blitzkrieg*, a form of warfare that relied heavily on speed and flexibility. Notably, this shift contributed to a diminished role for massed armies and the increased use of small, highly maneuverable forces—a trend that has continued into fourth-generation warfare.

In broad terms, this most recent modality of warfare is marked by its capaciousness. As Thomas Hammes (2005, 190) explains, fourth-generation warfare "uses all available networks—political, economic, social and military— to convince the enemy's political decision-makers that their strategic goals are either unachievable or too costly for the perceived benefit." Markedly, fourth-generation warfare does not attempt to win by defeating the enemy's military forces but instead victory is sought through a combination of guerrilla tactics, insurgencies, terrorist acts, civil disobedience, and other clandestine operations by targeting and undermining the enemy's political will (Hammes 2005, 190). In other words, war need no longer be based solely, or even primarily, on physical, armed confrontations but instead predicated on political, economic, ideological, and—increasingly—informational strategies (Durante 2014, 59). Most ominously, as Lind et al. (1989) caution, fourth-generation warfare seems likely to be widely dispersed and largely undefined; the distinction between war and peace will be blurred to a point of indistinction. As such, the subtext of Humpty Dumpty's rhetoric will only increase in importance: who will master the discourse of warfare to justify political, economic, and social objectives?

Certainly, military strategists will continue to develop more lethal ways to dominate the air, land, and sea. Increasingly, though, the most contested battlefield of the future will be the electronic theater of war. Cyberwar, for example, has and will continue to change the "nature" of war and, by extension, our understanding of peace. Traditional modalities of war—including both conventional and unconventional forms—are mostly conducted by human beings through the use of physical force: soldiers are sent into battle, armed with various weaponry and armaments. War, in its traditional guise, is inherently material, that is, warfare is elemental, fought on land, on sea, or in the skies. Cyberwarfare operates in a fundamentally different way by bringing the nonphysical domain to the fore (Taddeo 2014, 9). Consequently, cyberwar *expands* the terrains of possible conflict and, in the process, affords governments to wage war without seemingly being at war.

From nonnuclear high-explosives, precision-guided munitions, and stealth designs for aircraft, tanks, and ships, many governments continue to develop, refine, and deploy conventional weapons of war. Alongside these developments, however, are new electronics for intelligence-gathering, -interference, and -deception, and futuristic designs for automated and robotic warfare (Arquilla and Ronfeldt 1993, 142). An extension of fourth generational warfare, the use of information and communications technologies in warfare constitutes a military technological revolution of uncertain ends. As John Arquilla and David Ronfeldt (1993, 142) explain, "the future of war … will be shaped in part by how these technological advances are assessed and adopted." Of great concern, as Massimo Durante (2014, 71) cautions, is that cyberwar is essential covert and thus will raise questions of transparency and accountability.

From a technological standpoint, cyberwar comprises many different modalities, including, for example, the deployment of computer viruses to target an adversary's informational infrastructure. By way of illustration, consider so-called "smurf" attacks. A smurf attack, Maiarosaria Taddeo (2012, 107–108) explains, is a "distributed denial of service" (DDOS) attack, the aim of which is "to disrupt the functionality of a computer, a network or a website." In 2010, for example, Burma's military authorities allegedly launched a massive cyberattack against the country's electronic infrastructure prior to national elections (BBC News 2010). Other examples include repeated cyberattacks by Russia, China, and Iran on presidential elections in the United States (Burt 2020) and an Iranian attack against Albania (Gritten 2022).

From a political standpoint, cyberwar marks a turning point in the quest to master the discourse of war. As Taddeo (2014, 9) explains, cyberwar is "extremely appealing from both ethical and political perspectives." Promoted as a nonlethal form of combat, cyberwar *appears* to avoid bloodshed and human commitment and therefore liberates political authorities of the burden of justifying military actions to public opinion (Arquilla and Ronfeldt 1997; see also Taddeo 2014, 9). Consider, for example, a scenario whereby tensions between two states are heightened, compelling one government to launch a cyberattack against the other state's information infrastructure. Unlike the use of conventional weapons, cyberattacks cause no physical damage or destruction to the targeted state's physical infrastructure and no casualties are apparent. In this scenario, warfare does not necessarily involve physical violence and thus appears bloodless (Taddeo 2014, 10–11). However, concerted attacks on a country's digital infrastructure can induce considerable structural violence. As such, the concept of cyberwar cannot easily be associated with the traditional idea of peace and therefore requires us to revise not only the traditional idea of war but also of peace (Durante 2014, 64).

Regardless of organization, conventional or unconventional, regular or irregular, wars and warfare are costly endeavors—even for the supposed victors. As Sun Tzu (1994, 173) affirmed more than 2,500 years ago in *Art of War*, "No country has ever profited from protracted warfare." For this reason, if, following Clausewitz, war is the continuation of politics by other means, governments have long sought alternative means to achieve desired goals. Diplomacy, for example, is often employed and the corresponding efforts to comprise can lead to peaceful—that is, nonwarring—coexistence. However, compromises are generally accompanied with concessions, and, for many governments, this is unacceptable. Accordingly, governments have long sought to impose war by other means. To this end, international sanctions, but especially economic sanctions, have since World War Two been heralded as purportedly *peaceful* alternatives to war (Arya 2008; Mulder 2022). In fact, between 1945 and 2000, the United States has imposed sanctions, unilaterally or with other nations, far more frequently than any other government in the world; more than two-thirds of the sixty-plus sanctions during this time were initiated and maintained by the United States, and three-quarters of these cases involved unilateral U.S. action without significant participation by other countries (Gordon 1999, 387).

Simply put, sanctions largely avoid the scrutiny that military operations face, in the domains of both politics and ethics (Gordon 1999, 388). Framed as mild forms of punishment as opposed to acts of aggression, sanctions are touted as nonviolent and peaceful. The reality of sanctions is anything but peaceful, however. In 1991, for example, the United States and its allies, following the Persian Gulf War (1990–1991), imposed a series of draconian sanctions on Iraq's economy, resulting in a serious humanitarian crisis. As Abbas Alnasrawi (2001, 214) writes, the decline in nutritional standards, which was the direct result of the virtual disappearance of imports, led to a sharp rise in infant and child mortality rates. Also, economic sanctions contributed to the loss of more than two-thirds of Iraq's gross domestic product, the persistence of exorbitant prices, the collapse of private incomes, soaring unemployment, large-scale depletion of personal assets, massive school dropout rates, and a phenomenal rise in the number of skilled workers and professions leaving the country as economic refugees. Within ten years, international agencies estimated that upward of 1.5 million Iraqi people lost their lives because of economic sanctions, including more than 500,000 children (Alnasrawi 2001, 214). Remarkably, the deaths from sanctions far surpassed the number of Iraqis directly killed in the Persian war—an estimated 40,000 casualties, both military and civilian (Gordon 1999, 388). The gravity of this post-conflict "peace" is underscored when one considers that the heaviest toll of economic sanctions falls not on the government's leaders but on the civilian population, primarily women, children, the elderly, the sick, and the poor (Gordon 1999, 388).

In the end, as Denis Halliday (2000, 232) writes, "whatever the terminology used, the results [of economic sanctions] are indisputably contrary to the spirit and the world of numerous international humanitarian legal instruments." So conceived and implemented, economic sanctions constitute a form of sovereign violence and thus further obscure warfare in the guise of negative peace.

The terrain of war and peace in the twenty-first century

As we enter the third decade of the twenty-first century, over 200 conflicts of varying intensity are ongoing (HIICR 2022). These include approximately sixty armed conflicts—the highest number recorded since World War Two (Pettersson et al. 2021, 810). Russia's February 2022 invasion of Ukraine captured the headlines; however, armed conflicts raged across all continents and included all modalities of warfare: revolutions, coups, and civil wars, insurgencies and targeted killings of politicians, activists, and journalists, sanctions and cyberwars.[4] In Azerbaijan, for example, separatists clash over the Artsakh region; in Ethiopia, the ruling Addis Ababa regime clashes with the Tigray People's Liberation Front; and in Myanmar, myriad resistance groups challenge the illegitimate and brutal reign of a military junta (cf. Pettersson and Öberg 2020; Pettersson et al. 2021; Lay et al. 2022).

Spurred in part by armed conflict, over 103 million people have been forcibly displaced; these include 32.5 million refugees, 53.1 million internally displaced persons, and 4.9 million asylum-seekers. Of these, approximately one-third, 36 million, are children. Geographically, nearly three quarters, or 72 percent, of refugees originate from just five countries: Afghanistan, Myanmar, South Sudan, Syria, and Ukraine. Similarly, just five countries account for the vast majority of IDPs: Syria, Ethiopia, Afghanistan, Yemen, and Ukraine (UNOCHA 2022, 15–17).

In total, roughly one-quarter of humanity—two billion people—are subject to some form of physical violence. If one includes vulnerability to structural violence and electronic threats, few, if any, person on Earth is immune. Consider hunger. Currently, hundreds of millions of people are at risk of worsening hunger; and at least 222 million people across 53 countries are vulnerable to acute food insecurities. Indeed, starvation is a very real risk for 45 million people in 37 countries. Syria, for example, is home to 12 million food insecure people, equating to about 54 percent of the country's population (UNOCHA 2022, 22).

Conflict remains the key driver of acute food insecurity—more than 70 percent of people experiencing hunger live in areas affected by war and violence (UNOCHA 2022, 22). That said, economic sanctions and other forms of "nonlethal" warfare continue to harm people. Stemming from Russia's war in Ukraine, that country became the most-sanctioned country in the world. Other countries subject to intense sanctions include Iran, Syria, North Korea, Venezuela, Myanmar, and Cuba (Zandt 2022). Notably, many of those countries subject to structurally violent sanctions are not—legally—at war. Neither, though, are they experiencing peace in the sense of being free from grievous harm. Venezuela is a case in point. Since 2015, the U.S. government has, unilaterally and with key allies, imposed a variety of sanctions on Venezuela, contributing to a widespread humanitarian crisis that has disproportionately harmed its civilian population (Chowdhury and Lalla 2021; Hallak 2021). For example, due to lack of spare parts and new machineries, the country's oil refineries have fallen into disrepair. As such, Venezuela is unable to produce enough gasoline, diesel, and cooking gas to meet even its domestic demand. Consequently, fuel shortages have hindered the ability of people to go to work, to school, or even to the hospital; households struggle obtain sufficient resources simply to cook their meals. Venezuela's water infrastructure has also collapsed in many parts of the country: because of power outages, the distribution of water is rationed. As such, most households can obtain water for only a few hours every five to seven days—and that water must be boiled to make it potable because the government is unable to import chemicals needed for proper purification. In addition, impediments to food imports and high prices of available items have resulted in hunger and malnutrition. Overall, the devastation of Venezuela's economy has forced many millions of people to search for a better life elsewhere, whether as refugees or asylum-seekers; this outward movement has, in turn, been accompanied with a corresponding rise in family separations, violence, child labor, and prostitution (Chowdhury and Lalla 2021). The upshot is that economic sanctions have, in total, resulted in the premature death of tens of thousands of ordinary men, women, and children in a country *not at war*.

Beyond the bombs and bullets of armed conflict, and beyond the material deprivation spurred by economic sanctions, there remain the murky geographies of cyberwars and related actions. It is well documented, for example, that cyberwarfare and related uses in traditional military operations have been underway since at least the 1991 U.S.-led Persian Gulf War. And throughout the twenty-first century, drones and other robotic weapon systems have been widely deployed—again, predominantly but not exclusively by the United States. Likewise, several instances of cyberattacks have been documented, notably the aforementioned attacks on presidential elections in the United States. However, the scale, scope, and especially the intention of these incidents are not always clear. Jacquelyn Bulao (2023) provides a sense of the problem. Globally, there are more than 2,200 cyberattacks every day. In other words, every thirty-nine seconds there is a new attack somewhere. To be sure, not all these attacks are directed toward government institutions; financial, healthcare, and retail organizations are often prime targets. Many attacks are motivated by financial gain; in just the first six months of 2021 there were more than 300 million ransomware attacks. That said, coordinate ransomware attacks can wreak havoc on the information infrastructure of a country. In 2017 the North Korean ransomware known as *WannaCry* crippled logistics, communications, transportation, and government agencies in 150 countries (Bulao 2023). Did this attack constitute an act of war? At this point, both military and civilian security experts are hard-pressed to designate any incident as an act of war. What is clear is that the information revolution will profoundly transform our understanding of war and, by extension, peace. Increasingly, governments will leverage all means of state power to achieve desired ends, including both military and nonmilitary measures. To that end, a growing number of states and non-state actors will see "war as being something much more than military conflict" (Bartles 2016, 34).

That said, there are concerted efforts throughout the world to ameliorate human suffering and to promote positive peace. As Bernard Amadei (2019, 3) explains,

> ensuring the security of water, energy, land, food, shelter, and health resources through capacity building; improving access to skills and resources; promoting the rule of law, good governance, social justice, and equality and economic equity; and creating mechanisms for populations to address and recover from conflicts in a constructive manner are as crucial for positive peace as the absence of direct personal violence.

Throughout sub-Saharan Africa, for example, myriad efforts are underway to promote food security and sustainability (cf. Lind 2006; Shulika 2016). Notably, many of these efforts are propelled by women's organizations and, as such, effectively integrate myriad forms of oppression and exploitation prevalent in conflict and post-conflict societies.

Frequently, positive peace initiatives germinate at the local level in response to particular past and present conflicts. In Cambodia, for example, many grassroots efforts have emerged in an effort to promote awareness, understanding, and reconciliation in the aftermath of genocide (Tyner 2017). One such organization is *Youth for Peace* (YFP), a local nongovernmental organization that works to empower Cambodia's youths to become agents for positive change. Established in 1999, YFP promotes democracy, freedom, gender equality, human rights, and social justice through an assortment of community-building programs, for example, art workshops, public exhibitions, and peace-dialogue sessions. Not wanting to be defined by its violent past, YFP seeks to reconcile Cambodia's past violence by promoting a more hopeful post-violent society.

War is not inevitable but neither is peace. Too often, the horrific scenes of mangled bodies and bombed-out cities make the headlines, and, as such, we rarely see the more seemingly banal peacebuilding activities that actually improve people's lives instead of taking lives. Unfortunately, our ability to parse out peacebuilding from peace-destroying processes is obfuscated. Governments routinely profess peace rhetorically while promoting war materially. Both economic sanctions and cyber-security developments, as we've seen, are presenting as peaceful alternatives to war but, in reality, are often a continuation of war by other means. To fully appreciate and articulate positive peacebuilding efforts, it is necessary to walk side-by-side with those who commit themselves to the promotion of social justice and the overcoming of oppression and exploitation.

Concluding remarks

In this chapter I embarked on a journey to chart the ambiguous geographies of war and peace in our present moment. In so doing, I have tried to highlight the challenges of such a task. In this regard, the uncertainty that surrounds any navigation of the geographies of war and peace arises from the warning posed by Humpty Dumpty. Modalities of warfare change, and with these transformations so too does our understanding of peace change. At the most basic level, war remains as it always has been: to compel an enemy in such a way as to achieve a desired objective. To that end, belligerents use whatever means necessary or accessible: bombs or bullets, chemicals or computers. The history of modern warfare displays an uncomfortable truth, namely that the conduct of warfare has always, and will always, adapt and change in Darwinian fashion to changing political, economic, and social environments. But the motivations for war—the *politiks* of war—are stubbornly persistent.

Violence abounds in our present world. States and nations continue to fight over territory; resources remain coveted; and hubris and ideology still burn in the hearts of humanity. With rising death tolls in Ukraine, Syria, Yemen, Ethiopia, Myanmar, the list goes on, it is all but clear that humanity is far from exorcising the "scourge of war" from our world. The problem, though, is deeper, for even those places supposedly free from armed hostilities, is peace truly at hand? Negative peace may prevail; but negative peace can be equally war-like. If by peace we mean positive peace, that is, the active and unconditional pursuit of social justice, can we really say that any place in the world is free from war? Often, it is easy to become disillusioned. However, as the work of grassroots organizations such as Youth for Peace illustrate, peace can be at hand if we are willing to work positively toward a more just and equitable world.

Notes

1 Charter of the United Nations, available at https://www.un.org/en/about-us/un-charter/full-text. Accessed January 11, 2023.
2 Lewis Carroll, *Through the Looking Glass*, 94.
3 Pope Francis' speech is available at https://www.vatican.va/content/francesco/en/messages/urbi/documents/20221225-urbi-et-orbi-natale.html. Accessed January 9, 2023.
4 Real-time mappings of ongoing conflicts are provided by *Action on Armed Violence*, available at https://aoav.org.uk/; the Armed Conflict Location & Event Data Project, available at https://acleddata.com/; and the Uppsala Conflict Data Program, available at https://ucdp.uu.se/. Accessed January 10, 2023.

References

Alnasrawi, A. 2001. Iraq: Economic sanctions and consequences, 1990–2000. *Third World Quarterly* 22: 205–218.

Amadei, Bernard. 2019. Engineering for peace and diplomacy. *Sustainability* 19: 1–17.

Arquilla, J. and D. Ronfeldt. 1993. Cyberwar is coming! *Comparative Strategy* 12: 141–165.

Arquilla, J. and D. Ronfeldt. 1997. *In Athena's Camp: Preparing for Conflict in the Information Age*. RAND Corporation.

Arya, N. 2008. Economic sanctions: The kinder, gentler alternative? *Medicine, Conflict and Survival* 24: 25–41.

Bartles, C. K. 2016. Getting Gerasimov right. *Military Review* 96: 30–38.

BBC News. 2010. Burma hit by massive net attack ahead of election. *BBC News*, November 4. Accessed January 10, 2023. https://www.bbc.com/news/technology-11693214

Blattman, C. 2022. *Why we fight: The roots of war and the paths to peace*. Viking.

Bulao, J. 2023. How many cyber attacks happen per day in 2002? *Techjury.net*, January 5. Accessed January 10, 2023. https://techjury.net/blog/how-many-cyber-attacks-per-day/

Burt, T. 2020. New cyber attacks targeting U.S. elections. *Microsoft on the Issues*, September 10. Accessed January 10, 2023. https://blogs.microsoft.com/on-the-issues/2020/09/10/cyberattacks-us-elections-trump-biden/

Byers, M. 2005. *War Law: Understanding International Law and Armed Conflict*. Grove Press.

Carroll, L. 1946 [1872]. *Alice's Adventures in Wonderland* and *Through the Looking Glass*. Random House.

Chowdhury, S. and S. Lalla. 2021. US Sanctions on Venezuela violate the human rights that they claim to protect, says UN Special Rapporteur on sanctions. *Countercurrents.Org*, April 21. Accessed January 10, 2023. https://countercurrents.org/2021/04/us-sanctions-on-venezuela-violate-the-human-rights-that-they-claim-to-protect-says-un-special-rapporteur-on-sanctions/

Diehl, P. F. 2016. Exploring peace: Looking beyond war and negative peace. *International Studies Quarterly* 60: 1–10.

Durante, M. 2014. Violence, just cyber war and information. *Proceedings, First Workshop on Ethics of Cyber Conflict*, ed. L. Glorioso and A-M. Osula, 59–72. NATO Cooperative Cyber Defense Center of Excellence.

Galtung, J. 1969. Violence, peace and peace research. *Journal of Peace Research* 6: 167–191.

Gordon, J. 1999. Economic sanctions, just war doctrine, and the 'fearful spectacle of the civilian dead.' *Cross Currents* 49: 387–400.

Greenwood, C. 1987. The concept of war in modern international law. *The International and Comparative Law Quarterly* 36: 283–306.

Gritten, D. 2022. Albania severs diplomatic ties with Iran over cyber-attack. *BBC News*, September 7. Accessed January 10, 2023. https://www.bbc.com/news/world-europe-62821757

Hallak, C. 2021. The human cost of economic sanctions. *McGill Business Review*, March 25. Accessed January 10, 2023. https://mcgillbusinessreview.com/articles/the-human-cost-of-economic-sanctions

Halliday, D. J. 2000. The deadly and illegal consequences of economic sanctions on the people of Iraq. *The Brown Journal of World Affairs* 7: 229–233.

Hammes, T. X. 2005. War evolves into the fourth generation. *Contemporary Security Policy* 26: 189–221.

Heidelberg Institute for International Conflict Research (HIICR). 2022. Conflict monitor 2021. *Heidelberg Institute for International Conflict Research*. Accessed January 10, 2023. https://hiik.de/wp-content/uploads/2022/03/CoBa_01.pdf

Heyns, C. and W.H. Gravett. 2017. 'To save succeeding generations from the scourge of war': Jam Smuts and the ideological foundations of the United Nations. *Human Rights Quarterly* 39: 574–605.

Kaldor, M. 2013. *New and Old Wars: Organised Violence in a Global Era*. John Wiley & Sons.

Kappler, S. 2017. Positive peace. In *The SAGE Encyclopedia of Political Behavior*, ed. F. M. Moghaddam, 640–641. SAGE.

Keegan, J. 2011. *A History of Warfare*. Random House.

Lay, T., R. Kishi, C. Raleigh, and S. Jones. eds. 2022. *10 conflicts to worry about in 2022: Ethiopia, Yemen, the Sahel, Nigeria, Afghanistan, Lebanon, Sudan, Haiti, Colombia, and Myanmar*. Accessed January 10, 2023. https://acleddata.com/10-conflicts-to-worry-about-in-2022/

Lind, J. 2006. Supporting pastoralist livelihoods in eastern Africa through peace building. *Development* 49: 111–115.

Lind, W. S., K. Nightengale, J. F. Schmitt, J. W. Sutton, and G. I. Wilson. 1989. The changing face of war: Into the fourth generation. *Marine Corps Gazette* October: 22–26.

Lonsdale, D. J. 2008. Strategy. In *Understanding Modern Warfare*, eds. D. Jordan, J. D. Kiras, D. J. Lonsdale, I. Speller, C. Tuck, and C. D. Walton, 14–63. Cambridge University Press.

Muchembled, R. 2012. *A History of Violence: From the End of the Middle Ages to the Present*. Polity Press.

Mulder, N. 2022. *The Economic weapon: The Rise of Sanctions as a Tool of Modern War*. Yale University Press.

O'Connell, M. E. 2005–2006. When is a war not a war? The myth of the Global War on Terror. *ILS Journal of International and Comparative Law* 12: 535–539.

Olivius, E. and M. Åkebo. 2021. Exploring varieties of peace: Advancing the agenda. *Journal of Peacebuilding & Development* 16: 3–8.

Opotow, S. 2001. Reconciliation in a time of impunity: Challenges for social justice. *Social Justice Research* 14: 149–170.

Pettersson, T., G. Engström, M. Solenberg, S. Davies, N. Hawach, M. Öberg, A. Deniz, and S. Högbladh. 2021. Organized violence 1989–2020, with a special emphasis on Syria. *Journal of Peace Research* 58: 809–825.

Pettersson, T. and M. Öberg. 2020. Organized violence, 1989–2019. *Journal of Peace Research* 57: 597–613.

Pinker, S. 2011. *The Better Angels of Our Nature: Why Violence Has Declined*. Viking.

Schelling, T. C. 2008 [1966]. *Arms and Influence*. Yale University Press.

Shulika, L. S. 2016. Women and peace building: From historical to contemporary African perspectives. *Ubuntu: Journal of Conflict and Social Transformation*, 5: 7–31.

Slim, H. 2008. *Killing Civilians: Methods, Madness, and Morality in War*. Columbia University Press.

Smith, H. 2020. Clausewitz's definition of war and its limits. *Military Strategy Magazine* December: 9–14.

Speller, I. and C. Tuck. 2008. Introduction. In *Understanding Modern Warfare*, eds. D. Jordan, J. D. Kiras, D. J. Lonsdale, I. Speller, C. Tuck, and C. D. Walton, 1–13. Cambridge University Press.

Taddeo, M. 2012. Information warfare: A philosophical perspective. *Philosophy and Technology* 25: 105–120.

Taddeo, M. 2014. An analysis for a just cyber warfare. In *Proceedings, First Workshop on Ethics of Cyber Conflict*, ed. L. Glorioso and A-M. Osula, 8–16. NATO Cooperative Cyber Defense Center of Excellence.

Torreon, B. S. and S. Plagakis. 2022. Instances of use of United States Armed Forced abroad, 1798–2022. *Congressional Research Service*, March 8. Accessed January 8, 2023. https://crsreports.congress.gov/product/pdf/R/R42738/38

Turse, N. 2014. The startling size of US Special Forces. *Mother Jones*, January 8. Accessed January 8, 2023. https://www.motherjones.com/politics/2014/01/map-startling-size-us-special-forces/

Tyner, J.A. 2017. *Landscape, Memory, and Post-Violence in Cambodia*. Rowman & Littlefield.

Sun Tzu. 1994. *The art of war*, translated, with a historical introduction by R. D. Sawyer. Westview Press.

United Nations Office for the Coordination of Humanitarian Affairs (UNOCHA). 2022. *Global humanitarian overview 2023*. UNOCHA.

van Creveld, M. 2009. *Transformation of War*. Simon and Schuster.

von Clausewitz, K. 1956. *On War*. Routledge & Keegan Paul.

Walzer, M. 2000. *Just and Unjust Wars: A Moral Argument with Historical Illustrations*, 3rd edition. Basic Books.

Wolfendale, J. 2017. Defining war. In *Soft war: The ethics of armed conflict* ed. M. Walze, 16–32. Cambridge University Press.

Wright, Q. 1924. Changes in the conception of war. *The American Journal of International Law* 18: 755–767.

Zandt, F. 2022. The world's most-sanctioned countries. *Statista.Com*, March 9. Accessed January 10, 2023. https://www.statista.com/chart/27015/number-of-currently-active-sanctions-by-target-country/

6

FORGING SHARED SPACES FOR BUILDING PEACE

Kara E. Dempsey

Introduction

> Together we build peace and challenge injustice. At Corrymeela, our expression is "together is better" as we challenge ourselves and others to have the courage to combat sectarianism, hatred, bigotry, and homophobia. To recognize the dignity in others. We work to build a world that belongs to everyone. In that way, we build peace together.
>
> (interview with Corrymeela's leader, Alex Wimberly 10/17/22)

Ethnonational divisions in Northern Ireland fueled violence in the region, particularly during the tumultuous period known as "the Troubles" (1969–1998).[1] Despite the 1998 Good Friday/Belfast Peace Agreement, Northern Ireland remains a place of sporadic conflict, residential segregation, partition walls, and a fragile "negative peace." Significantly, the 1998 peace agreement acknowledged the presence of two separate communities (nationalists/republicans and unionists/loyalists) without providing a true path for a united, integrated society. Evidence suggests that many political leaders continue to bolster ethnonational and geopolitical divisions in order to maintain positions of power (Murphy and McDowell 2019; Wilford and Wilson 2006). These divisive forces underpin and reinforce the deeply entrenched binary perceptions of spatial belonging and identity in Northern Ireland (Dempsey 2022a; Gaffikin et al. 2016).

Residential and social enclaving, which intensified during the Troubles and continues today, reveal wider ethnographic divisions in society. Indeed, the "isolation and physical separation, partition walls, and dissonant cognitive maps of national enclaves often form the foundation from which the ethnonational

DOI: 10.4324/9781003345794-6

conflicts first emerged and continue to persist by reinforcing the non-inclusionary fissures between these communities" (Dempsey 2022b, 2). Thus, the 1998 Peace Agreement (hereafter the Agreement) fell short of producing a "positive peace" – that is, more than just the absence of war. As this book demonstrates, peace is a process forged by multiple agencies employing a myriad of spaces and sets of place-specific practices. Situated understandings of peace range from negative to positive. Positive peace is more than the absence of violence. It is the "absence of structural violence" (Galtung 1996), and includes the conditions and efforts to maintain a just and equitable society (Megoran 2011; Campbell 2020). In the case of Northern Ireland, ethnonational and structural violence negatively impacts nationalist/republicans ("Catholics") as well as many working-class unionist/loyalists. Thus, positive peacebuilding must include societal restructuring to safeguard equity for all inhabitants.

As many "top-down" peace efforts in Northern Ireland have proven unsuccessful (Hyde and Byrne 2015; Murphy and McDowell 2019), some grassroots organizations have made inroads in combating sectarianism. Considered Northern Ireland's oldest peacebuilding center, Corrymeela builds positive peace by creating "shared spaces" to bridge ethnonational divisions and foster societal change (BBC 2015). Presbyterian chaplain and former World War Two prisoner of war Ray Davey founded the organization in 1965. He believed that through meaningful cross-communal encounters in these shared spaces, Corrymeela would work to dismantle divisive ethnonationalism and address societal inequalities. While situated within the place-specific geopolitical environment of Northern Ireland, Corrymeela developed into an international peacebuilding center that has been recognized by the Dalai Lama, King Charles III, and peacebuilders in conflict areas such as Israel/Palestine, Afghanistan, and Bosnia (Tyler 2015).

This chapter examines how Corrymeela builds positive peace through spatial practices of cooperation, respect, and deep listening. The following is built from field and archival research in Northern Ireland (2014–2019), including observational research during site visits, and nine one-hour interviews (2014–2022) with Corrymeela instructors and administrators in Ballycastle and Belfast or over Zoom (Bryman 2016). This fieldwork contributes to discussions of how this peacebuilding organization forges positive peace in its local environment and beyond Northern Ireland's borders. The example of Corrymeela shows that peacebuilding can thrive through the construction of spaces that foster, and are fostered by, the search for positive peace. This stands in contrast to bounded spaces that are built by, and maintain, negative peace, at best, or conflict.

The chapter is divided into five sections. The first examines research on peacebuilding and "shared spaces" to situate Corrymeela within peace, reconciliation, and cross-community work. The second section provides a brief history of sectarianism and violence in Northern Ireland. The third section introduces

Corrymeela as an organization and its mission within Northern Ireland. The fourth section outlines two of Corrymeela's many peacebuilding programs (sectarianism and legacies of conflict) that foster positive cross-community interactions. This chapter concludes by contextualizing Corrymeela within global peacebuilding efforts.

Conceptualizing peace

Peace is an ongoing process; a doing and becoming of uneven and multifaceted range of political practices (Megoran et al. 2016; McConnell et al. 2014). Historically, however, political geography's analytical focus prioritized imperialist projects and war, with a narrow understanding of peace as merely the absence of war or violence (Williams and McConnell 2011). Within the last decade, peace geographies have challenged the subdiscipline to foster a greater inclusion of peace in research agendas (Macaspac and Moore 2022). Arguing that political geography too often prioritized projects of war and violence, critics faulted it for collaborating with a violence-based world order. Instead, they called for geographical theorizations of peace that "help shape an alternative, more just landscape" (Inwood and Tyner 2011, 443).

At the same time, geographic analyses of peace began to move away from privileging state-led conflict resolutions, such as an elite "liberal peace" standard, which frames peace as neoliberal (capitalist) democracies agreeing to avoid war (Campbell et al. 2021). In contrast, new perspectives highlight the agentive and intersectional elements of peace across various scales, including the local (e.g., Dempsey 2022a; Koopman 2019; Williams 2015). These "critical geographies of peace" investigate how peacebuilding is conceived, socially constructed, geographically implemented, and situated in specific places through various networks of diverse actors (Williams and McConnell 2011). They also highlight the banal, every day, and embodied dimensions of peace to examine how they are experienced differently across space and time (Koopman 2014; Woon 2017).

Conceptualizations of peace now distinguish between negative (the absence of war) and positive in which "individuals, groups, states and the international community address direct, cultural and structural violence" (Byrne and Senehi 2012, 37). For example, Kirsch and Flint (2016) highlighted negative peace when examining the disconnect between liberal peacebuilding and actual post-conflict reconstruction projects. Johan Galtung (1996), like Dr. Martin Luther King Jr. who described negative peace as the "absence of tension," defined it is the lack of direct violence, such as bodily harm, while positive peace is the "presence of justice." Megoran (2011) argues that positive peace works to transform the inequalities responsible for the violence in order to forge a just social order. Building on Addams' (1907/2007) concepts of peaceweaving, Shields and Soeters (2017) frame positive peace as creating collaborative relationships across divisions to fostering social justice and equity, collaboration, community engagement, and shared governance. This framework challenges us to examine how we can cultivate these positive forces through our theory and geographic practices (Megoran et al. 2016). We must examine the multiple forms of agency within spatial-temporal contexts to situate the complex, emancipatory, and multidimensional elements of positive peacebuilding (MacGinty 2014).

To ground positive peace in place, Mitchell and Kelly (2011, 307) posit that peacebuilding transpires in peaceful spaces, the "secure, manageable spaces that embody the norms of intervening actors, and which act as epicenters from which these strategies can be consolidated."

Indeed, space is central to both conflict and peacebuilding efforts to challenge the socio-spatial violence and segregation in Northern Ireland. More specifically, contestation over space is often the stimulus and indication of violence and segregation (Dempsey 2020; McDowell et al. 2017). Indeed, much of the geography of conflict and exclusion in Northern Ireland is a manifestation of individuals and groups competing to control space, reinforcing ethnoterritoriality along the lines of bounded delineation and marginalization (Dempsey 2022a). Even after the Agreement, widespread residential, commercial, and social segregation is common. Over 93 percent of students attend publicly funded ethnonationally segregated schools. Predictably, friendships among children in the region commonly align along these dividing lines (Dowler and Ranjbar 2018).

However, space that is specifically designed as a "shared" space underpins positive peacebuilding efforts to combat territorialization of violence in the region (Dempsey 2022b). These spaces are a critical element in conflict transformation as they can be utilized to promote integration and "teach peace" in divided societies (McDowell et al. 2017). They must be safe, nonpartisan, and adaptable in order to provide opportunities for peacebuilding and foster efforts to traverse the gulf between divided communities (Morrissey and Gaffikin 2006; Corrymeela 2018). Gaffikin et al. (2016) argue that these spaces, which must be managed and maintained, are best facilitated by local peace organizations. By establishing shared spaces that facilitate positive encounters across sectarian divides, organizations such as Corrymeela challenge ethnonationalism in Northern Ireland and forge peacebuilding.

Peace has different meanings across time, scale, and place.[2] To interrogate how peace is produced in situ, it is important to understand the spatial production and manipulation of peace (Macaspac and Moore 2022). As peace is place-specific, one must also examine power relations and the "larger geopolitical contexts within which they are situated … especially when external actors, such as the UN are part of the local situation" (Megoran and Dalby 2018, 253). In the case of Northern Ireland, various regional, state, and EU governments support peacebuilding projects. However, skilled grassroots organizations are arguably most familiar with local perceptions of peace and corresponding practices. They also are best positioned to address the everyday realities of a community in conflict.[3] Grassroots organizations such as Corrymeela implement peacebuilding by making space for positive encounters while working to address the structural violence present within local political institutions, social and resource inequities, and (in the case of Northern Ireland) various paramilitary organizations.

Northern Ireland: A region forged in strife

Northern Ireland was formed in 1921 as a constitute unit within the United Kingdom during a time of acute strife, particularly between "Irish Catholics" and "British Protestants." During this time, much of Northern Irish society normalized the employment of religious labels to describe an individual's cultural or ethnonational background. Despite the presence of various political ideologies, common categorizations divided the region into nationalist or unionist camps (with republican and loyalist affiliations traditionally willing to use violence to obtain their goals). Nationalists and republicans (commonly labeled as Catholic) called for the dissolution of Northern Ireland in favor of a united Ireland. In contrast, unionists and loyalists (commonly labeled as Protestant) wanted to maintain the region's constituency within the United Kingdom. Despite the ubiquitous nature of these categories, they are neither homogeneous nor universal (Dempsey 2022a). Indeed, there is increasing evidence to suggest that many refuse to identify or be categorized within these partisan classifications. Others reject the religious elements of the corresponding ethnonational communities (McDowell, Braniff and Murphy 2017).

Because Northern Ireland was created after the Irish War of Independence (1919–1921), during which Irish nationalists/republicans fought to end British control of the island, many within the newly formed Protestant-controlled Northern Irish regional parliament considered Catholics (nationalist/republicans) as the "enemy within." More specifically, they feared Irish-Catholic's desire to eradicate Northern Ireland in favor of a "Catholic-controlled" Ireland. The subsequent systematic discrimination against Catholics and sectarian narratives propagated by politicians fueled conflict in the region. Tensions erupted in the 1960s as "The Troubles" (1969–1998) unfolded as a period that witnessed ethnonational territorialization and sectarian violence, proliferation of paramilitary organizations, and widespread British security forces' intervention.[4] When rioting against civil rights marches to end Catholic discrimination turned violent, tensions within the region erupted. After the local police (the Royal Ulster Constabulary, RUC) proved unable to quell the violence, the British Armed Forces launched "Operation Banner" in 1969 to restore order in Northern Ireland. The occupational presence of the military only intensified violence within the region and prompted the relatively moribund Irish Republican Army (IRA) paramilitary organization to resurface. It initiated its guerrilla-style "Long War" against the military and loyalist paramilitary groups (Patterson 2013).

As a result of the violence during this period, local inhabitants and "peace keepers" erected "peace walls" to divide sectarian enclaves in contested locations, such as certain working-class neighborhoods in Belfast and Derry/ Londonderry. Rioting in convergence zones, or "interfaces," exacerbated sectarian residential segregation, driving inhabitants away into more homogeneous residential areas. Other bordering practices permeated the region as well. The border in Ireland was highly militarized during the Troubles, further intensifying sectarianism and retaliatory violence. The perceptual walls erected between the self-segregating communities were famously described in Northern Ireland's Nobel Laureate Seamus Heaney's poem *Whatever You Say, Say Nothing* (1975). In this work, he reflects on how children utilized name-heritage and school affiliations to discriminate and marginalize. His words shed light on the bigotry, evasiveness, and pervasive divisions present throughout the region.

Eventually, the long road to peace in Northern Ireland resulted in the 1998 Agreement (see Dempsey 2022a). While this accord may have reduced paramilitary and British military violence, it only succeeded in establishing a negative peace, with many describing Northern Irish society as apartheid (Grattan 2020). Indeed, the Agreement propagated and perhaps legitimized the perception of two distinct and mutually exclusive ethnic communities, with the state reinforcing and maintaining this siloed society. Subsequently, much of Northern Ireland remains highly divided along a bounded sense of belonging and ethnic territory (Gaffikin et al. 2016). For example, employment opportunities, education, and housing are commonly segregated, thereby eliminating many interactions and sustaining future divisions. Public transportation routes and shopping destinations are segregated, as are most banal, everyday experiences. Less than 7 percent of students attend "integrated" schools, interreligious marriage remains less than 10 percent of society, and the number of "peace walls" is now greater than during the Troubles (McAuley and Ferguson 2016; Dempsey 2022a). Evidence suggests that paramilitary attacks increased 60 percent from 2013 to 2017 (McDonald 2018) and fear regarding Brexit-related developments resulted in a number of casualties (Campbell 2021).

Corrymeela: Building positive peace in a divided community

Peacebuilding organizations like Corrymeela work to forge cooperation and collaboration in divided societies. For this, Corrymeela creates space for peace. Through its shared spaces, Corrymeela rejects negative peace as "good enough," instead working for social justice and the recognition of the dignity in all. Considered "Northern Ireland's oldest peace center," Presbyterian chaplain Ray Davey and some students from Queens University Belfast founded

Corrymeela in 1965 as an NGO and professional reconciliation center. For Davey, peace is "both positive and dynamic ... and must be built up and maintained due to the changing geographies of peacebuilding" (1993, 11). Recognizing that (positive) peace is a situated socio-spatial practice that is always in progress, this organization utilizes space to forge peace within divided society (Megoran et al. 2016).

Over time, the organization developed intentionally designed spaces they utilize to encourage a holistic approach to positive peacebuilding. With over 40 full-time staff and 150 contributing members, this NGO is ecumenical, but it is open to all – regardless of religion or atheistic beliefs. Reinforcing its message of "all our welcome," its founder pledged that

Corrymeela is officially linked with no church, but seeks to work with all. It originated among students and graduates of the University. While its beginning was entirely Protestant, it has long since passed from that position to open its doors to all.

(Davey 1971, 2)

Rooted in the desire to establish positive peace, its members strive to challenge violent and exclusionary ethnonational narratives, unjust power structures that underpin conflict, and lack of sustainable long-term plan for peace in the region. This includes advocating for local communities' needs, cross-community engagement, and respect for different perspectives on the conflict to help healing and peacebuilding through reconciliation.

From this foundation, Corrymeela's geographic focus broadened beyond Northern Ireland's sectarian strife. Today, as a progressive and inclusive center, it helps train international peace advocates and promotes reconciliation among all ethnicities, sexual orientations, creeds, physical abilities, and nationalities. This includes support for victims of abuse, sex trafficking, and refugees (Corrymeela 2022). Known as the "Open Village," its central facilities are located in the northwestern corner of Northern Ireland, on six acres of the coastline in Ballycastle.[5] Arguably, its rural, ocean view setting fosters a liminal space that, paired with its specifically designed peacebuilding programming, forges opportunities for positive change (Corrymeela 2018). As space is often contested in Northern Ireland, this secure "neutral" space is designed to foster connections that extend beyond its borders. Indeed, it has been described as a sanctuary and place of encounter where people collaborate to address social divisions and create a community of respect and inclusion (Corrymeela Strategic Plan 2013–2016). The Ballycastle site includes several residential areas, indoor and outdoor training facilities, conference rooms, and silent reflection spaces. The various residential facilities are divided by function, allowing Corrymeela simultaneously host a number of groups with different programming needs. For example, it offers rooms for participants in training sessions that are attached to or near the "main house," while more remote housing is available for victims of trauma and those who need more time before engaging with others.

While portions of its programming occur at its Ballycastle center, the organization coordinates continued programming after groups complete training or an organized experience at the center. Additionally, its facilitators work to instill the desire to maintain the cross-communal work established at Corrymeela when individuals return home. Indeed, its motto is "bring Corrymeela home with you when you leave." In this way, the organization fosters practices of peacebuilding through a network that transcends its local borders.

Corrymeela's vision of a peaceful and sustainable society is grounded in "social justice, positive relationships and respect for diversity" (Corrymeela 2022). Its intentionally designed programming emphasizes shared goals, embodied perspectives, the intersectionality of individual experiences with peace, and dignity of others. Corrymeela also strives to challenge systematic discrimination in housing, health services, education, and politics – including potential constitutional changes (McDonagh 2021). More specifically, its mission is:

> promoting an educational rationale for reconciliation, developing and supporting community relations from the periphery and facilitating difficult meetings. Indeed, in our post-conflict context [Northern Ireland], reconciliation centres should aim to be a meeting space open to all, a "community that creates diverse, and often unexpected, meetings across lines of difference", and a place that challenges societal institutions to promote trust building within their core structures and core business.
>
> *(Wilson 2012, 2)*

Through storytelling, "deep listening," experiential learning, peace training "toolkits," and international peacebuilding seminars, Corrymeela combats marginalization and multiscalar legacies of violence and sectarianism.[6]

The foundations of Corrymeela

When Corrymeela was first established, there was no pertinent blueprint for designing what its members envisioned for their peace and reconciliation center. Arguably, without a similar precedent, Corrymeela's contribution to Northern Ireland was

> its naïveté for creating an open and inclusive space to challenge society's rival narratives of "us" and "them," where new memories and more inclusive narratives of community were made possible. Within a society with widespread resentment and latent violence, the work of organizations like Corrymeela remains urgent.
>
> *(Tyler 2015, 364)*

One tenet shared by founding members was that this organization understood that peace and reconciliation had to be more than just *not* killing one another. Rejecting acceptance of a negative peace, Corrymeela endeavors to design a place that built positive peace (Morrow 2004). Since its formation, the organization worked locally to stop violence and work to make Northern Ireland "a community built on mutual respect and the political and social participation of all" (Callister 1990). To develop its unique programming, Corrymeela collaborated with Italian reconciliation organization Agape Centro to develop peacebuilding methods that could be adapted for Northern Ireland.

In 1966 Northern Ireland's Prime Minister Terrance O'Neill (in office 1964–1969) participated in Corrymeela's first public event, a conference entitled *Community 1966*. For this, Corrymeela invited community members and politicians to participate in challenging sectarian narratives and systematic social injustice throughout the region. During the conference, O'Neill pleaded for better community relations between Catholics and Protestants, and identified Corrymeela as the kind of organization that would forge a better future for the region. Despite the inclusive nature of the conference, many within the loyalist/unionist community perceived it as a threat to the "Protestant peace" in Northern Ireland. Some openly protested the peacebuilding conference and attacked O'Neill's car during his journey to speak in Ballycastle. However, the conference (and its media coverage) raised awareness of Corrymeela's mission. Soon volunteers and participants across the world started traveling to the center to learn peacebuilding.

After the Troubles began in Northern Ireland, members of Corrymeela rescued victims and "refugees" of paramilitary, British Armed Forces, and police violence and brought them to the Ballycastle Centre for rest and rehabilitation.[7] Corrymeela also hired conflict resolution mediators and peace studies specialists to help develop effective peacebuilding methodologies for its organization. As the organization grew, Davey hosted other international peacebuilding groups from Scandinavia, Africa, Oceania, and Central America to exchange ideas and practices. During these international interactions, Davey asked his members to consider ways to adapt the visiting organizations' peacebuilding strategies for Northern Ireland.[8]

Corrymeela today

Eventually, Corrymeela forged a dynamic system that imbues space with meaning and peacebuilding-purpose. Employing its shared space, it offers unique programming for various ages, educational backgrounds, and experience(s) with violence and trauma. This includes innovative positive peacebuilding and intersectionality methodologies such as storytelling, poetry, music, and art to foster both empathy and healing for all – those who perpetrated the violence (e.g., former combatants and paramilitaries), those who witnessed violence, and individuals actively experiencing trauma (Skarlato et al. 2013). Most of its training incorporates an in-residence experience with trained facilitators as part of a long-term program. During this residential element, visitors participate in training and peacebuilding exercises in neutral safe spaces. While programming is adapted for each group, all incorporate elements of spatial practices of cooperation and ritual (i.e., team-work tasks) to help build a sense of purpose, shared goals, and recognition of similarities.

The range of programming offered includes conflict intervention, interpersonal and inter-communal workshops on reconciliation in divided societies, storytelling for post-trauma healing, educational and peacebuilding training seminars, and tailored programming for fostering peaceful affiliations. These themes are then tailored for youth groups, families, teachers, government officials and agencies, churches, and international peacebuilding organizations.[9] Through these specially designed interactions, participants are afforded instruction amid several days of sustained contact with the "Other" that is designed to help individuals reflect, grow, and learn from one another.

For example, Corrymeela's programming on marginalization is designed to support members of the LGTBQ+ community, international forced migrants and refugees, former combatants and paramilitaries, and victims of gender-based violence, among others (Corrymeela 2022). Some of these workshops work specifically with refugees and forced migrants to foster their inclusion in society. They also challenge the forces that exclude or alienate those who may be labeled as "Other" (Corrymeela 2022). Other workshops paired openly homophobic Christian fundamentalists with members of the LGTBQ+ community for peacebuilding and empathy activities. At the culmination of one such workshop, a self-identified Christian fundamentalist apologized to the group in case his words ever hurt the other participants and the LGTBQ+ community as a whole community (personal observation 2017).

Unsurprisingly, participants may disagree. Indeed, the facilitators expect disagreements and contrasting opinions. However, Corrymeela's s strategy is to build a shared, safe space where constructive transformations transpire and "humanizing relationships" are forged or restored. By emphasizing the "sameness" of participants and skill building for "deep listening," Corrymeela intentionally exposes differences among individuals in a place that fosters respect, curiosity, and tolerance. Subsequently, Corrymeela has witnessed numerous examples of lasting and rich "cross communal relationships" (interview with Corrymeela's Leader Alex Wimberly 10/17/22).

Two avenues for peacebuilding: Combating sectarianism and legacies of conflict

While Corrymeela offers a variety of positive peacebuilding programs, this section investigates how it addresses sectarianism and legacies of conflict by focusing on the "4 Rs" Recognition (of inequalities), Reconciliation (for peace-building), Redistribution (of resources), and Representation (unequal partici-pation in decision making). For these programs, Corrymeela trains facilitators, mentors, and "youth camp counselors" who work with youth groups, univer-sity groups, individuals, families, community centers, and government officials. In a safe, shared space, it builds positive peace by addressing systematic issues through experiential learning, "shared goals," and establishing functional inte-gration among the participants.

Sectarianism

Corrymeela's anti-sectarian youth programs often begin with anti-bullying training through reciprocal storytelling to foster empathy. By listening to oth-ers' experiences, particularly from "the Other," facilitators work to build trust and an expectation of reciprocity among participants. Additional methods in-clude integrated sports or interactive games designed for collaborative work for success. Research demonstrates that specifically designed mixed sports can help forge a sense of camaraderie, community, and belonging that transcends sectarian and ethnonational identities (Koch 2016; Bleakney and Darby 2018; Giulianotti 2011). This may include integrated sports teams or handheld mazes that require everyone to contribute to successfully move the ball through the labyrinth. During a study abroad trip I led to Northern Ireland, my students participated in many of these integrated games and were surprised by how the design curtailed an individual's ability to exceed at any task. Instead, partici-pants needed to collaborate with all team members for a positive outcome. From this foundation, youth groups discuss sectarianism with facilitators who emphasize the "sameness" of all their participants.

Similar activities are provided to educators who visit Corrymeela. For ex-ample, educators can utilize Corrymeela's Educator Guides' UP SERIES that include short videos, storytelling, and discussion questions designed to make students reflect on the impact of sectarianism in their lives and society as a whole (see for example, Pettis 2013). The goal is to teach peace and peaceful actions that can disrupt sectarianism and asymmetrical power in society. Some accompanying activities may include workshops, such as a project on political flags in Northern Ireland. In this workshop, participants are first asked to share their perceptions of the flags. Then, they learn about the history of the flags and the political environment during which the flags were adopted. Even-tually, the group collaborates to design a new flag for Northern Ireland with shared and inclusive symbols (personal observation 2017).

There are also various programs designed for adults of all nationalities. Pro-gram offerings range from small groups to large community organizations, during which key practitioners discuss techniques and opportunities for inter-vention on sectarianism practices. Storytelling is commonly utilized with adult groups as a way to establish a foundation of trust, "shared experience," and respect for all participants. Corrymeela also offers anti-sectarian reading workshops that discuss a reading list designed to encourage participants to analyze sectarianism in the written work, their lives, and society. Other practices include integrated adult football teams organized by Corrymeela FC, which coordinates teams consisting of players from various social classes and a balance of "Irish" and "British" (or other creeds) on the teams. Because sports are commonly ethnonationally ascribed and divided in Northern Ireland (Dempsey 2022a), research suggests that cross-community sports teams are a successful form of positive peacebuilding in Northern Ireland. These experiences can help foster cross-community relations and challenge negative stereotypes providing a foundation for positive peacebuilding in society (Cardenas 2016).

When working with paramilitaries and other violent combatants, Corrymeela draws on research that demonstrates sustained, specifically designed programming with paramilitaries helped participants develop relationships with individuals from opposing organizations (e.g., Kapur, Campbell and Tutu 2018). In fact,

> sustained contact between individuals, sleeping, eating, socialising and engaging in communal activities together is considered a powerful force in the development of trust and relationships. The literature suggests that trust is more likely to develop where there are routine and frequent face-to-face interactions between different identity groups.
>
> *(Acheson et al. 2011, 24)*

For these groups, Corrymeela utilizes methods developed from collaborations with universities, such as "The Recovering from Violence Research Cluster" and "Moving from Violence to Peace" program, spearheaded by the Ohio State University's Mershon Center for International Security Studies (https:// www.corrymeela.org/resources/peace-and-conflict).

Corrymeela also designs workshops for international peace organizations or cross-community groups. For these, facilitators commonly ask participants to examine and discuss government documents such as "A Shared Future: Policy and Strategic Framework for Good Relations in Northern Ireland" (2005). This document calls for the establishment of "cross community" respect, equality, tolerance, inclusive communities, equality, trust-building, open dialogue, and shared political partnership and provides a productive foundation for the group to discuss how they can implement measures to produce the recommended changes. Other training packets focus on building just societies or peacebuilding in post-conflict society through shared grieving (see seedofsequoia.org).[10]

After these workshops, Corrymeela continues its partnerships with participants, providing additional training, resources, and encouragement to "do the necessary work for change at the local level and beyond" (Corrymeela 2022). In this way, participants engage in positive peacebuilding practices as they forge systemic changes to re-imagine sectarian narratives, actions, and larger societies.

Legacies of conflict

This program provides training and encouragement to enable individuals and institutions to forge a holistic agenda for a positive peaceful society. In Northern Ireland, like other post-conflict societies, many inhabitants continue to be affected by segregation, deep-rooted societal violence, and unrecognized trauma. As a result, it focuses "on training and facilitation that builds capacity across education and civil society, enabling individuals and institutions to be better equipped to develop a resilient, shared and democratic society" (Corrymeela 2022). Some workshops are designed for local citizens, while other are tailored for international participants such as refugees, Israeli and Palestinians, or genocide survivors. Its newest programming is designed to provide support for frontline workers (refugee advocates, Doctors Without Borders, etc.) to help them process "vicarious trauma" through counseling, physical activities such as boxing classes, and establishing a social care network for participants (interview with Corrymeela's Leader, Alex Wimberly 10/17/22).

Corrymeela also offers training workshops for volunteers, educators, politicians, and peace organizations that range from the local to the international (e.g., South Africa, racially segregated areas in southern United States, Israel/ Palestine, Sri Lanka, and the Balkans). Some of the topics include how to: break cycles of violence; integrate segregated societies and schools; disarm and reintegrate paramilitaries, genocide participants, or gang members; productively work with the police; address mental health concerns; and forge gender equality.

For example, when advocating for integration in violent societies, Corrymeela recognizes that processes are complex and require multiscalar practices for successful inclusion. At the local level, facilitators work with individuals to adjust to a society different than the pre-conflict period. Their main goal is to integrate individuals into present-day society and encourage participants to not revert to former actions or societal norms. They also identify the importance of addressing the psychological causes and consequences of violence by providing mental health support services and/or addiction treatment.

The intersectionality of this work recognizes the impact of gender roles and societal bias (e.g., violence is a "male behavior," so female combatants become societal pariahs), as well as limited economic opportunities for participants. For this, facilitators commonly work with local governments and potential employers to provide social and economic alternatives outside of violent organizations (e.g., paramilitaries or gangs). Providing alternative economic opportunities is a significant element of integration, not only for an individual, but also as a means to weaken the power of violent groups. Indeed, as these violent organizations are frequently entrenched within these communities and offer economic and social opportunities to its members and local inhabitants. Thus, these integration methods may challenge an armed group's standing and "pull factor" within a community, thereby reducing its attractiveness. Additionally, productive employment opportunities may replace the "sense of purpose" that many violent groups often offer to its members. Research suggests this is particularly true for marginalized individuals who frequently gain a sense of belonging and respect in these organizations that may not otherwise be attainable (DOJ 2022).

Ultimately, Corrymeela's peacebuilding facilitators recognize that connection was a fundamental element of successful re/integration. Subsequently, its members actively collaborate with other peacebuilding organizations to "build connection across programs, countries, cultures, and individuals to share research, policy, and practice knowledges to better address issues of re/integration and transition after violence" (Hooser and Billing 2021, 22). Through their work, participants strive to foster positive peace and meaningful transformations at the individual, local, and societal levels.

Conclusion

This chapter illustrates how Corrymeela conceives peacebuilding. Specifically, the example of Corrymeela shows how peace is socially constructed, geographically implemented, and situated in specific places with networks of diverse actors. In sum, this chapter highlights positive peacebuilding in a segregated society as a process in which peacebuilding agency and the making of spaces are inseparable and mutually constitutive. Through its intentionally designed "shared space" for peacebuilding, Corrymeela dismantles the spatial and social impacts of divisive ethnonationalism, segregation, and violence in Northern Irish society and beyond.[11] Because peace is a set of socio-spatial practices that are continuously produced, "shared spaces" can be a critical part of conflict transformation. They both ground and can facilitate positive peacebuilding efforts to transcend divisions in conflicted communities (Megoran et al. 2016; McDowell et al. 2017).

As many state-sponsored, "top-down" peace efforts failed to forge positive peace in Northern Irish society, this shortcoming has become the focus of many local organizations, such as Corrymeela. Over time, Corrymeela expanded its programming from the local and regional geopolitical environment of Northern Ireland to eventually include national and international peacebuilding practices and workshops. Indeed, as peace is a multilayered set of processes and political practices, this organization championed the agentive and intersectional elements of positive peace. It refused to accept negative peace (absence of conflict) as an acceptable societal standard and works to transform the inequalities responsible for the violence to create a just society. In this way, Corrymeela not only forges collaborative relationships across sectarian divisions, it works to end the cycle of violence that underpins such conflicts.

Indeed, Corrymeela's comprehensive message of positive peacebuilding includes its advocation for greater social justice and equity, community engagement, and shared governance within society. According to one of its public statements addressed to the Stormont Assembly, Corrymeela offers its "shared space" for peacebuilding and openly challenges leadership to do more for its citizens:

> All of the people living in this part of Ireland deserve leadership that is envisioned in its hope for a future that is to the betterment of all citizens; those who are Irish, or British, or both or neither. We call on politicians to speak of the achievements of peace and reconciliation in their electioneering, and ask them to honour those from within and without their party lines, national identities and community affiliations. Reconciliation is difficult, so is Leadership. At Corrymeela, we are willing to host gatherings – on and off the record – for community groups and political leaders who wish to meet each other in a spirit of cooperation, reconciliation and good will. A reconciling future will be built on a reconciling present. Let us share and make stories of reconciliation in this time rather than stoking fears and suspicion.
>
> *(Statement on the Stormont Assembly 2017)*

Corrymeela also recognizes and demonstrates that peacebuilding is not easy. Interactions between "oppositional" groups in this inclusive "shared space" are sometimes disquieting or disturbing, particularly for those with a sense of victimization or animosity for the "Other." However, despite protracted legacies of sectarianism, violence, and division, Corrymeela perseveres. It fosters the space and opportunity for transformational moments as well as the gradual, personal, and societal changes necessary for building and maintaining positive peace. As its peacebuilding programs offer guidance and avenues of hope for societies transitioning from conflicts, Corrymeela's influence extends far beyond the borders of Northern Ireland. As inter-communal violence endures in various geopolitical landscapes, the practices and spaces afforded by Corrymeela are both a source of healing and hope and a model for similar organizations that fight for peace.

Notes

1 Ethnonational violence in the region predates the establishment of Northern Ireland in 1921 (see Dempsey 2022a).
2 The intersectionality of peace also applies to gender. "United Nations' reports have identified that women and children are the most vulnerable populations within ethno-political conflict … and the UN unequivocally reaffirmed the importance of the role of women in peace-building and conflict resolution" (cited in Skarlato et al. 2013, 170).

3　For a hybrid (macro and local level actors) approach to peacebuilding see MacGinty 2014; Hyde and Byrne 2015.

4　For a history of the Troubles see Dempsey 2022a.

5　Corrymeela also has a Belfast location.

6　"Deep Listening" is when participants are trained to still their mind, prioritize listening over talking and avoid activity trying to prepare rebuttals; instead, aim to listen to understand, even if one does not agree (Corrymeela 2022).

7　Northern Irish inhabitants who fled sectarian violence in their residential communities were frequently described as "refugees" during the Troubles.

8　Today, Corrymeela now collaborates with groups and volunteers from the aforementioned locations as well as Southeast Asia, South Asia, and South America, among others.

9　For example, Corrymeela offers educators and youth groups training materials from many of its resource packs and "toolkits." One such example is its *Together: Building United Communities Good Relations* resource pack that offers 30 years' worth of materials that are government-endorsed guide to peacebuilding practices for students aged 11–19 (https://www.corrymeela.org/news/194/good-relations-week-goes-virtual).

10　As a result of the pandemic lockdown, Corrymeela now offers these international meetings as webinars as well.

11　For a summary of geographic scholarship of nationalism see Koch 2023.

References

Acheson, N., C. Milofsky, and M. Stringer. 2011. Understanding the role of non-aligned civil society in peacebuilding in Northern Ireland: Towards a Fresh Approach. In *Building Peace in Northern Ireland*, eds. Power, M, and M.C. Power. Liverpool University Press.

BBC. 2015. Prince Charles visits Northern Ireland's oldest peace centre Corrymeela. *British Broadcast Corporation*. Accessed May 22, 2015. https://www.bbc.com/news/uk-northern-ireland-32839220

Bleakney, J. and P. Darby. 2018. The pride of east Belfast: Glentoran Football Club and the (re) production of Ulster unionist identities in Northern Ireland. *International Review for the Sociology of Sport* 53: 975–996.

Bryman, A. 2016. *Social Research Methods*. Oxford University Press.

Byrne, S. and J. Senehi. 2012. *Violence: Analysis, intervention, and prevention*. Ohio University Press.

Callister, J. 1990. BBC Documentary *Ray Davey and the Corrymeela Community*. https://www.corrymeela.org/resources/media

Campbell, K. 2020. 'After the dust settles': The experiences of local peace and reconciliation organisations in post-Agreement Northern Ireland. A case study of the Corrymeela Community. PhD Diss. University of St. Andrews.

Campbell, L. 2021. Lyra McKee: two men charged with murder of Northern Irish journalist. *The Guardian*. September 16. https://www.theguardian.com/uk-news/2021/sep/16/lyra-mckee-two-men-charged-with-of-northern-irish-journalist

Campbell, S., D. Chandler, and M. Sabaratnam, eds. 2021. *A liberal peace? The problems and practices of peacebuilding*. Bloomsbury Publishing.

Cardenas, A. 2016. Sport and peace-building in divided societies: A case study on Colombia and Northern Ireland. *Peace and Conflict Studies* 23: 4–12.

Corrymeela 2018. Corrymeela community. Accessed May 2021. www.corrymeela.org

Corrymeela 2022. Accessed December 2022. www.corrymeela.org

Davey, R. 1971. Corrymeela is not an island. *Community Forum* 4.

Davey, R. 1993. *A channel of peace: the story of the Corrymeela community.* Marshall Pickering.

Dempsey, K.E. 2020. Spaces of violence: A typology of the political geography of violence against migrants seeking asylum in the EU. *Political geography* 79: 102157.

Dempsey, K.E. 2022a. *An Introduction to the Geopolitics of Conflict, Nationalism, and Reconciliation in Ireland.* Taylor & Francis.

Dempsey, K.E. 2022b. Fostering grassroots civic nationalism in an ethno-nationally divided community in Northern Ireland. *Geopolitics* 27: 292–308.

DOJ (Department of Justice, UK). 2022. Executive's tackling paramilitary activity, criminality & organised crime programme. Accessed May 15, 2022. https://www.endingtheharm.com/tag/northern-ireland/

Dowler, L. and A.M. Ranjbar. 2018. Praxis in the city: Care and (re) injury in Belfast and Orumiyeh. *Annals of the American Association of Geographers* 108: 434–444.

Gaffikin, F., C. Karlese, M. Morrisey, C. Mulholland, and K. Sterrett. 2016. *Making space for each other: Civic place making in a divided society.* Queens University Belfast Press.

Galtung, J. 1996. *Peace by peaceful means: Peace and conflict, development and civilization.* Sage.

Giulianotti, R. 2011. Sport, transnational peacemaking, and global civil society: Exploring the reflective discourses of "sport, development, and peace" project officials. *Journal of Sport and Social Issues* 35: 50–71.

Grattan, S. 2020. Northern Ireland still divided by peace walls 20 years after conflict. *The World.* https://theworld.org/stories/2020-01-14/northern-ireland-still-divided-peace-walls-20-years-after-conflict

Heaney, S. 1975. Whatever you say, say nothing. *North.* Faber and Faber. 53.

Hooser, K. and T. Billing. 2021. Moving from violence to peace. *The Recovering from Violence Research Cluster.* Mershon.

Hyde, J. and S. Byrne. 2015. Hybrid peacebuilding in Northern Ireland and the border counties: The Impact of the International Fund for Ireland and European Union's Peace III Fund. *IJCER* 3: 93.

Inwood, J. and Tyner, J., 2011. Geography's pro-peace agenda: An unfinished project. *ACME: International Journal for Critical Geographies* 10: 442–457.

Kapur, R., J. Campbell, and D. Tutu. 2018. *The troubled mind of Northern Ireland: An analysis of the emotional effects of the Troubles.* Routledge.

Kirsch, S. and Flint, C., 2016. *Reconstructing conflict: integrating war and post-war geographies.* Routledge.

Koch, N., ed. 2016. *Critical geographies of sport: Space, power and sport in global perspective.* Routledge.

Koch, N. 2023. Geographies of nationalism. *Human Geography* 16: 200–211.

Koopman, S. 2014. Making space for peace: International protective accompaniment in Colombia. In *Geographies of Peace*, eds. McConnell, F., N. Megoran, and P. Williams, 109–130. I.B.Tauris.

Koopman, S. 2019. Peace. In Antipode Editorial Collective, eds. T. Jazeel, A. Kent, K. McKittrick, N. Theodore, S. Chari, P. Chatterton, V. Gidwani, J. Peck, J. Pickerill, M. Werner, & M. W. Wright, *Keywords in radical geography: Antipode at 50*: 207–211.

Macaspac, N.V. and A. Moore. 2022. Peace geographies and the spatial turn in peace and conflict studies: Integrating parallel conversations through spatial practices. *Geography Compass* 16: e12614.

MacGinty, R. 2014. Everyday peace: Bottom-up and local agency in conflict-affected societies. *Security Dialogue* 45: 548–564.

McAuley, J.W. and N. Ferguson. 2016. 'Us' and 'Them': Ulster loyalist perspectives on the IRA and Irish republicanism. *Terrorism and Political Violence* 28: 561–575.

McConnell, F., N. Megoran, and P. Williams, eds. 2014. *Geographies of peace: New approaches to boundaries, diplomacy and conflict resolution.* Bloomsbury Publishing.

McDonagh, P. 2021. Shaking ideologies: A response to 'Pulpit to Public: Church leaders on a post-Brexit Island' by Gladys Ganiel. *Irish Studies in International Affairs* 32: 617–619.

McDonald, H. 2018. Northern Ireland 'punishment' attacks rise 60% in four years. *The Guardian.* https://www.theguardian.com/uk-news/2018/mar/12/northern-ireland-punishment-attacks-rise-60-in-four-years

McDowell, S., M. Braniff, and J. Murphy. 2017. Zero-sum politics in contested spaces: The unintended consequences of legislative peacebuilding in Northern Ireland. *Political Geography* 61: 193–202.

Megoran, N. 2011. War and peace? An agenda for peace research and practice in geography. *Political Geography* 30: 178–189.

Megoran, N. and S. Dalby. 2018. Geopolitics and peace: A century of change in the discipline of geography. *Geopolitics* 23: 251–276.

Megoran, N., F. McConnell, and P. Williams. 2016. Geography and peace. In *The Palgrave handbook of disciplinary and regional approaches to peace*, eds. Megoran, N., F. McConnell, and P. Williams, 123–138. Palgrave Macmillan.

Mitchell, A.L. and L. Kelly. 2011. Walking in North Belfast with Michel de Certeau: Strategies of peace building, everyday tactics and hybridization. In *Hybrid forms of peace: From the everyday to postliberalism*, eds. Richmond, O.P. and A. Mitchell. 277–292. Palgrave Macmillan.

Morrissey, M. and F. Gaffikin. 2006. Planning for peace in contested space. *International Journal of Urban and Regional Research* 30: 873–893.

Morrow, D. 2004. On the far side of revenge. *Speech at the Shared Future at Gluckman Ireland House.*

Murphy, J. and S. McDowell. 2019. Transitional optics: Exploring liminal spaces after conflict. *Urban Studies* 56: 2499–2514.

Patterson, H. 2013. *Ireland's violent frontier: The border and Anglo-Irish relations during the troubles.* Springer.

Pettis, S. 2013. ed. *Up Standing – Stories of courage from Northern Ireland.* Corrymeela Press.

Shields, P. M. and J. Soeters. 2017. Peaceweaving: Jane Addams, positive peace, and public administration. *The American Review of Public Administration* 47: 323–339.

Skarlato, O., S. Byrne, K. Ahmed, J.M. Hyde, and P. Karari. 2013. Grassroots peacebuilding in Northern Ireland and the border counties: Elements of an effective model. *Peace and Conflict Studies* 20: 4–26.

Statement on the Stormont Assembly 2017. *Corrymeela* 17 January corrymeela.org/news/69/statement-on-the-stormont-assembly

Tyler, A. 2015. Reconstituting Community in Divided Society: Faith-inspired Civil Organizations and the Case of Corrymeela. *Political Theology* 16: 346–366.

Wilford, R. and R. Wilson. 2006. The Peace Process. *Devolution Monitoring Programme 2008*. Queens University.

Williams, P. 2015. *Everyday peace?: politics, citizenship and Muslim lives in India*. John Wiley & Sons.

Williams, P. and F. McConnell. 2011. Critical geographies of peace. *Antipode* 43: 927–931.

Wilson, D. 2012. *Dún Laoghaire/rathdown comenius regio 'restorative approaches' programme 2010–2011 A formative evaluation*. University of Ulster.

Woon, C.Y. 2017. Children, critical geopolitics, and peace: Mapping and mobilizing children's hopes for peace in the Philippines. *Annals of the Association of American Geographers*, 107: 1–18.

7

THE VIOLENCE OF DEVELOPMENT AND THE PROSPECTS FOR PEACE

Colin Flint

Introduction

Development, peace, and violence. Which of these words, at first glance, does not seem to fit? I suspect that for readers in the wealthiest parts of the world the most likely choice would be the last one. For those in developed countries, visions of development are usually associated with an increase in wealth and wellbeing that is implicitly understood as nurturing peaceful circumstances and a growing sense of security, or what we know as positive peace (Galtung 1965, 233). However, throughout modern history ideas and projects that may fall under a broad umbrella term of development have been inseparable from conflict and violence.

To understand the geography of peace and conflict is to understand a global economy that generates gross inequalities of wealth and life-chances. This chapter will explain the generation and maintenance of inequality requires the use of many forms of violence. The connection between development and violence is inseparable, and uncomfortable for those who won the lottery of birth and enjoy relative wealth in a world of poverty. Hence, another outcome is that a lot of effort has been spent to deny or ignore the connection between violence and development. In this chapter we will pull back the curtain to expose development as inextricably violent. But the news is not all bad. There are many instances in history when the inequities and injustices of global economic processes provoked meaningful and productive peace movements. We may be at the cusp of another such moment.

Development is a word that is commonly used but often barely interrogated (Taylor 1989; Peet and Hardwick 2015). Specifically, developmentalism is an ideological framework created after World War Two and used as the basis for

DOI: 10.4324/9781003345794-7

a host of interventions in the poorest parts of the world (Taylor 1992). However, the idea that the rich countries of the world have a model for how the rest of the world should be, and a sense of what the poorer parts of the world should do to "develop," has a long history (Blaut 1993). Hence, we need to consider the *longue durée* of capitalism that has created and maintained an unequal world, and usually by using direct violence (Wallerstein 1979). In other words, a world of structural violence has been produced by direct violence. Though the idea of development or "civilizing" projects has been around for centuries, we also need to understand the different manifestation of ideas of development in different geopolitical epochs, especially as we seem to be in a period of intensifying geopolitical tension that involves different visions of development (Flint and Zhang 2019).

There are also signs that we are in a moment when the realities of global climate change could produce meaningful discussions about how our world should be, or a new vision of human security. Contemporary geopolitics of peace are simultaneously transnational and local as the outcomes of planetary catastrophe are experienced differently in different places (Koopman 2011; MacGinty 2021; Williams and McConnell 2011). The imperative of empathy for others in different places and the realization of connectivity at the global scale have catalyzed the agency of climate change activists (see O'Lear this volume). Yet this form of geopolitics is occurring at a time of interstate tensions that include Chinese competition with the United States, Russian threats to the West, and the rejection of Western modernity by religious fundamentalists challenging women's rights. This suite of antagonisms is also simultaneously local and global. We live in places situated within the context of a global economy, the threat of global war, and a changing global ecosystem and at a time when possibilities for peaceful interactions coexist with the growing shadows of war.

To explore the contemporary geopolitical moment of promise and concern we need to (i) situate contemporary geographic thought on the topic, (ii) understand the unequal structure of the global economy and why it has produced a geography of violence and sporadic steps toward peace, (iii) interpret the current moment of geopolitical tension as a competition between U.S.-style developmentalism and a new Chinese vision of global cooperation that is, nonetheless, likely to maintain existing patterns of inequality and violence, and (iv) explore why this seemingly depressing moment is also likely to be a time of peacebuilding.

A social science framework to explore the violence of development

An estimated 1.1 billion people live in these circumstances (Woodruff 2012). We tend not to consider this mass of humanity; it is an uncomfortable global fact. Instead, economic statistics tend to focus on national development.

Measures of economic growth and wealth are usually reported at the national scale; hence the term Gross National Product (GNP). Though it does not take much reflection to realize that there are disparities of wealth within countries, the picture of the world that is painted is one of trajectories of countries doing relatively well or poorly. Economists divide these periods of growth (or recession) into quarters or years. Furthermore, the good or bad economic fortunes of countries are sometimes situated in historic periods of varying lengths and specificity; "the age of imperialism," "postwar boom," and "globalization," (for example Agnew 2003, 85–86). From a person struggling to eat daily, through the annual fortunes of a national economy, to global epochs the experiences and circumstances of development span time from the immediate to eras; and involve scales from the body to the global.

The geography of inequality is common in mainstream approaches to development. If such inequality did not exist then there would be no point in advocating for, and attempting to achieve, development. Hence, we label countries as "developed," "developing," "less developed," "medium income," etc. in an ever-changing lexicon of terminology that essentially says the same thing: Some parts of the world are wealthy and others poor (Khan et al. 2022). The important points to consider about these labels are (i) that the unit of measurement is countries (development is seen as a national project), and (ii) that the trajectory of fortunes of a country is seemingly a function of the choices that country makes rather than its position in the unequal hierarchy of the world-economy.

The discipline of geography has long been focused on connecting geographic scale (Taylor 1981). A consideration of scale is particularly useful when we also categorize time from immediate- to long-term. The experiences of individuals occur within specific time-space settings. For example, you are reading this in a particular place (a house, a room, on an airplane, etc.) at a particular moment (an hour in a day in a year). However, to understand daily and place-specific experiences, whether they be extreme poverty or affluence, requires situating those moments in a broader picture. Development, and the violence associated with it, might best be understood from a global and long-term perspective to situate place-specific experiences.

What do we mean by long-term? World-systems analysis is a framework that takes the very long-term viewpoint (Wallerstein 2004). For our purposes we may start around 1450 (not a typo!) when a new form of economic and social relations emerged; a set of relations that we know as capitalism appeared in Europe. They were not the form of capitalism we know today. Over time, capitalism became global so that by the end of the nineteenth century every person in the world was living within a capitalist world-economy. Existing forms of society in the Americas and Africa, and feudalistic societies in China and Japan, were all destroyed by their engagement with global capitalism (Flint and Taylor 2018, 15). Note the word "destroyed" in the previous sentence. Mostly, existing forms of society outside of Europe were integrated into the capitalist world-economy through violence. Indigenous people were slaughtered, and their understandings of property, work, governance, and their relationship to the environment were deemed to be "primitive." Cannon-fire, rape, dispossession of property, and the prohibition of the use of ancient languages were forms of violence that ensured people now acted within the economic imperatives of global capitalism. This process is now known as racial capitalism (Robinson 1983).

These forms of violence were portrayed as "progress" within the countries of Europe (Said 1979, 12). We know the historical process as "colonialization" or "empire building." The discipline of geography was implicated through its construction of theories and concepts based on racism and environmental determinism that provided justification in physical and biological processes for socially constructed inequalities and colonial power relations (Ratzel 1896; Huntington 1915; Semple 1911; Blaut 1993). Slowly but surely the large-scale theft and violence involved are being acknowledged. Looted artifacts housed in museums of the wealthy countries are being returned. The United Kingdom is a reluctant participant; the much-acclaimed British Museum remains a warehouse containing stolen goods. Dispossession of statues, even the bodies of the buried (think of Egyptian "mummies"), and other reminders of a once thriving culture are a form of violence as it prevents acknowledgment of once-thriving and, for their time, advanced cultures.

Though the period of formal empires ended in the second half of the twentieth century, disparities of wealth across the globe remained. The world-systems perspective argues that such disparities are a necessary ingredient of the operation of the capitalist world-economy. This is a sobering thought as it means that the relative wealth and security of some people is a function of the poverty and insecurity of others. In other words, the peace of some is connected to the violence imposed upon others. To understand this relationship requires a brief excursion into the macro-geography of the capitalist world-economy.

The logic driving our global society is the relentless pursuit of profit that is exercised through the connection between, broadly speaking, two sets of processes (see Flint and Taylor 2018, 20-1). Core processes are the production of high value goods and services (such as robotics and banking) that result in large profits, that are undertaken by people earning high incomes and, therefore, able to consume at a high level. Peripheral processes are the production of low-value goods and services (such as some forms of agriculture and clothing manufacture) that result in low profits, that are undertaken by people earning low incomes and therefore limited to consume at a low, or even subsistence, level. Core and peripheral processes are located across the globe, but often they are clustered in different locations so that we can talk of core or peripheral places or countries. For example, the United States is a country in which core processes predominate, though there are many places in the country where we could easily find peripheral places (low-income jobs and poor communities).

The opposite can be said for a country like the Democratic Republic of Congo (DRC), where peripheral processes predominate but the ruling elite will profit and live well. Often the wealthy in poor countries gain through illicit practices related to the peripheral processes that dominate in the national economy. Cities and countries in which there is a relatively even balance of core and peripheral processes are called semi-peripheral. Though we often use the labels of a "core country" or "peripheral country" we should be wary of this shorthand and casual usage – cities and countries are always a mixture of the twin set of processes (Terlouw 1992, 143).

Importantly, these two sets of processes can only be understood in relation to each other (Taylor 1992). Economic activities that are considered core (making luxury cars, for example) require inputs that are considered peripheral (such as rubber for the tires or the extraction and processing of the minerals within the computer systems in the satellite navigation program). Hence, places across the world are connected. A town that is home to a factory making luxury cars is linked to those places harvesting rubber. The geography of the world-economy is one of connected inequalities.

Equally important is that the nature of these connections, and the fortunes of places and countries, changes over time. There are periods when the geography of core and periphery processes is largely stable, and there are moments when they are in flux (Flint and Taylor 2018, 21–25). We are in such a period of flux. We will describe the dynamism of periods of global economic change later in the chapter, but for now we can note that our framework connects the geography of daily experiences to periods of economic change within the *longue durée* of the global geography of structural inequalities.

The multiple and interconnected geographic scales and historic scopes help us understand the variation of people's experiences across space and time. These experiences are directly related to the idea that development is a form of human progress, and that the successful implementation of development projects will increase peace at the personal, national, and global scales. However, considering these intersecting scopes and scales provides a different picture: that historic projects of development are geopolitical projects that have necessarily and often violently imposed various forms economy and society on people (Power 2019; Essex 2013). Rather than a panacea for peace, development means peace for some in certain places and times, and violence and insecurity for others in different places and times.

Geographies of structural violence and cultural violence

A world-economy in which most of the world's population are poor, and many are very poor, is one in which a few are also extremely rich. To sustain such an unjust and unequal world requires the constant and pervasive use of violence.

Two forms of violence help us understand the inherently violent nature of the world-economy. The first is structural violence, or the harm imposed upon people through living in situations in which peripheral processes predominate (Galtung 1969, 170). This type of violence is often connected to environmental change with a differential impact upon children, women, and indigenous people (O'Lear 2021). Structural violence can be seen in the disparities in health and life-chances across the world. The average life expectancy of a male living in Chad is 51, compared to 81 years in Sweden (World Bank n.d.-a). This difference of 30 years is a structural loss of living years. If access to healthcare, nutrition, and housing in Sweden allows for an average lifespan of 81 years, then it is surely possible in other countries. The disparity is a form of premature death, just as we would think someone killed in a war or by gang-violence died young and unnecessarily. Structural violence can also be thought in terms of global disparities in life-chances, such as access to education. For example, only 44 per cent of girls have access to primary education in Equatorial Guinea compared to 99 per cent in France (World Bank n.d.-b), denying a pathway to self-realization more open to young men. Denying these forms of peace is a pervasive form of violence.

The persistence of structural violence is surprising in some ways. Wouldn't the people suffering from economic deprivation resist? And are people in the wealthier parts of the world not "good people" who would be appalled by these circumstances and wish to help? The lack of resistance of those in the periphery and the general apathy of the rich are explained by Galtung's structural relations of imperialism. Four sets of political relations connect the core and the periphery for economic interests while creating division between the sensibilities of non-elites across the globe.

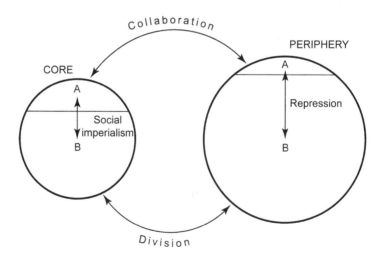

FIGURE 7.1 Galtung's structural relations of imperialism.

Source: Flint and Taylor 2018, 96.

Galtung's structural relations of imperialism simplify the political relations of the world into four dominant power relations (Galtung 1971). Elites in core and periphery countries cooperate with each other to ensure the flow of trade that, largely, benefits, the wealthy countries. Whether it be diamonds from the DRC or bananas from Guatemala or copper from Peru, the actual extraction process is through the payment of low wages, or even forms of unfree labor, while only a handful of people reap the profits. Elites in the core benefit from this trade by receiving cheap inputs into the higher-value production processes in the core. This relationship is dependent upon an exploited workforce in the periphery; and it is made supplicant through the use of violence. This violence can take many forms. For example, the suppression of labor organizations or political parties advocating for workers' rights. It may also take the form of unsafe working conditions and low wages that do not allow for the purchase of adequate nutrition or shelter. It may manifest itself as forms of indentured labor.

Why do most people in the wealthier countries do nothing about these forms of violence within the periphery? It is because of the relationship of division between workers in core and peripheral countries. Workers in the core are fearful of workers in the periphery in two ways. First, they worry that cheaper workers in the periphery will steal their jobs. Hence, there is a vested interest against development in the form of industrialization. Keeping peripheral workers employed in the extraction industries or low-skilled manufacturing is a way to prevent development into more skilled forms of manufacturing that could lead to unemployment in the core. For example, over the past decades we have seen employment in the automobile industry and call-centers offering consumer support be moved to what are labeled "developing countries." Another form of fear is that increased salaries for workers in the periphery will increase the price of products that workers in the core consume, whether it be bananas, T-shirts, or cell phones. Hence, there is no commonality of interest between workers in the core and the periphery. The result is that most people are unaware of, or turn a blind eye to, the forms of violence being suffered by most of the world's population.

How do we, in the wealthier parts of the world, live with ourselves given these circumstances? The answer is through the practice of cultural violence. This form of violence refers to the widely accepted labeling of people, or the process of "othering" (Spivak 1988). It is related to development as it is used to "blame the victim" for the structural violence they are experiencing, rather than considering their situation within the core-periphery hierarchy of the world-economy. Specifically, the othering of people in the periphery as people who are "undeserving" of the standard of living of those living in the core. In the age of imperialism othering was overtly racist – with people who were dispossessed deems to be savage and nonhuman (see Conrad (2018 [1902]) as a classic example). Today the same attitudes prevail but they are less overt – we often portray people in the periphery as lazy, corrupt, or violent. This process of dehumanization is a form of violence perpetuates structural relations. It is also a form of self-harm as we scapegoat and stereotype others, rather than understand their complexity as human beings in very difficult circumstances due to their geographic setting (Hobson 1902, 207–208).

Time-space contexts of development and violence

Galtung's (1971) model of imperial relations highlights the division of interests between workers in rich and poor countries. Despite this division, elites in the core always want to develop connections to the periphery. Mistakenly, this connection is often framed as primarily a search for new markets for the export of goods made in the core. Originally, colonialism was driven by the motives of plunder, then cheap labor for agricultural production and only later markets (Hayter 1981, 40–49). Though some export takes place, of course, the primary motive for the core to develop ties to other parts of the world is for economic interests – find places in the world where economic inputs can be obtained cheaply. It is ready access to economic inputs and not new markets that drives the ties. This leads to the conclusion that these networks are not about increasing the standard of living of people in the periphery but connecting them to the world-economy as areas of cheap labor.

From the mid-1400s through to the end of the nineteenth century, huge swathes of the world were forcibly integrated into the capitalist world-economy. Various forms of violence were used, and genocide was committed in a staggeringly broad and systematic fashion. The twin imperatives of economic integration and Christian proselytization led missionaries and merchants to the Americas where they killed, converted, and dispossessed. The flow of gold and silver in to Portuguese and Spanish coffers helped fuel wars, including those in the Mediterranean to challenge the Ottoman Empire and its desire to spread the geographic reach of Islam. As Britain joined the imperial act, it "found" what became Australia and New Zealand and claimed it to be "terra nullis" or empty territory: effectively making the aboriginal population "nonhuman" and legitimizing their killing as simply clearing the land for the spread of civilization.

In North America the nineteenth century ended with the slaughter of the "Indian Wars," much of this conflict under the much-celebrated President of peace, compassion, and democracy Abraham Lincoln. As with European countries before, the economic growth of the United States was fueled by the violence of the slave trade. The United States adopted the desire for empire with its annexation of Hawaii and imperial project in the Philippines. After Japan's war victory over Russia in 1905 (an outcome that shocked European sensibilities that presumed their racial superiority over Asians), the future of the newly industrialized country was also based on empire. A future in which Asian countries would be free of the yoke of European empire, but Japan would be dominant. Its cruel and bloody occupation of Korea (1910–1945) and subsequent military campaigns in Manchuria proved evidence that an Asian-led empire was little different from a European-led one.

The violence of empire will probably come as little surprise. Indeed, the promise of development was that imperial subjugation, the inequities of colonizer and colonized, and the economic, physical, and cultural violence necessary to maintain empire would be relegated to history. In its place was a new ideology, developmentalism. The overarching framework for the project of developmentalism was W.W. Rostow's ([1960] 2017) *The Stages of Economic Growth*. In it, Rostow laid out a step-by-step model in which poor countries, or what he called "traditional societies," could, if they followed a pathway defined by Rostow and other development experts, urbanize and industrialize and become paragons of "high mass-consumption." In other words, the whole world could obtain the lifestyle of a post–World War Two U.S. suburb. However, the idea of developmentalism was not separate from war. In fact, it was a mode of thought within a war. The subtitle of Rostow's seminal text was *A Non-Communist Manifesto*. Development was a weapon in the Cold War arsenal; a way to improve the standard of living of people to diminish the appeal of communism.

Rostow's vision was for "non-Communist literate elites" in what he called "transitional societies" to drive national projects that twinned democratization with economic development. In his words, "they have the right to expect the world of advanced democracies to help on an enlarged scale, with greater continuity; but it is they who must overcome the difficulties" that were the social, political, and economic challenges they faced ([1960] 2017, 144). Eerily, this vision was closely related to the practice of U.S. involvement in postcolonial wars, such as the Vietnam War. The U.S. strategy was to "bomb them [Vietnam] back to the stone age" (a statement usually attributed to Air Force general Curtis LeMay in his memoir, though it originally appeared in a 1967 column by the humorist Art Buchwald (Cullather n.d.)). Stunningly, the United States dropped over twice as much tonnage of bombs on Indochina in the Vietnam War than they had in the combined European and Asian theaters of World War Two (Young n.d.). Reducing a country to rubble was one way to wipe the slate clean, so to speak, and allow it to embark on the journey through the stages of growth unfettered from existing conditions – because the cities had been flattened, the crops poisoned by defoliants like Agent Orange, and the rice fields the new home for unexploded ordinance.

Furthermore, countries that did restructure their economies and societies along the model proposed by Rostow were often ruled by authoritarian regimes. These forces limited democratic freedoms, prevented the adoption of worker's rights, and suppressed segments of its population that were not on board with the national vision of "development" (Peet and Hardwick 2015, 97). Many of these countries, despite their violation of human rights, are erstwhile and consistent allies of the United States. In fact, the CIA initiated coups in Chile, Guatemala (bananas again), and Iran in the face of plans to nationalize the oil industry. President Jimmy Carter made a stark reference to the threat of war in his claim (made in 1980) that oil resources, vital at the time for the U.S. and the global economy, would be protected by any means necessary. Since then, the United States has embarked on two wars in the Persian Gulf region and continues to spar with Iran, a country with increasing security ties to China (Morrissey 2017).

A necessary part of U.S. developmentalism has been violence within peripheral countries. The United States has a training school for members of the secret police of countries that are keen to ensure that they are friendly to the interests of U.S. businesses and favorable trade relations. The program was initially called the International Military Education and Training Program and the main campus, known as the School of the Americas, was housed on a military base, Fort Benning, Georgia. It is now called the Western Hemisphere Institute for Security Cooperation. The program has graduated a series of torturers and killers for "undemocratic and brutal military regimes, such as Honduras, Haiti, Paraguay, Uruguay, Chile, Peru, Colombia, Panama, El Salvador, and Guatemala" (Flint 2022, 190). The geographic and historic extent of the School of the Americas exposes it as a purposeful and seemingly necessary aspect of U.S. foreign policy. It is in such a way that the United States has participated in the violence of Galtung's set of imperial relations. Repressing labor activists and preventing land reform in Guatemala, as just one example, allows us to buy affordable bananas in our grocery stores.

Despite the persistent and necessary structure of core–periphery relations, some countries experience economic growth and change in the structure of their economies. The United States is a good example, as over the course of three centuries it has changed from being a colonial possession, through a predominately agricultural economy and rural population, to a country of the world's greatest cities and largest and most advanced economy – plus being by far the world's most dominant military. In other words, the United States is a model for the type of transition Rostow ([1960] 2017) envisioned and advocated for. In the twentieth century other countries have also "developed." In the 1980s the label "Asian tigers" was given to Taiwan, South Korea, Singapore, and Hong Kong in recognition of their economic growth and role as key manufacturing countries (Gulati 1992). In the 2001 a new economic growth designation emerged – BRIC – referring to Brazil, Russia, India, China. Later, South Africa was included to complete the BRICS. Currently, Brazil, Russia, and South Africa are all facing their own political struggles, meaning that their economic fortunes seem a lot less rosy in 2023 than they were 20 years ago.

However, India's ongoing economic transformation is dramatic. Our interest is on China though, because it brings to light the curious contradiction of developmentalist thinking and its connection to geopolitics. The United States and its Western allies should be lauding China's economic accomplishments. Since the end of World War Two it has marched through Rostow's ([1960] 2017) stages of economic growth, increased the standard of living and levels of consumption for its people, urbanized to an impressive degree, and modernized its economy. Yet instead of saying "well done, China – you are the poster child of Rostowian development" the language coming from the United States is completely different. Instead, China is a "threat" to U.S. "interests." Its economic success is portrayed as a geopolitical risk that could lead to a third world war.

For its part, China is envisioning a new form of global economic relations. Not the "hegemony" of the United States and the postcolonial legacy of continuing core–periphery relations. Instead, China espouses visions of "South-South cooperation" that are "mutually beneficial."[1] China is creating a new set of relations with countries that are essential for the provision of energy (Russia and the Gulf countries), raw material inputs (Indonesia), food (Brazil), and inputs for its advanced manufacturing industries (such as computer chips for robotics and super-computers). It is also looking to Europe and the United States for markets for its goods (Flint and Waddoups 2021). All this sounds like economic relations; or a globalized world based on peaceful interaction. And yet we should be cautious. It was such economic integration that Norman Angell (1910) identified as making the prosecution of war "futile." He made the argument just four years before the onset of World War One.

China's presence on the African continent, its construction of ports and financial and management in Sri Lanka, Pakistan, and the Gulf would seem necessary and exemplary manifestations of development and the promotion of global trade. Instead, they are viewed with unbridled suspicion as means of putting poorer countries into debt (see Brautigam 2009 for a rebuttal) and building a network of military facilities to support its growing blue water navy (McDevitt 2020). Furthermore, China's attempt to develop its economy by growing its supercomputer industry is seen as a combined economic and military threat. As supercomputers are increasingly necessary for advanced weapons systems, the United States has imposed sanctions upon Chinese businesses and passed legislation promoting domestic production of computer technology. These actions can be interpreted as a step toward economic nationalism or protectionism.

In sum, violence and global economic relations have always been intrinsically linked: from the earliest times of European expansion across the globe, through the era of imperialism, into the US-led visions of development, to today's tensions between the United States and China as the latter develops its own connections to peripheral countries and its own vision of these relation-ships as better and different from those of the United States. Many forms of violence have been involved: genocide, support by wealthy countries for repres-sive governments that have protected external economic interests, destruction of property rights, and environmental degradation, among many others. In other words, the pursuit of "development" has required violence within and between states. Most sobering is the recognition that competition between the wealthier countries over control of parts of the periphery have often led to major wars, including the two world wars. The current tensions between China and the United States echo the words and actions between Britain, Japan, Germany, and the United States preceding World Wars One and Two. Devel-opment may be the pathway to war and not its panacea.

Development and opportunities for building peace

Just as violence and development have been intrinsically linked, the same can be said for peacebuilding and development. Development has been seen as the means to make both positive and negative peace. Though the results are mixed, at best, the idea that global politics has a purpose or imperative to "develop" has been a constant feature. However, the idea of what develop-ment is and who has the power to define and enact development policies oc-curs within the dynamics of global geopolitical competition and has altered the global landscape.

In the age of imperial competition peace was envisioned in two main ways. One was antislavery campaigns. This political movement connected the global scale of imperialism and the scale of national competition over imperial spheres of influence with the scale of the body; the essential humanity of every-one no matter their race. Though this sense of individual humanity may appear obvious to us now, it was a major rhetorical battle for antislavery campaigners to swim against imperial conceptions of a racial hierarchy. The academic disci-pline of geography, and other disciplines, was active in constructing knowl-edges of racial hierarchy that legitimated imperial control, labeled as the spread of "western civilization" (see Driver's 1991 discussion of British Victorian ex-peditions in Africa). These academic studies were the basis for environmental determinism that suggested inherent (ine)qualities depending on climate zones. In sum, European races were deemed superior and duty-bound to nurture or "develop" what were seen as inferior or "subhuman" races; broadly those in the tropics.

The intersection of academic thought, religious missionary work, and the search for economic interests was a thinly disguised excuse for empire. These joint enterprises were partially undermined by the antislavery movements' recognition of the humanity of all. For example, in 1787 British antislavery campaigners William Wilberforce and Thomas Clarkson established the experimental colony of Sierra Leone with the intention to "introduce civilization among the natives and to cultivate the soil by means of free labour" (quoted in James 1994, 185). British abolition of the slave trade in 1807 followed. However, the use of the word "civilization" indicates the cultural violence within the good sentiments of antislavery campaigners, and Freetown in Sierra Leone soon became one of many harbors enabling the global reach of the Royal navy (James 1994, 185–186).

The antislavery movement was a feature of the questioning of the legitimacy of empire. Anti-imperialism was, of course, a feature of nationalist movements in the periphery; those willing to be free of the shackles of imperial control. Anti-imperialism movements also sprouted in the colonizing countries. Liberals such as J.A. Hobson (1902, 51) saw imperialism as detrimental to workers in the core because it constrained national economic growth, while favoring specific business interests. In parallel, Marxists saw imperialism as the final or "highest stage of capitalism" that presaged the inevitable rise of socialism (as per Lenin's pamphlet first published in 1916; Lenin 1939). In their different ways, the liberals and the Marxists were challenging the set of relations identified by Galtung. The liberals were trying to ameliorate those relations; the Marxists were trying to destroy them. Both saw their political projects as means of generating a form of positive peace. The liberal agenda remains, but it has achieved limited success in changing global relations. Once the Marxists achieved power through the Bolshevik revolution in the Soviet Union, they managed to change standards of living in some countries (such as Cuba), while failing in others, such as Angola and the Republic of the Congo (Peet and Hardwick 2015, 178–180, 220–221). Most importantly, the Soviets maintained a paternalistic attitude of control over poorer countries. Ultimately, their project failed.

Significant geopolitical change was driven by movements within the colonized countries. The language of national self-determination and its association with vaguer ideas such as liberty and freedom were adopted across the periphery. Notable examples are the nationalism of Sun Yat Sen in China and the push away from Chinese feudalism through a goal of adopting Western liberal ideas and attaining independence (Cantile and Jones 1912). Ho Chi Minh shared the same vision for Vietnam, and observed the discussion resulting in the post–World War One Treaty of Versailles (Logevall 2012, 3–4). He was thrilled by the rhetoric that empire was dead, and that national freedom was achievable. Yet racism and control remained the dominant ideologies and practices of the West. In the end, Ho Chi Minh led the violent resistance to end French colonization of Indochina and, subsequently, the war against the U.S. attempts to maintain core control of this part of the periphery, what we call the Vietnam War.

Chinese, South-east Asian, and African projects of anti-imperialism were largely inspired by the words and actions of Mahatma Gandhi and the Indian movement against British imperialism. At the time, the idea that the use of nonviolence could end empire seemed ridiculous. Gandhi was ridiculed, especially by Winston Churchill (Herman 2008, 359). Yet his appeal to humanity against empire was a very successful use of geographical scale to change the nature of the debate. The bravery of his followers to stand up against imperial violence was astonishing. It blended peace with anti-imperialism and a sense of human development (the dignity and value of the individual). These ideas were subsequently put into practice by the civil rights movement in the United States under the leadership of Dr. Martin Luther King.

The slow, but steady, success of anti-imperial movements fostered a vast change in the practice of global violence: the end of formal imperialism. However, successes in achieving national self-determination occurred within another global geopolitical context: the Cold War. Newly independent countries, often in need of financial support, sought help from one side of the Cold War or another. Their independence was limited, and often peace was elusive as internal struggles became arenas for a proxy war between East and West, for example in Angola and Nicaragua (Halliday 1983, 92–96). But not all countries wanted to be pawns within the Cold War.

The Non-Aligned Movement (NAM) was a political project to organize Asian and African countries and be somewhat separate from the U.S.- and Soviet Union–led blocs that defined the geopolitical landscape of the Cold War. The leaders of India (Jawaharlal Nehru), Egypt (Gamal Abdel Nasser), and Yugoslavia (Josip Broz Tito) were the dominant personalities of the movement. The key moment in the formation of the NAM was a conference held in Bandung, Indonesia, in 1955. It notably engaged China as a partner. The West readily dismissed NAM's achievements, primarily because it seemed to interfere with the goal of the United States to lead a global order (Carafano et al. 2022). Seen simply through the lens of interstate competition, it could be seen as a futile endeavor through its inability to challenge the actions of the superpowers. However, the Movement published a set of principles at its 1955 Conference. And these are important to note for their attempt to seek pathways to positive and negative peace (The South Centre 2017). The call was for a respect for human rights, territorial sovereignty, and racial equality. These goals were to be achieved through an international order that did not enable arm-twisting by stronger external countries, even in the guise of collective security agreements. Disputes were to be resolved through international arbitration through, to quote principle nine in full, "Promotion of mutual interests and cooperation" (The South Centre 2017). In sum, the path to peace at the scale of the body and a nation's search for positive peace were to be found through a world order of equal countries acting in a legal and respectful manner.

The superpowers won. Many members of the NAM had little choice within the power dynamics of the Cold War but to align with one country or another. However, the goals of the Bandung principles are embedded within key contemporary UN documents, especially the Millennium Development Goals and their revision to address sustainability (Sustainable Development Goals Fund n.d.). The connection of peace and security at the scale of the body with the recognition that societal and global change is an imperative that combines negative and positive peace at scales from the body through the national to the global. Furthermore, the recognition that a just and equitable global scale (or in their words a "global partnership") is a required goal to enable peace and development at lower scales is evident. This commitment to a global perspective has been emphasized because of the challenges posed by global climate change, and the renewed blend of development goals with sustainability (Sustainable Development Goals Fund n.d.).

The pursuit of positive peace at multiple scales continues to evolve. Nongovernmental organizations, the UN, and countries all engage in development programs that see the scale of the body (in terms of physical health, well-being, and personal growth) within the vibrancy of communities and national stability (McCandless 2019). The stated intention may well be to ameliorate the conditions of the world's poorest and nurture contexts that will allow for personal security and national growth. However, foreign aid and the politics of development remains connected with geopolitical competition. China's emergence as a global player, especially through the construction of infrastructure has led to competition with the United States for influence within peripheral countries. The island nations of the Pacific region are a particularly acute arena (Medcalf 2020, 111–116). Despite the Western rhetoric of development and "rules-based" cooperation, and the Chinese rhetoric of "South-South cooperation" it is sadly easy to see the wealthiest countries acting in self-interest and with an eye to geopolitical competition rather than a primary interest in the well-being and future of the world's poorest (Flint and Waddoups 2021). The current move to economic nationalism was a feature of the 1930s and path toward World War Two.

Conclusion

In the wake of World War Two the United States promoted the ideology of developmentalism to rethink core–periphery relations. Postwar peace was meant to enable economic growth for a prosperous and democratic world that could be turned away from the appeals of global communism. Instead, the ideas of developmentalism went hand-in-hand with new forms of violence enacted by the core within the periphery. Within the context of the Cold War, the United States and its allies fostered coups and supported repressive regimes in the name of protecting "economic interests."

For all the faults of developmentalism I have identified, I agree with Rostow ([1960] 2017, 166) that "the fate of those of us who now live in the stage of high mass-consumption is going to be substantially determined by the nature of the preconditions process and the take-off in distant nations." In other words, the likely future of relations between the wealthiest and most powerful countries in the world will largely be a function of the relations between rich and poor countries, and the indirect relations between rich countries as they compete within poorer countries for global influence. In fact, we seem to be entering a new period of competitive developmentalism. The relative openness of a globalized economy is currently under challenge as China and the United States compete for control of the most profitable manufacturing industries, especially high-level computer chip production. Economic nationalism is going hand-in-hand with new visions of developmentalism, especially China's hope to access raw materials and manufactured goods for its domestic industry, and provides an outlet for exports. In turn, we are seeing a geography of trading blocs, sanctions, and tariffs that are mirrored in new security relations.

Just as Rostow's developmentalism was framed within the context of the global hostilities of the Cold War, we are in danger of seeing development only through the lens of geopolitics. And yet we are facing a planetary crisis in the form of global climate change, millions of people still live in abject poverty, and women and minorities face challenges to basic human rights. The impending global crisis is an opportunity for us to rethink the geographies of peace, violence, and global inequality. The impact of current and historic fossil fuel–driven economic growth in some countries upon bodies and communities in other parts of the world is an opportunity to expose the power relations of contemporary imperialism (Galtung 1971). The impending mass global migration driven by environmental change will make attempts to ignore the linkages between rich and poor places across the world impossible. When the destitute that have been forced to move through flood or heat knock on the doors of the rich, then connections will be made. Reflection on the impact of the bodies of those suffering famine catalyzed by global climate change caused by the economic activity in other countries will become stark. One reaction is to build walls, mental and physical, to deny these connections and prevent movement. The alternative is to reflect on what the geographic perspective has to offer: Scales and places are connected and peacebuilding that links bodies, places, countries, and the inequality of the global economy is imperative. The goals of development need to be defined and attained with an eye to positive peace at numerous scales for the sake of people and the planet. Currently, development remains a weapon in the drift to a new global military confrontation.

Note

1 China's rhetoric is aimed at increasing trade and investment ties with peripheral countries by claiming that these economic relations will not be exploitive, and in the process emphasizing that the history of U.S. and European relations have been based on imperialism in classic and contemporary forms.

References

Agnew, J. 2003. *Geopolitics: Re-visioning World Politics*. 2nd ed. Routledge.

Angell, N. 1910. *The Great Illusion*. G.P. Putnam's & Sons.

Blaut, J.M. 1993. *The Colonizer's Model of the World: Geographical Diffusionism and Eurocentric History*. The Guilford Press.

Brautigam, D. 2009. *The Dragon's Gift: The Real Story of China in Africa*. Oxford University Press.

Cantile, J. and C. Jones. 1912. *Sun Yat Sen and the Awakening of China*. Fleming H. Revell Company.

Carafano, J. J., M. Primorac, and D. Negrea. 2022. The new non-aligned movement. *The Heritage Foundation*. Accessed February 6, 2023. https://www.heritage.org/global-politics/commentary/the-new-non-aligned-movement

Conrad, J. 2018 [1902]. *The Heart of Darkness*. Suzeteo Enterprises.

Cullather, N. n.d. Bomb them back to the Stone Age: An etymology. *History News Network*. Accessed February 5, 2023. https://historynewsnetwork.org/article/30347

Driver, F. 1991. Henry Morton Stanley and his critics: Geography, exploration, and empire. *Past and Present* 133: 134–166.

Essex, J. 2013. *Development, Security, and Aid*. University of Georgia Press.

Flint, C. 2022. *Introduction to Geopolitics: Fourth Edition*. Routledge.

Flint, C. and P.J. Taylor. 2018. *Political Geography: World-Economy, Nation-State and Locality*. 7th ed. Routledge.

Flint, C. and M. Waddoups. 2021. 'South-South Cooperation' or core-periphery contention? Ghanaian and Zambian perceptions of economic relations with China. *Geopolitics* 26: 889–918.

Flint, C. and X. Zhang. 2019. Defining geopolitical context: China's dynamic foreign policy within global economic and hegemonic cycles. *Chinese Journal of International Politics* 12: 295–331.

Galtung, J. 1965. On the meaning of nonviolence. *Journal of Peace Research* 2: 228–257.

Galtung, J. 1969. Violence, peace, and peace research. *Journal of Peace Research* 6: 167–191.

Galtung, J. 1971. A structural theory of imperialism. *Journal of Peace Research* 8: 81–117.

Gulati, U.C. 1992. The foundations of rapid economic growth: The case of the four tigers. *American Journal of Economics and Sociology* 51: 161–172.

Halliday, F. 1983. *The Making of the Second Cold War*. Verso.

Hayter, T. 1981. *The Creation of World Poverty*. Pluto Press.

Herman, A. 2008. *Gandhi and Churchill: The Epic Rivalry That Destroyed an Empire and Forged Our Age*. Bantam.

Hobson, J. A. 1902. *Imperialism: A Study*. James Nisbet & Co.

Huntington, E. 1915. *Civilization and Climate*. Yale University Press.

James. L. 1994. *The Rise and Fall of the British Empire*. St. Martin's Griffin.

Khan, T., S. Abimbola, C. Kyobutungi, and M. Pai. 2022. How we classify countries and people – and why it matters. *BMJ Global Health* 7:e009707. doi:10.1136/bmjgh-2022-009704

Koopman, S. 2011. Let's take peace to pieces. *Political Geography* 30: 193–194.

Lenin, V.I. 1939. *Imperialism: The Highest Stage of Capitalism*. International Publishers.

Logevall, F. 2012. *Embers of War: The Fall of an Empire and the Making of America's Vietnam*. Random House.

MacGinty, R. 2021. *Everyday Peace: How So-called Ordinary People Can Disrupt Violent Conflict*. Oxford University Press.

McCandless, E. 2019. Beyond liberal and local peacebuilding – three critical framings to approach the complexity of conflict and fragility in Africa. *Africa Insight* 49. Accessed February 17, 2023. https://hdl.handle.net/10520/EJC-1d51ac91d0

McDevitt, M.A. 2020. *China as a Twenty First Century Naval Power*. Naval Institute Press.

Medcalf, R. 2020. *Indo-Pacific Empire: China, America and the Contest for the World's Pivotal Region*. Manchester University Press.

Morrissey, J. 2017. *The Long War: CENTCOM, Grand Strategy, and Global Security*. University of Georgia Press.

O'Lear, S. ed. 2021. *A Research Agenda for Geographies of Slow Violence*. Edward Elgar.

Peet, R. and E. Hardwick. 2015. *Theories of Development: Contentions, Arguments, and Alternatives*. 3rd ed. The Guilford Press.

Power, M. 2019. *Geopolitics and Development*. Routledge.

Ratzel, F. 1896. *The History of Mankind (Völkerkunde)*. Two volumes. Macmillan.

Robinson, C.J. 1983. *Black Marxism: The Making of the Black Radical Tradition*. Zed Press.

Rostow, W.W. [1960] 2017. *The Stages of Economic Growth: A Non-Communist Manifesto*. Martino Fine Books [Cambridge University Press].

Said, E. 1979. *Orientalism*. Vintage Books.

Semple, E. C. 1911. *Influences of Geographic Environment: On the Basis of Ratzel's System of Anthropo-Geography*. Henry Holt & Co.

Spivak. G. C. 1988. Can the subaltern speak? In *Marxism and the Interpretation of Culture*, ed. C. Nelson and L. Grossberg, 271–313. University of Illinois Press.

Sustainable Development Goals Fund. n.d. From MDGs to SDGs. Accessed February 6, 2023. https://www.sdgfund.org/mdgs-sdgs

Taylor, P.J. 1981. Geographical scales within the world-economy approach. *Review* 5: 3–11.

Taylor, P.J. 1989. The Error of Developmentalism. In *Horizons in Human Geography*, eds. D. Gregory and R. Walford, 303–319. Palgrave Macmillan.

Taylor, P.J. 1992. Understanding global inequalities: A world-systems approach. *Geography* 77: 10–21.

Terlouw, C.P. 1992. *The Regional Geography of the World-System*. Faculteit Ruimtelijke Wetenschappen Rijksuniversiteit.

The South Centre. 2017. Non aligned movement and Bandung principles as relevant today as ever: South Centre. Accessed February 6, 2023. https://www.southcentre.int/question/non-aligned-movement-and-bandung-principles-as-relevant-today-as-ever-south-centre/

Wallerstein, I. 1979. *The Capitalist World-Economy*. Cambridge University Press.

Wallerstein, I. 2004. *World-Systems Analysis: An Introduction*. Duke University Press.

Williams, P. and F. McConnell. 2011. Critical geographies of peace. *Antipode* 43: 927–931.

Woodruff, M. 2012. These guys found out the hard way what it's like to live on $1 a day. *Business Insider*, December 6. Accessed February 3, 2023. https://www.

businessinsider.com/what-its-like-to-live-on-1-a-day-documentary-chris-temple-zach-ingrasci-2012-12

World Bank. n.d.-a. Life expectancy at birth, male (years). Accessed 3 February 2023. https://data.worldbank.org/indicator/SP.DYN.LE00.MA.IN

World Bank n.d.-b School enrollment, primary (%). Accessed 3 February 2023. https://genderdata.worldbank.org/indicators/se-prm-enrr/?view=correlation

Young, M.B. n.d. No date. Bombing civilians: An American tradition. *History News Network*. Accessed February 5, 2023. https://historynewsnetwork.org/article/67717

8

POSTCOLONIAL CONFLICT IN SOUTHEAST ASIA

Rethinking the shatterbelt with colonial rupture in Asia's Cold War

Christian C. Lentz and Scott Kirsch

Introduction

Decolonization in Southeast Asia after the World War Two was deeply entangled in a series of violent global conflicts. During the war, an expansionist Japan swept aside European and American empires, opening oppositional spaces for nationalist mobilization during wartime occupation. Shortly after the Japanese surrender in 1945, nationalist leaders first in Indonesia and then in Vietnam declared independence, but the Netherlands and France refused, moving instead to reconquer their former colonies (Chandler et al. 2016). Resurrecting colonial ambitions sparked wars for self-determination marked by persistent civil strife. Fought in different insular and mainland contexts, these two struggles ultimately secured independence—Indonesia in 1949 and Vietnam in 1954—but not before ensnaring both in Asia's decades-long Cold War (1948–1990). Across the South China Sea in the Philippines, the postwar transition from Commonwealth to Republic (1946) was followed by establishment of extensive U.S. basing rights alongside enduring civil conflict on ideological and sectarian grounds. As we will see, what had begun as anti-colonial conflicts and postcolonial transitions became enmeshed in superpower rivalries, the consequences of which carried into postcolonial conflicts that would again roil the region from the 1960s onward.

To call this period "Asia's Cold War" is not to overlook local agency or the local stakes of conflict in favor of the conflict's global dimensions. Rather, it asserts the ways that conflicts occurring at multiple scales, among multiple and diverse agents, concatenated with a renewal of foreign intervention in the wake of Euro-American colonialism, initially along ideological lines of Cold War communism and anti-communism. In Southeast Asia, this brutal interplay of

DOI: 10.4324/9781003345794-8

internal conflict, regional fragmentation, and great power competition would leave millions dead, mostly noncombatants, in some of the century's most devastating conflicts. Asia's Cold War, in other words, was searingly hot, reflecting not the celebrated logic of deterrence but one of pervasive warfare in multiple contexts and forms (Chamberlin 2018).

As geographers sought to explain the persistence of conflict "after" World Wars One and Two, the classical geopolitical idea of the *shatterbelt*—a "region torn by internal conflicts whose fragmentation is increased by the intervention of external major powers" (Cohen 2015, 9; see also Kelly 1986; Hensel and Diehl 1994)—appeared to offer a useful syntax.

Though never widely adopted in other disciplines, the shatterbelt idea remains compelling as a spatial concept that integrates varied and competing agencies operating at multiple and intersecting spatial scales. As a lens onto postcolonial conflict, however, we argue that the concept might acquire greater efficacy if it is read through notions of *colonial rupture* (Lund 2020; Rasmussen and Lund 2017), as we discuss in the next section. Colonial rupture offers a historical sensibility that more adequately contextualizes, rather than naturalizes, the brutal violence of *Southeast* Asia's Cold War across multiple sites and scales. In subsequent sections, we engage three distinct national settings— Vietnam, Philippines, and Indonesia—which, while not meant to be representative of Southeast Asia, offer a means of "triangulating" the geopolitical moment (see Figure 8.1), highlighting the multi-scalar relations around which "Southeast Asia," as a distinctive regional setting for postcolonial conflict, took shape (Kratoska et al. 2005; Tyner 2007). Although our analysis pertains largely to the Cold War era, our approach remains relevant to understanding contemporary intersections of geographies of peace and conflict with local politics and global strategy, as discussed in the conclusion.

On shatterbelts, crush zones, and colonial rupture

Among the tasks for an emerging twentieth-century political geography, alongside planetary mappings of land and sea powers, insular crescents, and visions of civilizational conflict, was the work of identifying contested zones, belts, and regions across the earth's surface. The aim was to represent the world by "folding geographical difference into depluralized geopolitical categories" (Kearns 2009, 259), thereby developing a range of spatial concepts intended to provide stable—and actionable—geopolitical meaning for a world in flux. Conceptions of a world geographically *shattered*—dashed into fragments, "broken in pieces by a sudden blow or concussion," or left in ruinous damage (OED)—track with the efforts of classic geopoliticians, providing a simplified model for understanding, or explaining away, complex patterns of conflict and persistent violence across a shifting historical and geographical terrain. For Halford Mackinder (1904, 422), the outcome of the emerging "post-Columbian" age,

FIGURE 8.1 Southeast Asia.

a world without "new" territories for European empires to conquer, would be the rise of a "closed political system, and none the less that it will be one of worldwide scope." Henceforth, he insisted,

> Every explosion of social forces, instead of being dissipated in a surrounding circuit of unknown space and barbaric chaos, will be sharply re-echoed from the far side of the globe, and weak elements in the political and economic organism of the world will be shattered in consequence.
>
> *(Mackinder 1904, 422)*

Mackinder's ideas, notwithstanding their rampant imperialism and social Darwinism, have drawn reconsideration for their rendering of imminent geopolitical and geo-economic change (cf. Smith 2003). But what of the weak and shattered elements of the global whole that Mackinder's imagery evokes?

Some political geographers, like Isaiah Bowman (1931), in his mappings of frontier zones and "pioneer belts," would turn to colonialism's expansionist endgame, attempting to develop a "science of settlement" of marginal lands, both domestic (U.S.) and international, in anticipation of new global population pressures (see Smith 2003, 211–234). Others coming to grips with World War One and its aftermath turned to the idea of "bufferzones" between great powers, characterized by political fragmentation and violence exacerbated by external forces, to explore the implications of Mackinder's re-echoed explosions of social forces, albeit from a relatively narrow range of top-down perspectives. For example, James Fairgrieve (1915, 329–330) would describe the fraught position of a contiguous "crush zone" of small states, stretching around the Russian empire from the Baltic Sea to the Korean Peninsula, located broadly between the Russian "heartland" and surrounding "sea powers" as a region of buffer states which had "gradually come into existence" between the great powers. In his own reassessment of history's geographical pivot, Mackinder (1919) would similarly point to a band of Central European states extending from the Baltic to the Adriatic, occupying territories between Russia and Germany. Mackinder argued that the ethnic and economic diversity of these states and lack of regional cooperation resulted in outside intervention and dynamics of escalating conflict, as external powers competed for influence, alliances, and resources For John Frederick Unstead (1923, in Kelly 1986, 163), this "belt of political change" could similarly be understood as a source of wider conflicts, as "weak, antagonistic, dependent states" became caught up "within the interests of outside larger nations." The idea of the crush zone or shatterbelt, through which heterogeneity and weakness could be identified as a source of danger (and rationalization for intervention), thus performed a kind of meaning-making cultural work: It allowed conflict to be naturalized as the outcome of seemingly geological forces while shifting the blame for war to local patterns of difference rather than imperialist motives. Two decades on, Richard Hartshorne, in an essay addressing "The United States and the 'Shatter Zone' of Europe" (1944, 203), built on these earlier framings to point to the "thirteen countries situated between Scandinavia, Germany, and Italy on the one hand, and the Soviet Union and Turkey on the other … the 'shatter zone' of Europe, in the words of German propaganda of the interwar period." Hartshorne (1944, 203) argued that this was a geography Americans ought to know about, since "for the second time in a generation we are engaged in a war that originated in that belt."

While scholars thus framed Eastern Europe as a shatterbelt reflecting persistent dangers, even to the distant United States, the term was made geographically "modular" during and after World War Two with reference to the Middle East and Southeast Asia (Spykman 1944) and later with reference to sub-Saharan Africa (Cohen 1973). The physiographic geopolitical language of *shatterbelts* seemed well suited to the ideologically driven rivalries and new imperialisms of the Cold War. But the shatterbelt was a construction of region and scale that made sense from the outside, the top-down perspective of great power politics.

In his efforts develop "a more realistic classification of areas in terms of their political organization," Hartshorne (1941, 45) had been at pains to stress the post-imperial dimensions of Eastern Europe after the dismantling of Austria-Hungary in 1919. Associating persistent instability with promiscuous "zones of mixture," among varied cultural and religious groups in Eastern Europe, Hartshorne (1941, 50) lamented that "almost nowhere" could "sharp boundaries between the many different linguistic groups" be rendered satisfactorily.

Ironically, Hartshorne (1941) framed a response to this world of ambiguous boundaries by drawing a map of clearly defined regions and simple classifications. His analysis of the world of "potential states" comprised "pseudo-states," "repressed areas," and zones of "passive acceptance"—as distinguished from "organized state areas" (see Figure 8.2). He was less willing to extend political subjectivity to non-European peoples, to say nothing of the vast swaths of South America, Africa, Asia, and the far north which Hartshorne (1941, 46, 55) blithely deemed "permanently colonial" or "permanently dependent areas."

Ordinarily, a failure of prescience—in the work of a scholar writing amid a complex and overdetermined world-in-crisis—might be considered an unfair basis of critique. But Hartshorne's map fails so spectacularly to anticipate the possibility of postcolonial transition in many parts of the world that it serves to raise questions about the political geographer's deeply Euro-centric, decontextualized, and ahistorical understandings of colonialism and geographical difference. The map also allows us to explore the relations between colonialism and great power politics that concepts like the shatter-zone allowed political geographers to elide. Hartshorne's characterization of French Indochina among the world's spaces of "passive acceptance" was among the map's more regrettable geographical assertions, as recent history might have made clear to Hartshorne. Vietnamese peasants, workers, intellectuals, monarchists, and nationalists resisted French rule often and in organized fashion, including two large-scale revolts in 1930–1931, both bloodily suppressed, that convinced many that there could be no compromise with European empire, never mind "passive acceptance" of its French variant (Long 1973; Brocheux and Hemery 2009).

Accelerating after World War Two and unfolding through the Cold War, decolonization in Asia drove a world-historical transformation that reshaped a global geopolitical order but did not mark a clean break with its past.

Here, we introduce the idea of *rupture* to characterize decolonization as both a pivotal transition and an "open moment" of indeterminate spatiotemporal reordering (Lund 2020, 105). In our view, the process of decolonization

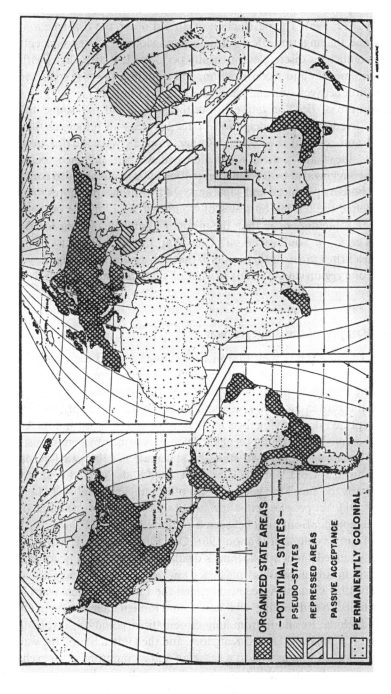

FIGURE 8.2 Richard Hartshorne's map of organized state areas, potential states, and permanently colonial areas.

in Asia resembles a frontier space where "the violent destruction of previous orders takes place and the territorialization of new orders begins" (Rasmussen and Lund 2017, 396). The destruction of an old order, the construction of a new one, and the lingering effects of associated violence resist a conventional focus on conflicts as discrete events neatly periodized. It also aligns with discussions linking war and reconstruction, and troubling any easy spatiotemporal distinction between war and peace (Kirsch and Flint 2011).

The rupture of decolonization in Southeast Asia did not sound the full liberation, freedom, or emancipation of its peoples—all principles that had motivated anti-colonial revolutions. Rather, empire's wake carried powerful institutions, ideas, and practices into independence that inflamed social tensions further aggravated by foreign intervention. The landscape also contained territorial borders produced and superimposed by European empires that continue to define the ten nation-states of Southeast Asia.[1] These postcolonial conflicts are something *more* than the inevitable outcome of their location in the "crush zone" of Southeast Asia's Cold War. Instead, they must be understood with a firmer grasp of their roots in the past, especially during the violent transition of a region caught up in the wake of shattered empires as well as new rivalries.

Vietnam: Origins of the 30 years' war

The artillery fire at Dien Bien Phu in 1954 echoed around the world. On 13 March, the People's Army of Vietnam (PAVN) opened fire on fortified positions of the French Expeditionary Forces. Within days, successive barrages destroyed the French airstrip, severing its main source of supply, and PAVN infantry captured the outlying forts in what, until then, France and its American ally had considered an impregnable defense. Alarmed by communist support for Vietnam's advance and stung by stalemate in bitter warfare on the Korean peninsula, Cold War hawks in the United States offered France three atomic bombs to "save" the garrison—an offer judiciously refused (Logevall 2012, 498–501). Yet peoples across the global south saw the same battle differently: as an epic confrontation between a colonized country fighting for independence from its colonizer (Fanon 1961; Oey 1961). Indeed, PAVN's victory on 7 May sounded French Indochina's death knell, catalyzing peace talks in Geneva that, in July, earned independence for Vietnam, Laos, and Cambodia.

The July 1954 Geneva Agreements ended the First Indochina War (1946–1954) but failed to end conflict, civil unrest, and foreign intervention in former French Indochina. Resulting from long diplomatic negotiations involving the Soviet Union, China, the United Kingdom, and the United States, the warring states at Dien Bien Phu agreed to a ceasefire and a compromise. In exchange for France's withdrawal and independence for its colonies, Vietnam would be divided at the 17th Parallel, roughly midway between south and north, until elections on reunification in 1956. The elections were never held, however. The temporary partition of Vietnam became, instead, a political border between a U.S.-allied Republic of Vietnam (RVN) in the south and a Soviet and Chinese-backed Democratic Republic (DRV) in the north.[2] The peace ostensibly achieved at Geneva contained the seeds that would later grow into the Second Indochina War (1960–1975), known to Americans as "The Vietnam War." In all, armed conflict in and over Vietnam lasted 30 years (Young 1991). Ending on 30 April 1975, conflict between northern and southern Vietnams—each armed by superpowers—had already spilled into the former French colonies of Laos and Cambodia, where decolonization had likewise become violently entangled in Cold War rivalries and local disputes.

Vietnam's war of independence against France was not simply a prologue to war against the United States, as conventional American understandings would have it. Nor was an already-unified Vietnam simply fighting foreign powers, French and/or American, as official Vietnamese historiography portrays it (Pelley 2002). Rather, great power intervention in the early 1950s transformed a colonial war into an enduring Cold War hotspot (Lawrence and Logevall 2007; Chamberlin 2018), or, indeed, a shatterbelt. But understanding why the early period was formative, we argue, requires not just a "top-down" geopolitical perspective on warfare and diplomatic negotiations but also a contextual analysis of local disputes, civil strife, and colonial racism. In other words, incorporating the regional and local dimensions of conflict commensurate with decolonization helps explain why violent social dynamics lasted far into the postcolonial era and spread beyond Vietnam into Laos, Cambodia, and Southeast Asia more broadly.

The First Indochina War was embattled by multiple armed forces, competing factions, and preexisting social formations, features that scholars have increasingly recognized as civil war (Chapman 2013; Lentz 2019; McHale 2021; Goscha 2022). Certainly, war against an imperial France intent on reclaiming its Indochinese colony after World War Two was the main conflict, beginning with guerilla warfare in Hanoi in 1946 and ending with fixed-position combat in Dien Bien Phu eight years later. But Vietnamese society over this period was far from unified. Nor was its emerging state a unitary or fully representative political actor. Rather, colonial ethno-racial categories, long-standing regional-cultural differences, and sharp political-economic inequality complicated nation-building and vexed state formation.

Out of this fissiparous mix, a disciplined group of Marxist-Leninist activists rose in the 1940s to mobilize the masses, develop a conventional army, and centralize control of the DRV in a single-party state (Goscha 2022). The communist party secretly steered the Viet Minh, a unified front for national liberation that led armed resistance and popular mobilization early in the First Indochina War (Huynh 1982). Foremost among others, President Ho Chi Minh, Party Secretary Truong Chinh, and General Vo Nguyen Giap adapted a communist model of statecraft to Vietnam, including leveraging nationalism to mobilize a peasantry against a foreign power while putting its labor and capital to work in service to their rule (Lentz 2019). And yet DRV state formation was marked by regional variation and contextual situations that rebounded in lasting, cross-border conflict. Competing sovereignties during war meant that the DRV operated in "archipelago-like territorial formations" (Goscha 2022, 8) where, initially, its power was limited and highly contested. Fragmented in space and fluctuating over time, isolated DRV political formations strove to build support among the masses, connect with one another, and gather strength. The communist revolutionaries also fought off rival Vietnamese parties, eliminated factions within their ranks, and developed a powerful coercive force, the PAVN, founded in 1950 with ongoing support from the People's Republic of China (PRC).

An ostensibly unified front against colonialism was thus prone to violent splits that pitted Kinh/Viet peoples against other Kinh/Viet, Tai, Khmer, or other peoples and groups seeking political domination in unsettled spaces. In the Mekong Delta, pervasive violence split anti-colonial resistance and drove civil conflict for years to come. First, the Viet Minh largely dissolved in 1947 when one wing allied with communists opposed to France split from another wing willing to collaborate with France for eventual independence. Simultaneously, violence between Khmer and Kinh/Viet peoples broke out, in part because of differences over whose nation, Cambodia or Vietnam, represented their political aspirations. The ethnic violence spilled back and forth across the porous Cambodian border. "This messy double fracture shaped southern politics up to 1975 and beyond" (McHale 2021, 6).

Like the Mekong Delta, political actors elsewhere on Vietnam's ethnolinguistically varied terrain encountered DRV cadres, soldiers, and guerillas aiming to negotiate power relations, incorporate preexisting arrangements, and defeat rivals. The town of Dien Bien Phu, site of Vietnam's celebrated 1954 victory, was riven by racialized political-economic tensions stemming, in part, from France's colonial policy of divide and rule. The battle took place in the Black River region, a multicultural borderland with Laos and China where Kinh/Viet peoples, Vietnam's numerical majority, were a minority compared with Tai, Hmong, Khmu, Dao, and other ethnolinguistic groups (Cam Trong 1978; Le Failler 2014). Based on a long-standing "policy of races" that aimed to box the Kinh/Viet population in the Red River Delta, French administrators privileged Tai elites on this montane periphery with lighter taxes, official positions, arms and military training, and control of the lucrative opium crop (Brocheux and Hemery 2009, 62–63). As a result, the French-allied Tai Federation was the region's dominant political force at its establishment in 1949. Alienated by this colonial arrangement, rival Tai elites, Hmong opium cultivators, and poor peasants found common cause with revolutionaries hailing from the Red River Delta (McAlister 1967; Lentz 2017).

DRV cadres and Viet Minh organizers in the Black River region struggled initially less with France than the Tai Federation. Even after its defeat in 1953, civil conflict did not cease. Through combat at Dien Bien Phu and across northern Vietnam, PAVN targeted opponents of DRV rule while cadres mobilized civilian resources to feed combatants. These actions were part of a longer process of securing DRV territory and asserting its hegemonic sovereignty (Lentz 2019). After the battle, exhausted civilians protested that national liberation had delivered neither on promises of prosperity nor respite from state claims on labor and capital. Worried also about atomic bombs, many fled for Laos (Lentz 2021). Social unrest in the Black River region drove a millenarian movement, crushed by the DRV in 1957, that resonated with local *religio-cultural traditions* and aspirations for political autonomy. These tensions endured in Laos during the so-called "Secret War" (1964–1973) when, in the Second Indochina War, the CIA armed Hmong militias to fight border-crossing PAVN troops and their local allies, including other Hmong political groups (McCoy 2002; Lee 2015).

Just as violence stemming from the First Indochina War would spill into Cambodia and Laos for years to come, so did great power intervention ramp up in response to its conclusion. Two months after negotiations in Geneva, in September 1954, diplomats from the United States, the United Kingdom, France, the Philippines, Thailand, New Zealand, Australia, and Pakistan committed in Manila to defending one another from "communist aggression" (Avalon 1954). The pact created the Southeast Asia Treaty Organization (SEATO) based in Bangkok, part of U.S.-led efforts to contain communism in Asia. Meanwhile, as France withdrew from Vietnam, the United States turned its military and economic aid toward southern Vietnam, where power was fractured among heavily armed factions competing violently for supremacy (McHale 2021). In 1955, after he had defeated his rivals in Saigon and the Mekong Delta, Ngo Dinh Diem emerged as America's "Miracle Man" to lead the nascent RVN (Chapman 2013). A fervent anti-communist, he would also steer the country toward confrontation with the DRV, eventually engulfing Vietnam in conflict all over again. What the 1954 Geneva Agreements had intended as a temporary partition at the 17th Parallel had, in fact, created two Vietnams armed by Cold War rivals, amplifying a civil war that spread, in the wake of empire, through the shatterzone of Southeast Asia.

The Philippines: Subic foothold

If the wreckage of French Indochina—the 30 years' war, its violent entanglements in Laos, and the chaos and genocide it helped to spawn in Cambodia—constituted the brutal core of Southeast Asia's Cold War, it was by no means the only site of social upheaval and mass violence in the region. Nor did it contain the only settings to have made such conflicts possible. As historian

Alfred McCoy observes (2016, 975), the remarkable coincidence of forces and events that swept across Asia at the close of World War Two arrived together with "a discordant element," that is, "the decolonization that marked the end of European empires." In this "comingled, even contradictory mix of war and revolution," nationalism and communism marked the onset of the Cold War era "across a vast swath of Asia—from India to Burma, Malaya, Indonesia, the Philippines, all the way to Vietnam, China, and Korea." Including Japan and its empire, he adds, "virtually the whole of Asia felt the impact of this imperial transition" (McCoy 2016, 976). This continental-scale rupture was experienced differently in the Philippines, where the former colonizer was a rising global hegemon, at once an agent of European decolonization and an active participant in military efforts to control the Asian "Rimland" in a new Pax Americana (see Smith 2003, 347–373).

For McCoy (2016, 977), persistent U.S. military presence after independence made the Philippines "the awkward exception" in Asia's imperial transition. Having already gained U.S. congressional authorization for independence in 1935 (concurrent with the establishment of the Commonwealth), Filipino leaders rapidly secured independence after the war in 1946 in an "orderly transition" of sovereign power. Though it is questionable whether *orderly* can be applied to a nation invaded twice in four years (by Japan in 1941 and the United States in 1944–1945) and cantering toward a new war of peasant resistance with its roots in *Hukbalahap* (or *Huk*) resistance to the Japanese occupation (Kerkvliet 1977). When independence came, McCoy argues (2016, 977), "it was soon circumscribed by massive U.S. military bases that remained throughout the forty years of the Cold War," interrupting the regional pattern of Japanese occupation and subsequent independence. As sites geared to project military force to a wider area, the impacts of the bases were geographically extensive, as well as pervasive in Philippine politics and society. Meanwhile, the archipelago's regional situation in U.S. geopolitics shifted from one of overstretched imperial vulnerability to offshore staging ground, rear base, and logistics hub. But basing agreements also offered resources for Filipino agents working toward their own ends. The reconstruction of the Philippines as a bulwark against communism can be seen as the outcome of relations between a Filipino political class intent on building power and gaining resources *through* anti-communism (Woods 2020), and U.S. leadership committed to an ongoing alliance with its former colony and to its Subic "foothold" on the South China Sea.

The classical idea of a shatterbelt describes a "two-tiered structure of conflict" between a local layer, characterized by political turmoil, social fragmentation, and economic malaise and an "international" layer distinguished by "great power competition for footholds among various states of the region" (Kelly 1986, 163). If there were two "tiers" of conflict in the Philippines, then they collapsed into one another, or became generalized, in particular settings. Even before independence, Commonwealth President Sergio Osmeña, keen to guarantee funding for the reconstruction of Manila, had agreed to a statement of general principles guaranteeing continued use of American military bases. These bases had already been conceived by U.S. leaders as "springboards" rather than defensive positions in Asia's emerging "postwar" order (McCoy 2016). Hence reconstruction, alongside decolonization, served the United States as a "hegemonic strategy" (Kirsch and Flint 2015, 39), helping to make other states want what it wanted. A 1947 treaty, ratified unanimously by the Philippine Senate and broadly supported in a national plebiscite, granted the United States a 99-year lease on 23 military installations in the archipelago, including then little-used Subic Naval base, initially under U.S. military jurisdiction (including the neighboring city of Olongapo), without limits on their use for offensive operations. In turn, the Philippines received $620 million for war reconstruction (Berry 1989, 36), and the continuing support of its former colonizer around an agenda that linked anti-communism, reconstruction, and development both inside and beyond the Republic's territorial borders.

Hosting the formation of SEATO in Manila in 1954 was, of course, no geographical accident. By the time the new member states pledged their mutual defense in the face of communist aggression, following U.S. bilateral treaties with Japan (1951), Philippines (1951), South Korea (1953), and Taiwan (1954), plans for what would become a $170 million renovation of Subic Bay were underway (McCoy 2016, 996). Turning Subic into the "service station" of the Seventh Fleet, including the "relocation" of the barrio of Banicain (and leveling of a mountain) to produce new land for an airstrip, was completed in 1956. Along with nearby Clark Air Force Base, headquarters of the Thirteenth Air Force, Subic would be among the largest U.S. overseas bases for the next four decades. No longer a reflection of earlier—historically justified—naval concerns over U.S. overstretch in the Philippines (see Kirsch 2023, 23–28), Subic was positioned as a rear base in a network of bases, basing rights, and circulating forces surrounding ("containing") Asia from Japan, Okinawa, to the Philippines, backed by Guam, the enduring U.S. military colony in the Western Pacific within B-52 Bomber range of Cambodia (Rimmer 1997; McCoy 2016).

While the terms of the lease would be made more palatable to Filipinos in subsequent negotiations regarding local sovereignty issues, local opposition was initially tamped down, paving the way for the use of the base as a launching point for interventions in Indonesia (1958) and the Taiwan Straits (1958), Thailand (1962), and after the Gulf of Tonkin incident in August 1964, as a key staging area and logistics hub during the Second Indochina War (Rimmer 1997, 328). Indeed, U.S. intervention in Vietnam can crudely be traced through the circulation of people and things through Subic. Monthly ship visits increased from an average of 98 in 1964 to 215 by 1967, for example, with an astonishing 4.2 million sailor visits over the year. The Naval Supply Depot distributed more than four million barrels of fuel oil per month to meet the needs of Subic and Clark, the latter connected by a 41-mile pipeline (Rimmer 1997).

At a local scale, the accumulation of such geopolitical "rounds of investment" would have profound impacts. The ship repair facility included three dry docks that, by the 1960s, regularly employed 15,000 Filipino workers (McCoy 2016; Rimmer 1997). The military complex of Subic and Clark also provided a training grounds and rear base for U.S. Marines in Vietnam and Northeast Asia. Not surprisingly, the city of Olongapo, which did not gain independence from U.S. jurisdiction until 1959, grew rapidly, gaining in population from 45,000 in 1959 to perhaps 125,000 permanent residents during the 1960s—along with an estimated transient population of 50,000 (Rimmer 1997). Hence, in addition to the industry of the port, Olongapo was built around the boom-and-bust economy of ship visits which encouraged specific sectors, including sex work (largely decriminalized), rental housing, drug-running, bars, discos, and massage parlors at sites including Olongapo's Barrio Baretta (Rimmer 1997; Enloe 2000; Reyes 2019).

As Cynthia Enloe (2000) has described the pervasive links between military basing and sex work in Southeast Asia (including Subic and Clark in particular), the bases must also be understood as part of a wider network of migration and remittance flows. Thousands of (primarily) young women migrated from predominantly rural homes to sexualized entertainment businesses near military bases, even as many were themselves cut off from returning home due to social taboos. The war economy—and peace dividend—thus offered economic rewards for some alongside disruptive social transformations around the military enclaves.

The impacts of basing agreements spread broadly within Philippine borders, chiefly in providing support and military assistance to the new Philippine state. Along with postwar rebuilding, the total estimated $700 million in U.S. support for the Philippines from 1946 to 1950 was used to finance the new government and its armed forces (Kerkvleit 1977). Initiated under U.S. postwar occupation and carrying forward through independence, efforts to disarm, harass, and arrest former anti-Japanese allied *Hukbalahap* guerillas after the war contributed to the rise and fall of the insurgency over the next decade. Largely a war of peasant resistance, Filipino leaders reframed the *Huk* insurgency in more ideological terms as a means of attracting U.S. funding, including extensive CIA assistance (Kerkvleit 1977; McCoy 2009). U.S. support poured in to the tune of $500 million from 1951 to 1956 (about $5.5 billion in present US dollars), including $117 million designated for Philippine military assistance (Kerkvleit 1977, 244). U.S. support under the mutual security and training program also distributed surplus military gear and weaponry: gunboats, aircraft, tanks, guns, ammunition, mortars, bombs, jeeps, communication technology, and spare parts—materiel that would help the Philippine government crush the rebellion by the mid-1950s (Kerkvleit 1977). The CIA played an active role in Philippine politics, from supporting status quo leaders in electoral campaigns to planting a labor organization led by an anti-communist

American priest during an era of "lead-pipe diplomacy" on Manila's docks (Hawkins 2022, 150–199). The geography of conflict in the Philippines thus reflected the multi-scalar and distributed dimensions of postcolonial warfare (and collapsing of distinctions between "levels") across multiple social, political, and economic sites.

Alongside other U.S. allies, Filipino leaders would continue to benefit from transforming "local political struggles into sites of global communist revolution and international warfare" (Woods 2020, 3) in Asia's Cold War, and by leveraging new terms for maintaining use of the bases alongside the increased traffic of the Second Indochina War. Under President Ferdinand Marcos in 1966, the 99-year lease on military bases was reduced to 25 additional years, allowing Marcos to appeal to nationalist dissatisfaction across the political spectrum while increasing his own bargaining power. Unsurprisingly, when Marcos declared martial law in 1972, consolidating his power around a generalized state of warfare in response to student and worker protests over electoral corruption, along with growing Maoist and sectarian insurgencies, the United States did not abandon its "man in Asia." Indeed, even after the United States conceded the war in Vietnam, Subic and Clark remained important links in the chain of bases that would extend to the atoll of Diego Garcia and the "Indo-Pacific." Marcos continued to leverage the bases as a means of securing U.S. support to help legitimize his authoritarian rule (McCoy 2016), until the regime was brought down by internal contradictions and People Power in 1986. In 1991, against the wishes of Philippine President Corazon Aquino, the Philippine Senate by a vote of 13–12 rejected an extension of the Military Bases Agreement, effectively terminating an American foothold which had endured for the length of Asia's Cold War.

Indonesia: Revolution and counterrevolution

From October 1965 through 1966, a half-million to one million alleged communists in Indonesia were murdered by their countrymen, sometimes mobs and neighbors, but more often paramilitary organizations, street gangs, or Islamic youth groups. Many more were jailed for years inhumanely in what remains "one of the largest and swiftest yet least examined instances of mass killing and incarceration in the 20th century" (Robinson 2018, 3). The army and police stood by, sometimes assisting with transport, detention, and logistics. Political violence strayed into communal violence targeting ethnic Chinese, and opportunists took advantage of the disorder to settle local scores. The scale of the killing itself, as well as the incarceration and ongoing intimidation of survivors, amounts to a profound erasure of historical memory in Indonesia.

In addition to the brute fact of loss and its lasting effects on Indonesian society, the Cold War raised the ideological stakes motivating the killers and justifying their actions ever after. Because the violence ushered a shift in national leadership—the fall of President Sukarno and rise of General Suharto—the perpetrators, allied with the latter, were able to displace responsibility, impugn their victims, and taint the historical record. The political and communal bloodbath came on the heels of the abduction and killing of six anti-communist generals by leftist military officers on 1 October 1965. Known as the "September 30th Movement," the military crisis became an opportunity seized by Suharto—head of the army's strategic reserve—to counterattack, regain control, and, eventually, usurp the presidency. Suharto and the army quickly blamed the movement on the home-grown Communist Party of Indonesia (PKI) and on the PRC, accusing the latter of backing the PKI in a plot to spark a national uprising and turn Indonesia communist. Thus did the general come to command the armed forces and an ideological narrative, helping him sideline Sukarno, butcher the political opposition, scapegoat Chinese at home and abroad, and win U.S. support. Suharto would go on to rule Indonesia for decades, entrenching myths around his rise that have outlasted his fall in 1998 and persist into an era of democracy for Indonesia and renewed scholarly inquiry into a pivotal Cold War episode (Roosa 2006; Zhou 2019).

To locate Indonesia's mass killings in a shatterzone, as we do, brings our case studies of Southeast Asia's mid-twentieth-century rupture into world-historical comparison, one that examines the region's national, anti-colonial revolutions in relation to the global geopolitical shift from European empire to U.S. hegemony. Like Vietnam, Indonesia's declaration of independence after Japanese occupation in World War Two had destabilized European domination and opened spaces for oppositional organizing (McCoy 2016).[3] Like France, the Netherlands at first benefited from British troops on the ground and then sought to reconquer its former colony, precipitating a colonial war, euphemized by the Dutch as a "police action." Violence in the resulting Indonesian Revolution (1945–1949), however, was not directed simply against foreign foes, as nationalist historiography would have it, but strayed also into civil strife between rival political factions, ethnicized animosities, and sectarian movements (Sidel 2021). As in the Philippines and southern Vietnam, the United States initially helped deliver Indonesia's national independence but ultimately circumscribed its sovereignty in pursuit of heating Cold War anti-communism and containment of China and northern Vietnam (Ngoei 2020). Most of all, the violence that characterized Indonesia's break with Dutch colonialism, like that in Vietnam with the French, escalated steadily as the Cold War heated up.

By analyzing the mass killings in 1965–1966 as an instance of postcolonial conflict, we understand this episode, including its antecedents and aftermath, in relation to Indonesia's prolonged and incomplete rupture with colonial domination. Indonesia's "revolution" was less an egalitarian social transformation led by a disciplined party, as in Vietnam, than an anti-colonial movement cobbled out of contending forces and disparate regions that came together nationally to topple an old regime and achieve independence (Reid 2011; Goscha 2022). Certainly, there was long-standing, widespread enthusiasm for radical change among the Indonesian masses as well as a vibrant communist party, the PKI, that was legal and electorally competitive until Suharto banned it in 1966. But, notwithstanding the replacement of colonial officials in the revolution, the social order that emerged out of Dutch rule assumed a national form without significantly altering the hierarchies that Dutch colonialism had produced (Reid 2011, 186; Zhou 2019). To the contrary, popular action for radical social change, especially land reform, prompted counterrevolutionary backlash in 1948 and again in 1965 (Reid 2011, 180, 185).

The Madiun rebellion of 1948, the backlash against it, and the international response all took place during the anti-colonial war, helping facilitate Indonesian independence. It began when a Soviet-oriented group of Indonesian communists, including youth groups, leftist military factions, and unionists, led a revolt in eastern Java against the Republic and began to mobilize the peasantry there (Kahin and Kahin 1995, 31). Declared illegal by President Sukarno, the Madiun uprising was crushed by an Indonesian military that acted in hand with local Muslim militias "whose holy war was as much against the left as against the Dutch," leaving as many as 25,000 dead (Vickers 2013, 111–112). That conservative forces rallied violently to preserve a social order forged under colonial rule made it "unlikely that any fundamental restructuring of Indonesian society would take place" (Reid 2011, 36). Internationally, however, the Republic's firm response and swift military action convinced U.S. policymakers of Indonesia's anti-communist credentials. The countermovement thus contributed to a decisive shift in U.S. foreign policy, from backing Dutch reconquest to promoting Indonesian independence, that increased pressure on the Netherlands at the United Nations to yield sovereignty, which they did in December 1949 (Kahin and Kahin 1995).

The sequence of attempted revolution, concerted counterrevolution, and Cold War intervention would repeat in 1965 but at a greater scale of violence. Over the intervening period, Sukarno became more authoritarian, declaring an era of "Guided Democracy" in 1959 that ended parliamentary democracy, nationalized Dutch industry, and increased military power (Vickers 2013, 148). To balance military might, he relied increasingly on the PKI as a mass political base, which, in turn, agitated for action against landed elites, calling for redistribution of land held by mosques and local officials that stoked sectarian conflicts and class tensions (Robinson 2018). Economic conditions steadily deteriorated, and inflation swelled year on year through 1965. Internationally, Sukarno hosted the Asia-Africa Summit in Bandung in 1955 and formed a strategic relationship with China adhering to nonalignment in the Cold War (Zhou 2019). He thus challenged SEATO and earned the ire of the Eisenhower administration, which, in 1957–1959, deployed the U.S. Navy's Seventh Fleet, headquartered at Subic Bay, to back separatist rebellions in Sumatra and Sulawesi that destabilized the Republic (Kahin and Kahin 1995). Meanwhile, in the "Malayan Emergency" (1948–1960), Britain led a brutal counterinsurgency targeting its communist party but disproportionately affecting its ethnic Chinese population, convincing Sukarno that the colony's 1963 bid for independence was a "British neo-colonial plot" (Ngoei 2020, 135). His policy of Confrontation (1963–1966), or armed opposition against Malaya, including the British base in Singapore, received support from the PRC. Yet fighting a border war against Indonesia's neighbor alienated the army, amplified economic instability, and drove the Anglo-American alliance to seek Sukarno's ouster clandestinely. Although their role in the presidential transition and the mass killings remains a closely guarded secret, the anti-communist allies applauded the outcome and resolutely backed Suharto's New Order through the Cold War (Robinson 2018).

Conclusion

Too often overlooked in narrow analyses of "the Vietnam War," the three cases presented here—the First Indochina War in Vietnam, militarization of the Philippines, and mass killings in Indonesia—illustrate deeper historical causes and a broader field of conflict that help explain the brutal production of a postcolonial shatterzone in Southeast Asia throughout much of Asia's Cold War. Or reproduction. Although developed to investigate a specific moment in the twentieth century, our approach to postcolonial conflict is also relevant to understanding geostrategic and political-economic relations that endure to this day. On 2 February 2023, the United States and the Philippines announced a base agreement that allows the U.S. military to operate out of nine locations, marking a significant return to the archipelago 30 years after the end of the Cold War.[4] The U.S. Secretary of Defense declared that the agreement would increase "interoperability" within striking distance of Taiwan to counter an assertive PRC there, in the South China Sea, and the Indo-Pacific more broadly. President Ferdinand Marcos Jr. announced that he could not "see the Philippines in the future without having the United States as a partner" (Wee 2023). Breaking with his predecessor who threatened to scrap the Visiting Forces Agreement, Marcos the younger has thus followed in the footsteps of his dictatorial father, raising alarms about the cozy and opaque relationship between Filipino political elites and American geopolitical projects (Apostol 2023).

This new geopolitical alliance bears the hallmarks not only of an incomplete rupture with old forms of colonial rule but also of geopolitical strategies held over from the Cold War.

In this chapter, by extending our historical imaginations to think about the *longue duree*, our approach has been to reframe the intersection of the Cold War and postcolonial conflict in Southeast Asia. We argued for understanding

Cold War violence in relation to the region's political transition out of empires even as military and political links forged between Vietnam, the Philippines, and Indonesia continued to remake the region as a strategic theatre in the Cold War. Our approach demonstrates that conflicts from the late 1940s onward emerged out of multiple ruptures with empires and the region's subsequent entry into a rising, bipolar world order. After World War Two, in other words, preexisting contests over decolonization collided with an ideological, heavily weaponized conflict to generate hot wars, mass violence, and civil strife. As these conflicts intertwined and grew in scale, they forged connections across new nation-states that gave shape to Southeast Asia, helping produce a region where local power struggles intersect with global geopolitics to this day.

Notes

1 These borders still generate conflict, as in contested claims on the South China Sea or in their unfinished demarcation between Vietnam and Cambodia (McCoy 2016; Heng 2022).
2 Because they had not ratified the 1954 Agreements, the United States and RVN claimed that they were not bound to the elections. For more on Geneva, see Asselin (2013) and Nguyen (2012).
3 Unlike Vietnam, where the Viet Minh resisted Japanese occupation, nationalist leaders in Indonesia initially welcomed the Japanese military due to its defeat of the reviled Dutch state.
4 Periodic U.S. ship visits to Subic Bay had continued, particularly after the 2012 Enhanced Defense Cooperation Agreement.

References

Apostol, G. 2023. Dancing with America has been a curse for the Philippines. *The New York Times*, February 7. Accessed February 10, 2023. https://www.nytimes.com/2023/02/07/opinion/us-philippines-china-military-curse.html

Asselin, P. 2013. *Hanoi's Road to the Vietnam War, 1954–1965*. University of California Press.

Avalon Project, Yale University. 1954. Indochina: Final declaration of the Geneva conference on the problem of restoring peace in Indo-China, July 21, 1954. Accessed January 25, 2023. https://avalon.law.yale.edu/20th_century/inch005.asp

Berry, Jr., W.E. 1989. *U.S. Bases in the Philippines: The Evolution of the Special Relationship*. Westview Press.

Bowman, I. 1931. *The Pioneer Fringe*. American Geographical Society, Special Publication No. 13.

Brocheux, P. and D. Hemery. 2009. *Indochina: An Ambiguous Colonization, 1858–1954*. Translated by L. Dill-Klein with E. Jennings, N. Taylor, and N. Tousignant. University of California Press.

Chamberlin, P. 2018. *The Cold War's Killing Fields: Rethinking the Long Peace*. HarperCollins Publishers.

Chandler, D., R. Cribb, and L. Narangoa, eds. 2016. *End of Empire: 100 Days in 1945 that Changed Asia and the World*. NIAS Press.

Chapman, J. 2013. *Cauldron of Resistance: Ngo Dinh Diem, the United States, and 1950s Southern Vietnam*. Cornell University Press.

Cohen, S.B. 1973. *Geography and Politics in a World Divided*, 2nd edn. Oxford University Press.

Cohen, S.B. 2015. *Geopolitics: The Geography of International Relations*. Rowman & Littlefield.

Enloe, C. 2000. *Maneuvers: The International Politics of Militarizing Women's Lives*. University of California Press.

Fairgrieve, J. 1915. *Geography and World Power*. Kings University Press.

Fanon, F. 1961. *The Wretched of the Earth*. Translated by C. Farrington. Grove.

Goscha, C. 2022. *The Road to Dien Bien Phu*. Princeton University Press.

Hartshorne, R.J. 1941. The politico-geographic pattern of the world. *The Annals of the American Academy of Political and Social Science* 218: 45–57.

Hartshorne, R.J. 1944. The United States and the 'Shatter Zone' of Europe. In *Compass of the World: A Symposium on Political Geography*, eds. H.W. Weigert and V. Stefansson, 203–214. Macmillan.

Hawkins, M.B. 2022. From colonial cargo to global containers: An episodic historical geography of Manila's waterfront. Unpublished PhD Dissertation, Department of Geography, UNC-Chapel Hill.

Heng, K. 2022, Apr 12. Cambodia-Vietnam relations: Key issues and the way forward. *ISEAS Perspective* 2022 36: 1–1.

Hensel, P.R., and P. F. Diehl. 1994. Testing empirical propositions about shatterbelts, 1945–1976. *Political Geography* 13: 33–51.

Huynh Kim Khanh. 1982. *Vietnamese Communism, 1925–1945*. Cornell University Press.

Kahin, A. and G. Kahin. 1995. *Subversion as Foreign Policy: The Secret Eisenhower and Dulles Debacle in Indonesia*. New Press.

Kearns, G. 2009. *Geopolitics and Empire: The Legacy of Halford Mackinder*. Oxford University Press.

Kelly, P.L. 1986. Escalation of regional conflict: Testing the Shatterbelt concept. *Political Geography Quarterly* 5: 161–180.

Kerkvliet, B.J. 1977. *The Huk Rebellion: A Study of Peasant Revolt in the Philippines*. University of California Press.

Kirsch, S. 2023. *American Colonial Spaces in the Philippines: Insular Empire*. Routledge.

Kirsch, S. and C. Flint, eds. 2011. *Reconstructing Conflict: Integrating War and Post-War Geographies*. Ashgate.

Kirsch, S. and C. Flint. 2015. Geographies of reconstruction: Re-thinking Post-war Spaces. In *The Politics of International Intervention: The Tyranny of Peace*, eds. M. Turner and F.P. Kühn, 39–58. Routledge.

Kratoska, P., R. Ruben, and H. Nordholt. 2005. *Locating Southeast Asia: Geographies of Knowledge and Politics of Space*. National University of Singapore Press.

Lawrence, M. and F. Logevall, eds. 2007. *The First Vietnam War: Colonial Conflict and Cold War Crisis*. Harvard University Press.

Le Failler, P. 2014. *La Riviere Noire: L'Integration d'une Marche Frontiere au Vietnam*. CNRS.

Lee, M. 2015. *Dreams of the Hmong Kingdom: The Quest for Legitimation in French Indochina, 1850–1960*. University of Wisconsin Press.

Lentz, C. 2017. Cultivating subjects: Opium and rule in post-colonial Vietnam. *Modern Asian Studies* 51: 879–918.

Lentz, C. 2019. *Contested Territory: Dien Bien Phu and the Making of Northwest Vietnam*. Yale University Press.

Lentz, C. 2021. The crucible of Dien Bien Phu: Making Vietnam in the First Indochina War. In *Political Violence in Southeast Asia since 1945*, eds. E. Zucker and B. Kiernan, 117–133. Routledge.

Logevall, F. 2012. *Embers of War: The Fall of an Empire and the Making of America's Vietnam*. Random House.

Lund, C. 2020. *Nine-Tenths of the Law: Enduring Dispossession in Indonesia*. Yale University Press.

Mackinder, H.J. 1904. The geographical pivot of history. *The Geographical Journal* 23: 421–437.

Mackinder, H.J. 1919. *Democratic Ideals and Reality: A Study of the Politics of Reconstruction*. Constable and Company.

McAlister, J. 1967. Mountain minorities and the Viet Minh: A key to the Indochina War. In *Southeast Asian Tribes, Minorities, Nations*, ed. P. Kunstadter, 771–844. Princeton University Press.

McCoy, A. 2002. America's secret war in Laos, 1955–75. In *A Companion to the Vietnam War*, eds. M. Young and R. Buzzanco, 283–313. Blackwell.

McCoy, A. 2009. *Policing America's Empire: The United States, the Philippines, and the Rise of the Surveillance State*. University of Wisconsin Press.

McCoy, A. 2016. Circles of steel, castles of vanity: The geopolitics of military bases on the South China Sea. *Journal of Asian Studies* 75: 975–1017.

McHale, S. 2021. *The First Vietnam War: Violence, Sovereignty, and the Fracture of the South, 1945–1956*. Cambridge University Press.

Ngoei, W.Q. 2020. *Arc of Containment: Britain, the United States, and Anti-Communism in Southeast Asia*. Cornell University Press.

Nguyen, L-H. 2012. *Hanoi's War: An International History of the War for Peace in Vietnam*. University of North Carolina Press.

Oey, H. 1961. *Asia Menang di Dien Bien Phu*. A.-A.

Pelley, P. 2002. *Postcolonial Vietnam: New Histories of the National Past*. Duke University Press.

Rasmussen, M. and C. Lund. 2017. Reconfiguring frontier spaces: The territorialization of resource control. *World Development* 101: 388–399.

Reid, A. 2011. *To Nation by Revolution: Indonesia in the 20th Century*. NUS Press.

Reyes, V. 2019. *Global Borderlands: Fantasy, Violence, and Empire in Subic Bay, Philippines*. Stanford University Press.

Rimmer, P.J. 1997. US Western Pacific geostrategy: Subic Bay before and after withdrawal. *Marine Policy* 2: 325–344.

Robinson, G. 2018. *The Killing Season: A History of the Indonesian Massacres, 1965–66*. Princeton University Press.

Roosa, J. 2006. *Pretext for Mass Murder: The September 30th Movement and Suharto's Coup d'Etat in Indonesia*. University of Wisconsin Press.

Sidel, J. 2021. *Republicanism, Communism, Islam: Cosmopolitan Origins of Revolution in Southeast Asia*. Cornell University Press.

Smith, N. 2003. *American Empire: Roosevelt's Geographer and the Prelude to Globalization*. University of California Press.

Spykman, N.J. 1944. *The Geography of the Peace*. Harcourt, Brace.

Cam Trong. 1978. *Nguoi Thai o Tay Bac Viet Nam*. NXB Khoa Hoc Xa Hoi.

Tyner, J.A. 2007. *America's Strategy in Southeast Asia: From the Cold War to the Terror War*. Rowman & Littlefield.

Unstead, J.F. 1923. The belt of political change in Europe. *Scottish Geographical Magazine* 39: 183–192.

Vickers, A. 2013. *A History of Modern Indonesia*. Cambridge University Press.

Wee, S. L. 2023. U.S. to boost military role in the Philippines in push to counter China. *The New York Times*. February 3. Accessed February 10, 2023. https://www.nytimes.com/2023/02/01/world/asia/philippines-united-states-military-bases.html?smid=em-share

Woods, C. 2020. *Freedom Incorporated: Anticommunism and Philippine Independence in the Age of Decolonization*. Cornell University Press.

Young, M. 1991. *The Vietnam Wars, 1945–1990*. HarperPerennial.

Zhou, T. 2019. *Migration in the Time of Revolution: China, Indonesia, and the Cold War*. Cornell University Press.

9

FEMINIST GEOPOLITICS AND EMPATHETIC ENCOUNTERS WITH THE UNSEEN

Reconsidering *Black Hawk Down* twenty years later

Orhon Myadar and Tony Colella

Introduction

The film *Black Hawk Down* (2001) is based on the Battle of Mogadishu (also known as the "Black Hawk Down" incident), a U.S. military intervention in the 1993 Somali Civil War. The film is one of many successful Hollywood films based on distant, foreign affairs in which the American military is portrayed heroically. Recent blockbusters with similar subject matter include *12 Strong* (2018), *Zero Dark Thirty* (2013), and *American Sniper* (2014), but situating *Black Hawk Down* as both a response to the geopolitical events that inspired it and the film's approach to the peoples represented therein provides a particularly salient examination regarding the valorization of American soldiers and the Othering and erasure of all else (Nguyen 2013).

Black Hawk Down is an adaption of Mark Bowden's book *Black Hawk Down: A Story of Modern War* (1999). The film won two Academy Awards, marking its success within the entertainment industry. However, while the film is popular entertainment, it is also a global artifact through which a shift in U.S. foreign policy was articulated. Under the Truman Doctrine of the Cold War period, the United States aimed to provide economic, military, and political "aid" to countries that were seen as being threatened by the spread of communism. In the post–Cold War era, as demonstrated in *Black Hawk Down*, the U.S. actions began to be framed in a more humanitarian light. In the film, American soldiers are depicted as devoted, loyal, and selfless heroes in their mission to help Somalia and its people to the extent that the filmmakers excised one character out of the storyline because he did not fit into the carefully curated image of the U.S. military in the film. To be sure we do not question

DOI: 10.4324/9781003345794-9

the bravery, heroism, and selfless acts of individual soldiers who were involved in the mission. Nor do we question the humanitarian intent behind the mission. We rather probe deliberate framing of the mission's devastating (however unintended) human and material tolls on the Somali side through the prism of the carefully curated image of the U.S. military action. Even the notion of "humanitarian aid" is fraught with contradiction and political performance (e.g., Autesserre 2014; Bachmann 2014; Henry and Higate 2009; Higate and Henry 2010). Higate and Henry (2010, 38), for example, analyze the production of "peacekeeping," and argue specifically that the meaning attached to the peacekeepers' practices "cannot be disaggregated from broader-level political processes that feed into audience expectation."

This curating begins with the case of John Grimes, one of the film's main protagonists, who was given a different name than his real-world counterpart, John Stebbins, at the behest of the Pentagon. After his return from the Battle of Mogadishu, Stebbins was convicted of rape and sodomy of a child under the age of twelve (United States Army Court 2005). Erasing this individual from the film's storyline reveals as much as it aims to obscure. Even as Stebbins was transformed into Grimes, an all-American everyman hero, the film transformed Somali civilians into American caricatures of "foreigners" in distant places, effectively faceless subjects whose lives are not equally valued. In other words, the film effectively "others" these civilians within its narrative construction of the event which makes it easy for the viewers not to celebrate their lives or mourn their losses.

In this chapter, we explore this othering process in *Black Hawk Down* and other war blockbusters, and suggest how viewers can critically oppose this othering with an empathetic encounter. Although geographers have engaged with *Black Hawk Down* within critical geopolitical scholarship, the role of the empathetic encounter has not been explored in this otherwise important body of work (Carter and McCormack 2006; Dalby 2008; Dodds 2008). Our goal here is to provide the means by which viewers can empathize with othered subjects and in so doing intentionally warp the explicit intention of the film to present its American soldiers as unequivocal, unquestionable heroes. In so doing, we embrace how Pedwell (2012, 280) defines empathy—"an affective portal to different spaces and times of social justice"—as a way to unsettle *Black Hawk Down*'s dominant narratives.

Our focus on empathy with those who are not fully represented in policy discourses as well as media representation of such discourse is rooted in the assumption that empathy is central to any conflict resolution and peacebuilding efforts. Holmes and Yarhi-Milo (2017, 107), for instance, argue that empathy is "required for overcoming long-standing hostilities" and without empathy any efforts resolving conflicts would fail. To prevent disastrous consequences of warfare, the public must appreciate the full horrors of warfare and the impact it has on diverse parties, including those marked as the Other. By only valuing each life equally, can we unsettle the dominant narratives in political discourses as well as in war films that divide people us *vs.* them.

In understanding what is invoked and silenced in *Black Hawk Down*, we start with a background section to situate the political and social landscape when the film was released. The theoretical consideration that follows engages a body of critical feminist scholarship that intersects with our main conceptual framework. We call primarily on critical feminist geopolitics that exposes invisible violence (e.g., Dowler and Sharp 2001; Fluri 2011; Gilmartin and Kofman 2004; Hyndman 2007; Mountz and Hyndman 2006), as well as Judith Butler's concept of "grievability" which allocates which bodies, peoples, and cultures are viewers are "allowed" with which to empathize. We then analyze the film through the lens of empathetic encounter via what is seen and what is unseen in the film. We then offer our concluding thoughts on the impact of the violence of silencing of competing voices in *Black Hawk Down*.

Situating *Black Hawk Down*

In the 1990s, brutal civil war, collapsed government, and widespread famine ravaged the small East African nation of Somalia (Myadar 2022). However, the Bush administration considered the crisis in Somalia a "third-tier issue" within the strategic priorities of U.S. foreign policy (DiPrizio 2002). Thousands of Somalians perished during this time and thousands of others fled the country (Myadar 2022). Those who remained faced catastrophic famine, which only began to attract public attention in America beginning in mid-1992. In July 1992, *The New York Times*' Jane Perlez published a story on the gravity of the famine:

> In the damp, gray dawn in this remote Somali bush town, 25,000 men, women, children, their rib cages protruding, their eyes listless, shuffled with their last bit of strength today toward outdoor kitchens for a scoop of food. Hundreds, too feeble to eat, died while they waited.

The article featured photos of children and women as they waited for food. Other U.S. news networks also began to report on the crisis, showing vivid images of starving people, especially children. The media impact, or the "CNN factor," as the pundits called it, was said to have had a profound influence on both U.S. foreign policy decisions and the public opinion pertaining to the crisis (Kennan 1993). On December 4, less than five months after Perlez reported on the famine, President George H. W. Bush announced that the United States would be supporting the Operation Restore Hope mission in Somalia. Five days later, the first U.S. Marines landed on the shores of Mogadishu. The Somalian intervention, championed by the media and sanctioned by the public, materialized in the days before President Bush's term was complete and he left office.

In his detailed survey of media coverage leading up to the intervention, Jonathan Mermin (1997) illuminates a complex, dialectical relationship between the media and government policies. Mermin argues that rather than the media independently deciding what topics are reported, and when and how, the government plays a critical role in setting and staging the news agenda. In the case of Somalia, Mermin suggests it was not the media who pushed the government to act upon the crisis; rather, stories on Somalia appeared *after* the decision for intervention had been set in Washington (ibid.). The media covered events in Somalia to the extent it did only because there was a political will on the part of Washington to induce a U.S. response (Mermin (1997)).

Operation Restore Hope, as a U.S.-led humanitarian mission, aimed to provide food and aid to those affected by the civil war and famine. The mission involved a large military presence and was intended to stabilize Somalia and establish a secure environment for humanitarian efforts. The Battle of Mogadishu, also known as the "Black Hawk Down" incident, was a military operation carried out by U.S. Special Forces in October 1993 as a part of Operation Restore Hope. The specific mission that led to the Battle of Mogadishu aimed to capture key lieutenants of the warlord Mohamed Farrah Aidid, but quickly turned into a prolonged urban firefight, resulting in the deaths of eighteen U.S. soldiers and hundreds of Somalians after two Black Hawk helicopters were shot down. The incident subsequently led to a U.S. decision to withdraw from Somalia. More largely, the Battle of Mogadishu illustrated starkly the shortcomings of militaristic "peacekeeping" missions and interventions on behalf of "failed" states.

Produced nearly ten years after the George H. W. Bush administration made the decision to send the U.S. troops to Somalia, Ridley Scott's film depicted events during the battle. Despite the film's temporal distance from the actual event, the timing of the film's release was politically significant. The film was originally scheduled for release in March 2002. However, to capitalize on the fomenting sense of patriotism and fervent nationalism that surged throughout much of the United States during the post 9/11 period, the film's release accelerated to December 28, 2001 (limited release) and a broad release on January 18, 2002. The timing of the film was considered important because it depicts a charged experience of combat that resonated with the 9/11 affects of popular geopolitics in the United States (Carter and McCormack 2006, 239). "That glean of patriotism within *Black Hawk Down* is not because of the tragedy that struck, but the acclaim that followed does tie itself to the feeling that swept America" (Gleadow 2022).

Furthermore, *Black Hawk Down* was an illustrative case of how Hollywood supported Bush administration's War on Terror—an American-led military campaign launched to fight against terrorism (Lisle and Pepper 2005). Hollywood executives and the Bush administration agreed to foster Americans' support of U.S. troops and frame the War on Terror as the United States fighting "evil" (e.g., Boggs 2017; Davies and Wells 2002; Dodds 2008; Kumar and Kundnani 2014). For instance, Karl Rove, who served as senior adviser to President Bush, met with Hollywood's top executives to explore how the entertainment industry could play a role in the administration's war efforts (New York Times 2001). During this meeting, Rove proposed several ideas, including how Hollywood could reassure American children and families during the uncertain times (ibid.). Rove's concern to allay American families and children's concerns highlights the absence of his regard for Somali families and children.[1] Needless to stay, Rove and his ilk received what they hoped to achieve: a curated representation of the U.S. military. And the film industry received what it needed in return: logistical, material, and technical support from the Pentagon and Department of Defense including "two C-5 transport planes, four Black Hawk choppers, four 'Little Bird' helicopters, pilots from the 160th Special Operations Aviation Regiment (SOAR), and more than 100 Army Rangers" (Lisle and Pepper 2005). The film also employed military advisers to ensure that the portrayal of military protocol was accurately represented.

However, the relationship between Hollywood and the U.S. military predated the 9/11 attacks. In the case of *Black Hawk Down*, the Pentagon and the Department of Defense (DoD) had been involved in the film's production for over a year (Lisle and Pepper 2005; Hoglund and Willander 2017). In this way, the DoD was able to shape how the film's story was told, deviating quite substantially from Bowden's original account. The Pentagon's intervention can be seen as the "forcible framing" of the U.S. role in Somalia (Butler 2005, 826). Together the Pentagon and the DoD rewrote the disastrous and botched armed intervention as an "epitaph of bravery, commitment and selflessness" (Lisle and Pepper 2005, 173). As such, *Black Hawk Down* does more than providing entertainment and cinematic experience. It presents a particular political message as government propaganda, and, by doing so, shapes social texts surrounding U.S. military engagement or "social form of communication where ideas and discourses can be created" (Schrager 2017, 156).

As a political statement, the film illustrates the shifting geopolitical landscape of post–Cold War American foreign policy. More specifically, rather than justifying warfare with ideological goals (as during the Vietnam War), the film depicts the U.S. military's engagement in Somalia as driven by humanitarian goals. As Flint and Falah (2004) suggest, U.S. geopolitical interests at the time were not defined by the defense of national borders. Rather, the Bush administration's strategy was "extraterritorial" or "located within other and 'othered' sovereign spaces" (Flint and Falah 2004, 1380). The pretext predicated upon concerns for human rights allowed the United States to engage in warfare across the globe "under a beacon of being able to define and deliver life for all" (Flint and Falah 2004, 1389).

The social text undergirding the political message is that the U.S. military engagement in Somalia (overseas in general) is just and benevolent, and therefore worthy of public support. This social text is critical, especially when the film depicts urban warfare where the distinction between combatants and noncombatants is blurred and where civilian neighborhoods are turned into battlefields (Coward 2008; Graham 2008; Myadar and Colella 2022; Shaw 2000). In this bloody, messy, and violent engagement of warfare, the film's version whitewashes the realities of war and sanitizes the U.S. military actions not only in Somalia but also across the world. In doing so, it constructs a social narrative that supports the U.S. extraterritorial military in a way calculated to sway popular opinions regarding armed interventions (Myadar and Colella 2022).

At a glance, *Black Hawk Down* is a war movie based on a story about bravery, camaraderie, and the complex reality of war. The story highlights the selfless and heroic acts of an elite group of American Rangers and Delta Force soldiers whose mission is to capture the warlords who were terrorizing the nation. The film powerfully depicts the sacrifice the U.S. servicemen paid in their mission to help the poor people of war-torn Somalia, cultivating empathy and care for these soldiers. The audience is left with renewed appreciation and respect for the ultimate sacrifice of the U.S. servicemen (Gleadow and Down 2022). The "good war was central not only to Hollywood portrayals of combat but also to the production and reproduction of discourses of American national identity and national purpose rooted in more general and nebulous ideas about the defense of freedom" (Carter and McCormack 2006, 235).

The film, based on "a true story," represents actual events and real people, yet like other films that invoke historical conflicts and events, *Black Hawk Down* can be read as a political strategy with deliberate meanings and nuances (Dixon 2008). The affective power of the film is to bring the drama of the battlefield to viewers' gaze and show that the characters presented onscreen are heroes and protectors (Carter and McCormack 2006, 235). *Black Hawk Down* succeeds in doing so. It is easy to empathize with the onscreen soldiers and root for their survival and triumph. However, what is missing here is the same type of affective engagement and empathetic encounter with Somali civilian causalities. In their analysis of historical empathy, Muetterties and Bronstein (2020) demonstrate how critical empathy is in understanding different ways individuals are portrayed in textbooks and how important it is for students to consider the complex contexts that inform these portrayals as well as their own interpretations. Likewise, unless viewers are deliberate in seeing beyond what is represented within the cinematic frame, they become passive recipients of the particular version the makers of *Black Hawk Down*, in concert with the U.S. military, presented to them.

Theoretical considerations: The empathetic encounter

To probe the tensions between what is told and what is untold in *Black Hawk Down*, we rely on critical feminist geopolitics to examine the *overlooked* within the film's storyline via the concept of empathetic encounter. We define the empathetic encounter as a deliberate form of inquiry that is centered on a collective care and empathetic engagement with those whose stories are often silenced and unseen by war films. We use the empathetic encounter as a way to respond to Winkler's call to "break through the control of the state-media nexus and produce accessible information that exposes the insidious 'feminized'—and consequently, invisiblized—face of warfare waged by the world's superpower" (2002, 427). We specifically focus on the collusion between the U.S. military with the entertainment industry to whitewash the violent realities of the U.S.-led military action in Somalia. Critical feminist geopolitics in particular provides a way to render the invisible acts of violence visited upon unsuspecting people visible (Dowler and Sharp 2001; Fluri 2011; Hyndman 2007; Mountz and Hyndman 2006). As Hyndman reminds us, "feminist geopolitics blurs conventional borders between civilian and military" and urges us to consider the civilian subjects as equally valued political subjects both in Somalia and in the United States (2019, 8).

The concept is built upon our sensible encounter concept which we used previously to express a way of resisting the dominant narratives in a film by encouraging the viewer to become an active agent in being attentive to and unsettle hidden narratives (Myadar and Colella 2022; cf. Hole 2016; Marks 2000; McHugh 2015). We used the concept *sensible encounter* to critique another war film, *American Sniper* (Myadar and Colella 2022). The concept combines Laura Marks's (2000) concept of sensible cinema and Kristin Hole's (2016) concept of the encounter with cinema. The sensible encounter is particularly useful in the context of geopolitical imaginaries in film because it allows for a critical reexamination of the distance between the United States and foreign lands in terms of not only geography but also symbolism, epistemology, and intimate politics. Moreover, the sensible encounter unsettles the dominant narratives of a film by insisting that the viewer embodies individuals, not states, and so goes against the standard symbolism of war films in which key characters represent states and powers (ibid.). With the sensible encounter, individuals represent human beings, with all their messy, multidimensional lives that are often abrogated for the exigency of a dominant power's political message.

We refine the sensible encounter as the *empathetic encounter*, based in part upon Kozloff's (2013) *cinema of engagement* concept. Kozloff's concept advocates for, among other elements, shrinking the distance between viewers and events, analyzing power structures, locating films in real events, and inspiring work for social justice. In particular, in viewing films such as *Black Hawk Down*, we refocus our attention on the notion of empathy, which can be understood as "a process in which one person imaginatively enters into experiential world of another" (Bondi 2003, 71). In condensed retelling of stories such as in films, empathy is the key to effectively invoking compassionate responses in viewers (Keen 2011). As Bondi suggests, empathy can create "interpersonal and intrapsychic spaces in which similarities and differences can be mobilized, expressed and explored" (ibid.).

To situate *Black Hawk Down* in a broader spatiotemporal frame, we draw from films set during different wars. For example, Davison applies this approach to the film *Gallipoli* (Weir 1981), which is focused on the Gallipoli campaign of World War One spatially, but not temporally, similar to *Black Hawk Down*, and in so doing provides the means to both bring an empathetic encounter to the film and extract from it an empathetic encounter with the characters and contexts represented. Davison (2017) proposes a pathway to building *historical empathy*, a process by which an audience approaches a historical record with open-mindedness, care, and imagination. They explore evidence and builds contextual knowledge based on the historical record. This approach also allows the audience to consider multiple perspectives and become aware of how past beliefs may differ from those in the present, and then critically judge the present (2017, 151).

Films are never neutral, and both Horton and Clausen (2015) and Marcus and Stoddard (2007) investigate competing narratives in war films (for World War Two and the American Civil War, respectively) for both the historical knowledge they can carry and how viewers can empathetically encounter them. Horton and Clausen use spatiotemporally diverse films to establish "a sophisticated and nuanced sense of what it means to be the victor, the vanquished, and the occupied" (2015, 336). This contrasting approach provides another way to engage with films, as films from diverse sources can be used to establish empathetic encounters in the spaces of what they may share, what one may say that another may not, or what is lost entirely. Marcus and Stoddard (2007) warn against the danger of competing narratives, especially where filmmakers may obscure fact, such as John Stebbins becoming John Grimes—however, like Davison (2017), we emphasize the importance of the introduction of new perspectives. At the same time, as Susanne Urban (2008) reminds us, embodying a victim's perspective can be traumatic. Despite this discomfort, we recognize that films can allow viewers to inhabit and encounter diverse perspectives and identities in many contexts, not just educational or historical.

In the following section, we examine *Black Hawk Down* with empathetic encounter as our lens to see beyond what is scripted within the cinematic frame. We extend our empathetic encounter to the voices and stories of those whom are untold and unheard.

What is seen

Black Hawk Down opens with a scene that sets the stage for the viewers to encounter Somalia in a particular way—as an apocalyptic land with bodies of starving children and women, some are barely alive. These haunting images evoke pathos from viewers in a representation of human need integral to viewers' affective response. The opening scene thus sets the stage, and perhaps justification, for the humanitarian goals of the U.S. military in Somalia, while simultaneously framing the distance between Somalia and the United States.

The audience learns of a famine of a biblical scale that was ravaging Somalia. The subsequent scene shows a Red Cross food distribution center being attacked by Aidid militia forces from the vantage point of Staff Sergeant Matthew Eversmann (portrayed by Josh Harnett). The bird's-eye view of this carnage from Eversmann's helicopter cockpit signals the panoptic and elevated position of the U.S. soldiers while introducing their moral urgency to justify their presence. As unarmed civilians are attacked, Eversmann is tormented that the U.S. soldiers are not allowed to intervene because the UN mandate prevented them from engaging with the militia unless they are fired at first. The scene sets the stage for viewers' impassioned engagement: they would want the soldiers to do something, to help the needy and helpless victims and defeat the forces of evil.

Black Hawk Down presents its soldiers as surrogates for an audience that wishes to think of itself as moral and heroic. In this all male and mostly white ensemble of actors depicting U.S. servicemen, the individuality of each character is carefully constructed. It is easy to relate to and feel for the characters, and their lives are given deeply affective precedence.

> [B]y focusing on the individual soldier, and by suggesting that the US soldiers are willing to experience the violent crucible of war-even to sacrifice themselves for their nation and to make the world a safer place—the film depicted the presence of US military on foreign soil as natural.
>
> *(Hoglund and Willander 2017, 376)*

Eversmann is presented as empathetic as any of the soldiers can be in that moment, and invites viewers into his decision between doing and watching— because the film would rather its audience believe in the soldiers' actions, even as the audience's current action (watching) is held up as equivalent to the less heroic action (doing) presented.

Once the U.S. presence in Somalia is justified cinematically, the needy Somalis drop into the background for much of the film. The "evil" Somalis are used as a part of the storyline to enhance the need for violence. The film then shifts to focus on the terror inflicted upon the U.S. soldiers and the ultimate sacrifice of the eighteen soldiers who lost their lives during the operation. The mission was supposed to be a snatch-and-grab operation to enter the capital city of Mogadishu and capture two of Aidid's lieutenants from a house they were believed to be attending in a meeting. But instead, it ended up becoming a violent eighteen-hour urban firefight during which two U.S. choppers were gunned down killing hundreds of people and injuring many others.

The "evil" Somalis are used as a part of the storyline to enhance the need for this rampage and violence.

The militarization process relies on the perceived virtue and integrity of the U.S. military industrial complex. Hollywood has done its part in constructing and promoting the glorified version of military operations around the world. Dalby (2008) draws our attention to the figure of the warrior in Western filmic warfare. Using three Ridley Scott's films, Dalby examines the warrior figure in various combat landscapes spanning different time periods. In his discussion of *Black Hawk Down*, Dalby critiques the warrior code that is at the heart of the film. One of the important warrior codes is "leaving no one behind," whether alive or dead. The Rangers in the film live by this code even when it puts their lives at risk. Dalby reminds us that the dying soldier's supposed wish to be remembered as having fought well is not only a deliberate rearticulation of the warrior code but also an attempt more broadly to justify and glorify U.S. military engagement in distant lands. But that is precisely the point, that these soldiers are brave heroes, willing to sacrifice their lives not only for each other, but for the Somalis. Because the soldiers were in Somalia to provide relief supplies to hunger-stricken Somalis, the film effectively articulates the selflessness of these soldiers. In the good warrior fashion, Eversmann presents the soldiers with two options: "We can help, or we can sit back and watch a country destroy itself on CNN." As such they assume the role of saviors (white and male) (e.g., Hughey 2014).

These differences, between both Americans and Somalians, and "bad" and "good" Somalians, are reinforced by the titular Black Hawk helicopter, which not only provides narrative distance between the United States and Somalia but also as a material device through which "bad" Somalis are extinguished and "good" Somalis are rescued. The spectacular rendering of its demise along with the image of the bodies of American soldiers being dragged by a mob of Somalis down the streets of Mogadishu forms a cinematic climax that draws viewers' affective response and empathetic encounter with what is seen within the film's frame.

What is unseen

Beyond what is seen in the film, however, there is much that is unseen. For instance, the transformation of John Stebbins into John Grimes reveals a complex relationship between cinema and geopolitics. More specifically, how the military-entertainment industry functions not only as a site of economic exchange and an entertainment medium but also as a broader ideological tool of the state. The inconspicuous detail of the erasure of Stebbins's real name in the film, despite his character remaining a central part of the film, is a manifestation of the conduct of the John Stebbins himself. After the Battle of Mogadishu, Stebbins was found guilty of sexually assaulting his six-year-old daughter, and is serving a thirty-year prison term (United States Army Court 2005). It is one of the unseen and untold examples of violent realities of warfare.

If empathy allows us to enter into "the experiential world of an other" (Bondi 2003, 71), how can viewers experience an encounter with invisible war causalities whose names are not named and whose individuality is removed? Such a perspective requires attention for the erasure of those voices whose stories that are not told or immediately apparent. While the American Rangers' bravery is thoughtfully constructed, this façade obscures systemic, structural, and interpersonal violence depicted throughout the film. The film was sanitized of any negative impacts of the U.S. mission on Somalian civilians.

One of the untold contexts surrounding the U.S. role in Somalia is the U.S. contribution to Somalia's current state of affairs. During the Cold War, Somalia had been of significant geopolitical interest to both the Soviet and U.S. governments because of the country's strategic location and natural harbor, and with each party providing military aid and weapons at different points in history (Myadar 2022). American oil companies actively supported Siad Barre, whose regime ruled the country from 1969 to 1991. However, by the 1990s, the Cold War ended and Somalia's strategic importance to the United States waned. When the country disintegrated and entered a state of grave humanitarian crisis in 1991, the proliferation of weapons in Somalia supplied by the Soviet and U.S. governments made the fate of post–Cold War Somalia especially precarious with heavily armed factions fighting for the control of the country.

Beyond the untold story of the broader geopolitical role that the United States had played in Somalia, the lives of countless Somalians who perished or were injured during the battle are not told or grieved. While the original accounts of Bowden feature various incidents of unnecessary killings of civilians by U.S. soldiers, the film only shows killings that were "necessary" from the vantage point of a U.S. soldier. For example, in a scene where a woman is shot by U.S. troops, the soldier does so only after she begins to fire at the troops using her dead husband's rifle. But the mission was carried out in Mogadishu, a densely populated city with a population that had swollen to 1.5 million due to the influx of refugees, taking the lives of many innocent civilians (Department of Defense 1994). While names of the Americans appear onscreen, along with information about the honors they received, the massive number of Somalian civilian casualties during the mission was yet another violence of silencing in the film. It diminishes the carnage of war and desensitizes violent loss of human lives, robbing viewers of the opportunity to fully appreciate the lives of those who were cut short but whose stories are not fully told. The lives of Somalian civilians within the cinematic frames are represented as "precarious," in Butler's (2016) words, as their lives are never lived or lost in the full sense thus are not "grievable" (Butler 2016). As such neither their lives nor their deaths can evoke fully realized, empathetic reactions from the viewers.

Moreover, the individuals are treated as worth in terms of their respective nationalities at the time of their deaths, and render them as either American ("grievable") or Somalian ("ungrievable").

An empathetic encounter provides one way not only to experience Somalia outside of its apocalyptic frame, but to embody and grieve, if necessary, its people who appear unnamed, powerless, and unseen. Our recognition of Somalian and American lives as equal, while the filmmakers' intention to render them as unequal by taking refuge in the sheer number of Somalian lives loss, can help make those unseen names and unwritten lives "grievable," and therefore can encourage an empathetic encounter with the film's absences.

Furthermore, the film does not allow viewers to experience an emotional connection with Somalians, as they are mostly depicted in the film as background noise rather than affective, individual people. If they are featured, it is as the antagonists who help accentuate the film's narrative and the benevolence of the U.S. soldiers. For example, local civilians are depicted as chaotic assemblages of barbaric and untrustworthy aliens whom the soldiers refer as the 'skinnies.' In one scene, for instance, Staff Sergeant Eversmann can be heard telling others to "watch out for skinnies. They are all over the rooftop." This term is a form of discursive violence toward the very people whom the mission had been mandated to protect, and whom indeed were practically living "skeletons wrapped in skin" because of the harrowing famine (Times Daily 1993).

Throughout the film, the Somalians are reduced into what *The New York Times* called "a pack of snarling dark-skinned beasts, gleefully pulling the Americans from their downed aircraft and stripping them. Intended or not, it reeks of glumly staged racism" (Mitchell 2001). "Set against the well-meaning and disciplined US forces are hordes of ill-disciplined, gun-toting Somalians cast in none-too-subtle terms as marauding savages" (Lisle and Pepper 2005, 174). This is a form of racialized violence targeted against those whom the mission was intended to help.

Omar Jamal, a Somalian-American activist, extends the critique in arguing that the Somalian people were portrayed as uniformly barbaric and savage. "We don't know what Americans will think of us Somalis after they watch this movie," he wrote (Hassan 2002). For example, the film's chief antagonist is a man referred as Mo'alim. He is depicted as the epitome of senseless violence. However, in his account, Bowden, who interviewed Mo'alim, reveals a troubled young man who was caught in the violent world that surrounded him ever since he was a teenager. As Bowden writes, Mo'alim found a live American soldier in one of the choppers that been shot down. Bowden (2010) describes how Mo'alim protected the ranger from an angry mob of Somalians and this fact is intentionally omitted in the film version. Furthermore, according to Bowden, the film does not feature any Somalian actors, nor were any Somalians hired for consultations to ensure accuracy (Hassan 2002). The lack of representation robs Somalians of their agency to tell their own stories, to embody their struggles and to authentically represent themselves.

In addition to its racialized violence against Somalians, the film suffers from the gendered exclusion of women from the film's storyline, and no female character was included in the cast ensemble. In one of the poignant conversations in the film between Sergeants "Hoot" Gibson and Eversmann, Hoot explains his motivation (finding himself in battlefields) as being "about the men next to you." The film's racialized and gendered undertone is concealed by the deliberate attempt to make the events of the Mogadishu as a benevolent act of the U.S. servicemen.

Finally, the film silences the voices of those who are hurt by the U.S. war efforts outside the actual battlefield both spatially and temporally. The fact that Stebbins, one of the leading characters, committed a heinous crime against his child is deliberately concealed when the Pentagon and filmmakers changed Stebbins name to avoid any controversy. That is a violence in itself as it removes an opportunity to shed light on other consequences of the violence of the war, which extends beyond actual warfare. For example, studies have shown that many war veterans suffer from various posttraumatic disorder (Trautman et al. 2015). Incidents of child abuse involving military families either leaving or just returning from deployment had risen 30% since 2001 (ibid.). By veiling Stebbins's story behind a fictional (glorified) character, the military-entertainment complex silenced the screams of the victims of warfare including Stebbins's own child.

If films presented true realities of wars, without privileging one life over other, they could be used as a platform to cultivate empathy for war victims. If viewers are provided an opportunity to learn stories of different people, communities and groups without othering them across superficial constructions of difference, films can be used as a tool for peacebuilding.

Conclusion

The 2001 film *Black Hawk Down* was not a mere entertainment production or a simple commodity solely aimed at profit making. From its production to its release date, the film reflects the tangled relationship between the entertainment industry and the U.S. military complex. It was produced in such a way to reframe the U.S. military actions during the botched Battle of Mogadishu in a single-dimensionally benevolent light.

From the gaze of starving Somalian children through the barbarian mob of people, *Black Hawk Down* constructs a particular narrative that justifies the intervention and violence that followed during the U.S. military operation in Mogadishu. Eighteen American soldiers were killed and seventy others wounded during the mission. The bodies of American soldiers being dragged through the streets of Mogadishu by a mob of Somalians became the embodiment of the

U.S. sacrifice in its humanitarian goals. The lives of these servicemen are celebrated and mourned in the film while the civilian casualties on the Somalian side are left unmarked and unmourned as if "all lives are not equally valued" (Hyndman 2007, 39). By relying on the idea of feminist care and our conceptual tool empathetic encounter, we draw attention to the silent erasure of Somalian casualties from the cinematic frame. We also shed light on some of the violence of warfare that transcends the actual battles spatially and temporally.

Although we acknowledge that bravery and selfless acts of the U.S. soldiers for the mission they believed in, we argue that understanding the full scale of violence of wars and how they impact the lives of innocent people is necessary to pushing against the narrative that minimizes violent realities of warfare. By only deliberately focusing on the silenced, overlooked and "Orientalized" voices, can we empathetically and imaginatively begin to enter experiential world of those who the film deliberately excluded. If empathy is critical to conflict resolution and peacebuilding, then films too can be used as a way to foster affective and empathetic engagement with the victims of warfare, and in so doing films can be used as a tool to nurture peacekeeping efforts.

Note

1 It is noteworthy that in *Black Hawk Down*, all the heroic soldiers are men while the Somalian social landscape is depicted as a helpless mass of mostly women and children.

References

Autesserre, S. 2014. *Peaceland: Conflict resolution and the everyday politics of international intervention*. Cambridge University Press.

Bachmann, J. 2014. Policing Africa: The US military and visions of crafting 'good order'. *Security Dialogue* 45: 119–136.

Boggs, C. 2017. *The Hollywood war machine: US militarism and popular culture*. Routledge.

Bondi, L., 2003. Empathy and identification: Conceptual resources for feminist fieldwork. *ACME: An International Journal for Critical Geographies* 2: 64–76.

Bowden, M., 2010. *Black Hawk down: A story of modern war*. Atlantic, Inc.

Butler, J. 2005. Photography, war, outrage. *PMLA* 120: 822–827.

Butler, J. 2016. *Frames of war: When is life grievable?*: Verso Books.

Carter, S. and D.P. McCormack. 2006. Film, geopolitics and the affective logics of intervention. *Political Geography*, 25: 228–245.

Coward, M. 2008. *Urbicide: The politics of urban destruction*. Routledge.

Dalby, S. 2008. Imperialism, domination, culture: The continued relevance of critical geopolitics. *Geopolitics*, 13: 413–436.

Davies, P.J. and P. Wells. 2002. Life is Not Like a Box of Chocolates. In P.J. Davies and P. Wells, eds., *American film and politics from Reagan to Bush Jr.* Manchester University Press.

Davison, M., 2017. Teaching about the First World War today: Historical empathy and participatory citizenship. *Citizenship, Social and Economics Education* 16: 148–156.

Department of Defense. 1994. Report of the defense science board task force on military operations in built-up areas, Washington, DC: *Office of the Undersecretary of Defense for Acquisition and Technology*, November, 19.

DiPrizio, R. C. 2002. *Armed humanitarians: US interventions from northern Iraq to Kosovo.* JHU Press.

Dixon, D. 2008. "Independent" documentary in the United States: The politics of personal passions. *The geography of cinema–A cinematic world.* Franz Steiner Verlag, 65–83.

Dodds, K. 2008. Hollywood and the popular geopolitics of the war on terror. *Third World Quarterly* 29: 1621–1637.

Dowler, L. and J. Sharp. 2001. A feminist geopolitics?. *Space and Polity* 5: 165–176.

Flint, C. and G.W. Falah. 2004. How the United States justified its war on terrorism: prime morality and the construction of a 'just war'. *Third World Quarterly* 25: 1379–1399.

Fluri, J. 2011. Armored peacocks and proxy bodies: Gender geopolitics in aid/development spaces of Afghanistan. *Gender, Place & Culture* 18: 519–536.

Gilmartin, M. and Kofman, E., 2004. Critically feminist geopolitics. In *Mapping women, making politics: Feminist perspectives on political geography*, ed. L. Staeheli, E. Kofman and L. Peake. Routledge, 113–125.

Gleadow, E. Black Hawk Down Review. https://cultfollowing.co.uk/2022/01/09/black-hawk-down-review/. January 9, 2022. Accessed 9/3/2022

Graham, S., ed. 2008. *Cities, war, and terrorism: towards an urban geopolitics*: John Wiley & Sons.

Hassan, O. 2002. Somalis differ on worth of 'Black Hawk'. http://the.honoluluadvertiser.com/article/2002/Jan/25/il/il03a.html. Accessed 8/09/2018

Henry, M. and P. Higate. 2009. *Insecure spaces: peacekeeping, power and performance in Haiti, Kosovo and Liberia.* Bloomsbury Publishing.

Higate, P. and M. Henry. 2010. Space, performance and everyday security in the peacekeeping context. *International Peacekeeping, 17*, 32–48.

Hoglund, J. and M. Willander. 2017. Black Hawk-down: Adaptation and the military-entertainment complex. *Culture Unbound* 9: 365–389.

Hole, K. 2016. *Towards a feminist cinematic ethics: Claire Denis, Emmanuel Levinas and Jean-Luc Nancy.* Edinburgh University Press.

Holmes, M. and K. Yarhi-Milo. 2017. The psychological logic of peace summits: How empathy shapes outcomes of diplomatic negotiations. *International Studies Quarterly* 61: 107–122.

Horton, T.A. and K. Clausen. 2015. Extending the history curriculum: Exploring World War II victors, vanquished, and occupied using European film. *The History Teacher* 48: 321–338.

Hughey, M. 2014. *The White savior film: Content, critics, and consumption*: Temple University Press.

Hughey, M. 2007. Feminist geopolitics revisited: Body counts in Iraq. *The Professional Geographer* 59: 35–46.

Hyndman, J. 2019. Unsettling feminist geopolitics: forging feminist political geographies of violence and displacement. *Gender, Place & Culture* 26: 3–29.

Keen, S. 2011. Fast tracks to narrative empathy: Anthropomorphism and dehumanization in graphic narratives. *SubStance* 40: 135–155.

Kennan, G. F. 1993. Somalia, Through a Glass Darkly. Published: September 30, 1993. A, p. 25. http://www.nytimes.com/1993/09/30/opinion/somalia-through-a-glass-darkly.html?pagewanted=all. Accessed 8/8/2018.

Kozloff, S. 2013. Empathy and the cinema of engagement: Reevaluating the politics of film. *Projections* 7: 1–40.

Kumar, D., & Kundnani, A. 2014. Imagining national security: The CIA, Hollywood, and the war on terror. *Democratic Communiqué, 26*, 5.

Lisle, D. and A. Pepper. 2005. The new face of global Hollywood: Black Hawk down and the politics of meta-sovereignty. *Cultural Politics*, 165–192.

Marcus, A.S. and J.D. Stoddard. 2007. Tinsel town as teacher: Hollywood film in the high school classroom. *The History Teacher* 40: 303–330.

Marks, L. U. 2000. *The skin of the film: Intercultural cinema, embodiment, and the senses.* Duke University Press.

McHugh, K. E. 2015. Touch at a distance: toward a phenomenology of film. *GeoJournal* 80: 839–851.

Mermin, J. 1997. Television news and American intervention in Somalia: The myth of a media-driven foreign policy. *Political Science Quarterly* 112: 385–403.

Mitchell, E. 2001. *Mission Of Mercy Goes Bad In Africa*. December 28, 2001, E, p. 1.

Mountz, A. and J. Hyndman. 2006. Feminist approaches to the global intimate. *Women's Studies Quarterly* 34: 446–463.

Muetterties, C.C. and E.A. Bronstein. 2020. Scoundrel or freedom fighter? Creating historical empathy inquiries. *Social Studies Research and Practice* 15: 155–165.

Myadar, O. 2022. Place, displacement and belonging: The story of Abdi. *Geopolitics* 27: 462–477.

Myadar, O. and T. Colella. 2022. Feminist geopolitics, cinema, and the sensible encounter in American Sniper. *GeoJournal* 87: 133–140.

Nguyen, V.T., 2013. Just memory: War and the ethics of remembrance. *American Literary History* 25: 144–163.

Pedwell, C. 2012. Economies of empathy: Obama, neoliberalism, and social justice. *Environment and Planning D: Society and Space* 30: 280–297.

Perlez, J. 1992. Deaths in Somalia outpace delivery of food. July 19, 1992. *The New York Times*.

Schrager, B. 2017. A dialogic approach to alien movies. *Yearbook of the Association of Pacific Coast Geographers*, 153–167.

Shaw, M. 2000. *New wars of the city: 'Urbicide' and 'genocide.'* University of Sussex.

The United States Army Court. 2005. http://www.armfor.uscourts.gov/newcaaf/opinions/2005Term/03-0678.htm. Accessed 09/09/2018.

Times Daily. 1993. A day of suffering. Children's stories help explain Operation Restore Hope. Charles Hanley. 9A.

Trautmann, J., J. Alhusen and D. Gross. 2015. Impact of deployment on military families with young children: A systematic review. *Nursing Outlook* 63: 656–679.

Urban, S.Y. 2008. At issue: Representations of the Holocaust in today's Germany: Between justification and empathy. *Jewish Political Studies Review*, 79–90.

Weir, P., 1981. Gallipoli: I felt somehow I was really touching history. *Literature/Film Quarterly* 9: 213.

Winkler, P. 2002. (Feminist) Activism Post 11 September: Protesting Black Hawk down. *International Feminist Journal of Politics* 4: 415–430.

10

THE SPATIALITIES OF NONVIOLENT PEACE ACTIVISM IN THE MIDST OF WAR

From Colombia to Ukraine

Sara Koopman

How can one work nonviolently to build peace in the middle of an armed conflict zone? Grab a protest sign and stand in front of a tank? People have indeed done that, but too often the only form of nonviolent action people know or think of is a street protest. In a war, a street rally is neither safe nor necessarily effective. There is a wide range of other nonviolent action tactics that do not involve coming together in a public place and that use other socio-spatial strategies instead to build peace.

Geographers have long argued for understanding space as more than just a container for society, a grid on which we live (Lefebvre 1992). Space is shaped by what we as society do to it, not just in terms of buildings and roads but also what we do in spaces day to day, and how we imagine spaces, and who can be there, and what can be done there. The relationship between society and space goes both ways. Society is also shaped by space. Those roads, for example, do shape how easily one can move, and who one interacts with daily and how, and who we understand to be neighbors. To say that space and society each shape the other is a bit misleading because they never exist apart.

As such, space is constantly shaping peace and the ways that we work to make it in society. So too those ways shape the spaces of peace that are built. In this chapter, I focus particularly on the socio-spatial dynamics of various ways nonviolent action has been used to build peace. First, I discuss different meanings of peace, and how geographers have contributed to these understandings. Next, I cover different ways of understanding what is considered nonviolent action for peace. Then, I look at some of the socio-spatial dynamics of such action in two countries in the midst of war: Colombia and Ukraine. The key point here is that peace(s), and peace activism(s), are shaped in and

DOI: 10.4324/9781003345794-10

through the spaces and times in which they are made, as they too shape those spaces. I look for ways that space and spatialities create openings for nonviolent action, but also ways that they constrain them. I also explore ways that such actions create different sorts of peace spaces. I look for connections across contexts and argue that geographers have a role to play in weaving this sort of web. First, let me begin with what peace and nonviolent action can mean, and what space has to do with it.

Can peace be built in the midst of war?

Peace is often thought of as the opposite of war. In Peace Studies, this is known as negative peace or the absence of war. This sort of peace is seen as an end point – once you get to peace, you are done (until war breaks out again, and you have to start over working for peace). But in real life, war and peace are never that black and white. In lectures beginning in 1899, Jane Addams, co-founder of Hull House and of the Women's International League for Peace and Freedom, first talked about what she called "*positive* ideals of peace," contrasting them to *negative* peace (Carroll and Fink 2007). In her book based on the lectures she changed to calling them newer ideals of peace, but she continued to contrast them to the term negative peace (simply not war). Her newer ideals with "a sense of justice no longer outraged" (1907, 8) are clearly what we now think of as peace with justice, or positive peace. As opposed to a negative peace, a positive peace with, for example, enough to eat and warm and dry housing is more clearly an ongoing process, something that we build together and keep building. Addams saw that this was an uneven process that would require different sorts of peacebuilding efforts, and that these would change across time and space. For example, Addams was a leader in the settlement house movement, building home and education spaces for new immigrants as part of her peace activism. Today, activists at the U.S.-Mexico border work to make the immigration process slightly more humane by offering water and rest sites to immigrants crossing the desert (and are sometimes arrested for doing so, like geographer Scott Warren) (Warren 2019).

The term "positive peace" as such was famously used by Martin Luther King (1963) in his letter from a Birmingham jail where he writes about "negative peace which is the absence of tension" and "positive peace which is the presence of justice." The term is often misattributed to Galtung (1969), who cited neither Addams nor King, but built on the term by offering a framework whereby negative peace is the absence of direct violence and positive peace is the absence of structural violence. He would later go on to add that it was also the absence of cultural violence. Ironically, he thereby defines positive peace by what it lacks, rather than by what it has, as Addams and King did.

Geographers continue to build on these ideas of positive peace, and argue that it is an ongoing process that varies across time, place, and also scale

(McConnell, Megoran, and Williams 2014; Bregazzi and Jackson 2016). Peace is too often thought of as something kept only at the national scale, as in when a president or prime minister talks about maintaining national security. But you can build peace inside you, in your family, in your neighborhood, in your city, and in your region (Koopman 2017). These scales are obviously not disconnected. Pain and Smith (2008) see the global and intimate as intertwined, like a DNA helix with multiple connectors, and constantly shaping each other. More recently, Pain (2015) argued that they are so tightly entwined that violences across scale function as "intimacy-geopolitics." Her work on domestic violence as "intimate war," both inside and shaping a purported peace, was extended by Brickell (2015), who looked at the violences of attempts to build "intimate peace" in the purported post-conflict zone of Cambodia. She illustrates how ideas of peace at what may appear to be a micro level are shaped by and sustain both ideologies and practices at larger scales.

Peace is always situated, meaning it is always built some*where* and at some *scale* for some *people* (Koopman 2017). Peace is built in different ways in part because it means different things to different groups even in the same time and place (Koopman 2011a). If, as has been happening in the invasion of Ukraine, your city being bombed has set off large fires in the area and you have asthma, the most important step toward peace for you right now might be to stop the fires so that you have clean air to breathe – while if instead you have low blood pressure and the bombing has also taken out the electricity and thus the heating in winter, actually building another small fire in a wood stove might be the more urgent thing.[1] Peace is multifaceted, precarious, and built and rebuilt every day by every one of us. We can make spaces and moments of peace even in the midst of armed conflict.

One of the powerful ways we build peace is through nonviolent action. This is more than political action that is not violent – it is action that works *against* violence (of various sorts, including direct, structural, and cultural), without using violence.[2] It can be understood as engaging in *un*armed conflict to address a power imbalance (Sheehan 2014). But to put it that way is to again focus on what we are working against, rather than what we are working for – just as Galtung does with his definition of positive peace. This is a common emphasis, in both activism and activist-academic work. Yet nonviolent action, like peace, can be understood as both negative and positive. We can build peace not just by working to stop war and injustice (obstructive nonviolent action) but also by "being the change" and building alternatives. For example, community gardens are a form of constructive nonviolent action.

Mahatma Gandhi was a driving force behind India's challenge to an empire, and an innovator of nonviolent action. He called obstructive and constructive action the two wings of the bird of nonviolence, and emphasized the latter, particularly at the end of his life (1948). Some campaigns, such as Gandhi's call to refuse to pay the British occupier's tax on salt and instead make your own from the sea, can be both obstructive and constructive. Like Gandhi, Kropotkin, an early anarchist geographer, also argued that instead of spending energy fighting what we do not want in the world, we should focus more on building the world we want. In particular he argued for doing so through what he called mutual aid. He lamented that there were so many descriptions of acts of violence, but little chronicling of the everyday ways "the masses" built peace, even while war was being waged (1902, 116–17).

Peace exists inside war, and war inside peace. Seeing them as all-or-nothing opposites is a paradigm that can discourage us from trying to achieve peace. Or the tendency can be to focus solely on studying conflict, believing all it takes for peace is to end war (Koopman 2011a). If we understand peace not as an opposite or as an endpoint, but as a *doing* (through purposeful action), then it becomes something we all can do, and indeed need to do, and do again and again. Such an approach makes the term "peacebuilding" repetitive. The building aspect is inherent and ongoing. Peace is always being built; it is never done.

Because it is a socio-spatial process, how we use nonviolent action to build peace will always vary. We have much to learn from different "situated knowledges of peace," as described by Williams and McConnell (2011). Because human geography looks at the relationship between society and space, and how each is constantly shaping the other, geographers are particularly well placed to connect the various experiences and experiments in peace and justice making around the world through what Haraway (1992) calls "non-innocent conversations," meaning ones that recognize that we each have situated knowledge, based on our location and experience. Ideally these conversations acknowledge that peace(s) and peacebuilding both shape and are shaped by the spaces they are made in, but also weave the peace(s) together and foster solidarity – helping movements grow and learn from and connect with each other (Koopman 2011a). I aim to do a bit of this here, and I turn now to nonviolent activism for peace in two armed conflicts: Colombia and Ukraine.

Making space for peace in Colombia

Colombia has had one of the longest internal armed conflicts in the world. There have been various guerrilla groups fighting the government since at least the 1960s (some argue there was no break from the fighting in the 1940s). Either way, it started long before the drug trade and has primarily been a fight about unequal access to land – but drug money has been fuel on the fire, funding arms and fostering violence. But in the midst of the war, some brave grassroots groups have been innovating ways of building peace. In this section, I describe the nonviolent action strategy of accompaniment as it is used in Colombia.

The slogan of Peace Brigades International (PBI) is "making space for peace." I conducted research with PBI and other accompaniment organizations in Colombia and discussed what that slogan meant to them and how they understood space, peace, and how they were making it (Koopman 2011b). In most conflict zones, some people are more likely to be attacked than others. In particular, certain outsiders tend to be left alone. Accompaniment uses that relative safety to provide protection, by putting people who are less at risk next to local civilians who are under threat, as a sort of "unarmed bodyguard" (Mahony and Eguren 1997). This is sometimes done 24/7 when the risk is very high, but often done only when people need to travel through various sorts of particularly risky spaces. This might mean riding with them on a jeep or a canoe, or walking on a stretch of path in the rainforest.

Why does simply walking with someone build peace? Because usually, it is not just any body at risk that is being accompanied, but the bodies of those struggling to build peace and justice in the midst of conflict. Often this simply means the peace of being able to stay on the land you are farming, in a context where over one in six people have been violently displaced, most of them small farmers.[3] And usually, it is not just any body that is doing the accompanying, but a privileged outsider body that is less likely to be killed (Koopman 2011b). Who one is, where one acts, and how one acts all shape possibilities for nonviolent action for peace.

Different political actions and spatialities intersect. Accompaniers rely on and leverage systems that make *certain* lives matter more – paradoxically to support local peace organizing efforts that are instead working to make *everyone's* lives safer, but particularly the lives of peace organizers under attack. The United States sent $13.2 billion in aid, most of it military, to Colombia between 2000 and 2022 (Isacson 2022). Ironically being a major funder of war makes it more possible for grassroots activists to work for peace by making a U.S. passport more powerful in that context. If a person is stopped at a Colombian military (or even paramilitary) checkpoint with a U.S. passport that will get much more attention than, say, one from Burkina Faso. Accompaniment takes much more than just being a foreigner walking alongside a Colombian activist under threat. First, specific location matters. A U.S. passport has historically been no deterrent at all to the Colombian guerrillas – indeed U.S. passport holders were at higher risk of being kidnapped by them. As such, accompaniment has traditionally been used in areas where the greatest risk was from the paramilitaries, death squads loosely linked to the official Colombian army that receives all that U.S. aid. But second, even in those areas, being a lone actor with a U.S. passport would not make much of a difference.

The spatiality of networks is key to how accompaniment works, as it is done by PBI and other organizations like Peace Presence and Community Peacemaker Teams. They rely on networks with the ability to pressure chains of political and military influence in other spaces/times. These can raise the stakes of an attack by making it clear it would have heavy consequences.

For this to work two things must happen. The local paramilitary leader needs to know that the accompanier is there, and the local paramilitary leader must have heard from the local military commander to leave the accompanier alone. The commander generally needs to have received that message from their superiors, who were pressured from the U.S. embassy, who were pressured from members of the U.S. Congress, who were pressured from constituents, some of whom heard about the situation from the parents of the accompanier, who spoke about it at a dinner event in the basement of their church. Obviously, this is an oversimplification of how these chains work. Accompaniers also often meet with actors all along the chain. Once these pressure chains are established, they do not need to be activated each time to work.

Today these pressure chains are more likely to start by someone hearing about a threat on social media than in the church basement mentioned above. New information and communication technologies (ICTs) foster new and more rapid connections across both distance and various sorts of social difference, such as age. Many of us have access to a video camera on our phones and can quickly upload or live-stream short videos that can provide an intimate sense of others' daily lives and struggles (like a paramilitary checkpoint or death threats). Unfortunately, many groups have now come to over rely on social media, and grassroots organizing across those basements has weakened (particularly so after COVID, but this overreliance has been increasing for years). In the face of growing algorithm friction on various sites, and increased outright censorship on Twitter, this is a worrisome trend for the future of peace activism.

Not all accompaniers in Colombia are from the United States. Many are from Europe and Canada, which have also been major aid funders. Canada is also home to powerful mining companies working in Colombia. While accompaniers are commonly from these more powerful countries, some are from other Latin American countries. A few teams, like Peace Presence and Community Peacemaker Teams, also have Colombian accompaniers on their teams. Latina accompaniers are often able to strengthen a group's security analysis by being better able to read nuances in communication and build stronger community connections.

It is hard to untangle passport privilege from racial and class privilege. In Colombia, people from the United States are widely imagined to look a certain way, and have certain resources. Because they did not fit that stereotype, Latina accompaniers from the United States told me they would, for example, make a point of speaking English at a military checkpoint, even if it was only into their phone – sometimes on a fake phone call even! They might also simply get their blue passport out in advance and have it very visibly in hand.

When hearing about accompaniment some imagine that it primarily works through racial privilege, and that any lone white person walking alongside a Colombian under threat will provide protection. Whiteness is, undoubtedly, part of the privilege that is leveraged – but without any of those chains of influence to turn to in case of emergency, it would not go far. Ideally whiteness and other forms of privilege are being used against themselves by accompaniers, who use them to support organizing projects in Colombia that in the long run dismantle systems that make certain lives matter more.

It is also interesting that there are generally far more women accompaniers. Women are more socialized for care work, but it is also true that they are more widely imagined as civilians and less likely to be seen as a threat. Of course, other social locations such as age and physical ability shape how accompaniers are seen and treated. Accompaniers generally work in pairs to accompany a person or group under threat as they move through a more dangerous area. Teams try to ensure that there is a combination of passports (and other strengths) on any particular trip.

Accompaniment has kept safe some of the bravest organizers working for peace and justice under incredibly threatening conditions. In the process, it has built many ties between organizers in the global North and South that continue after the accompanier returns home, weaving a web that is both for, and in a sense of, peace. Colombia is now the country with the largest number of international organizations providing accompaniment, but this tactic began in Central America in the 1980s and has since spread around the world. The webs of solidarity created by accompaniers and their networks of support build different and more positive relationships between places, beyond just working to stop acts of violence.

Accompaniers travel to different countries, and travel alongside peace activists under threat when they travel through dangerous areas inside those countries, but the key spatiality they tap in to as they do that work is networks, and the chains of influence across other places that they can activate in case of a threat or an attack. Chains ending in Washington, DC and Brussels, the seat of the EU, are particularly key. But as they walk alongside peace activists in danger, accompaniers are not only pushing back against hegemonic policies and practices of (in)security that they do *not* want (antigeopolitics). They are also nurturing other types of nonviolent security in and through connection that they *do* want, what I have called alter-geopolitics (Koopman 2011b).

Geopolitics is widely imagined as something done just by politicians, for (usually military) security. Alter-geopolitics is instead about all different sorts of people coming together not just for security, to stay alive – but to live *well*. Likewise positive peace is about well-being, which requires security of food and housing and health and more. The organizations that are accompanied in Colombia are building positive peace by doing seed exchanges, collectively drying their cacao, and more. Far too often though, their first priority has to be physical safety. It is hard to organize your community for access to food if you are afraid that you could be killed for speaking out about paramilitaries taking the land.

Using space for peace in Ukraine

If Colombia is one of the oldest armed conflicts in the world, Ukraine, as of this writing, is both one of the newest and most intense. Russian troops crossed the border between the two countries and began invading Ukraine in February of 2022. Cities far from the front lines are regularly bombed by Russia. Electricity and water infrastructure is particularly targeted, making it hard for Ukrainians to safely stay in their homes. Even amid bombing and full-out trench warfare though, nonviolent action is possible and has been widespread in Ukraine, as I will describe in this section. Given the intensity of the current war it is not surprising that much of it has been focused on building negative rather than positive peace.

Different settings and spatialities require, and make possible, different forms of nonviolent action. Any tactic will always need to be adapted to a particular space, to particular threats. For example, accompaniment in Ukraine happens in less formal ways, and not always through established organizations. In Colombia, accompaniers help people who have chosen to live and work together to avoid displacement. In Ukraine, informal accompaniment networks have come together to help people displaced, and support those who have been forced to flee bombing and occupation. Throughout Europe local people arrived at train stations with signs offering housing to arriving refugees and asylum seekers. Because most men have not been allowed to leave Ukraine, this has been a flood of women and children. The UN recorded 7.8 million Ukrainians in other parts of Europe by December 2022 (with 2.85 million of those in Russia, many forcibly taken there) (UNHCR 2022). Numbers are surely higher, due to both underreporting and refugees in the Americas, but even so that is nearly one-third of all Ukrainians. Yet many more are displaced within the country.

One of the ongoing dangers faced by women refugees is pressure for sexual favors or outright sexual assault – particularly when staying with people who have not been screened in any way. One of the ways that women have built alternative security through solidarity has been by offering support from networks of volunteers on Telegram and WhatsApp chat groups that offer tips and referrals (Daza Sierra 2022a). These have included connections to car rides and trains as well as short-term places to stay along the way as women flee inside Ukraine and across the continent. People have also used Facebook to find friends of friends (of friends) who could host. These social networks move back and forth between offline and online connections that help women find not peace, but at least a more peaceful refuge from war. In the process, networks have been strengthened and connections woven across distance and difference that may help to build peace across Europe in other ways as people go on to use these connections for other organizing campaigns, such as for more clean energy.

Women are in some contexts more at risk, but they have also used patriarchal structures and imaginaries (using them against those very structures in a sense) to build more safety. Women are widely not imagined to be combatants. Indeed, as Cynthia Enloe (2000) puts it, they are lumped into "womenandchildren" as the quintessential civilians. As such, in conflict zones around the world, women are more able to get through military checkpoints. As mentioned above, international accompaniment teams in Colombia have historically had more women on them. In Ukraine too, Nina Potarska, the coordinator of the Women's International League for Peace and Freedom, reports that "thanks to the invisibility that women gain in war," women have been key in helping with evacuation from regions being invaded (Daza Sierra, 2022a, 23). Peace is not the same across time and space, but even in the same space, what peace means and how we build it plays out differently for different people.

Some have not been able to flee the bombing in Ukraine. In particular many elderly and people with disabilities have been physically unable to manage the hardships of the trip, and some have been reduced to living in the root cellars of their homes that have been destroyed by bombing (Petrasyuk 2023). Many have refused to move, even to nearby shelters, wanting to protect what is left of their homes but also afraid to go somewhere new where they do not know people, or how to get around. One of the hardest parts of displacement is losing your social fabric – what neighbor can help you fix this thing, or whose niece knows that person you need to get through to. This extends to place-specific knowledge gained through years of living in a particular place. For example, knowing where there are trees with fruit in the summer, and where one can forage chanterelle mushrooms in the spring.[4] Losing this knowledge is even harder for the elderly that rely more on these social and ecological networks and are generally less able to use online networks for support.

Some have found a bit more space for peace at home in the midst of war through a constructive nonviolent action campaign that involves literal construction. A network of churches has been funding the building of, and offering volunteer labor to help with the placement of, prefabricated tiny homes right next to destroyed homes. These have primarily been offered to *babushkas* (grandmas, elderly women). They are called "hope homes" and offer those who receive them the ability to slowly work on rebuilding their own homes while maintaining the social and land connections that are all the more important as we age. I heard about this project through the sort of networks that have made this work possible. My aunt (in the United States) forwarded an email from a distant cousin, Aaron Gibbs, who was doing pastoral service in Slovakia when Russia invaded. He reached out to pastors in Ukraine, and together they created this project. Aaron's emails have been forwarded through different church and friend and family networks and helped to fund the project.[5] In the process, again, the networks that make this peace work possible in themselves can also weave other forms of peace. For example, in my case, they have built stronger connections across my family.

Making it safer to stay, making it safer to go, both have been important nonviolent action tactics for peace in the context of war in Ukraine. Making it harder for the invaders to come in has also been an important tactic for resisting occupation. Slowing down the invasion has given people time to flee. Astoundingly, many of the Russian troops were given no functioning GPS. Their personal cell phones do not work on Ukrainian signals. There have literally been columns of Russian tanks coming down roads but not sure where to turn. In this context, the simple act of sabotaging road signs has been a widespread tactic of resistance. Many were simply removed or erased, some were covered with "f---off," some signs pointing to particular towns were turned around, and some signs were changed so that all directions pointed to The Hague (a reference to the eventuality of being tried for war crimes and one of many ways that Russian conscripts are pressured to stop killing) (see Figure 10.1) (Daza Sierra 2022a, 2022b). This tactic is a reprise of one used by Czechs to resist Soviet invasion in 1968 (Lakey 2022). The Ukrainian government's road

FIGURE 10.1   Ukrainian road sign modified to tell Russian invading troops that all roads lead to The Hague (the location of the international criminal court).

Source: https://commons.wikimedia.org/wiki/File:FM2CGiVXEAMSDTY.jpg

agency actually encouraged people to remove or change the signs, first by sharing on social media a photoshopped version of a sign changed to "f---off," then by sharing photos of offline actions that, when shared online, shaped other offline actions in other places (Bella 2022). Locals, of course, did not need the signs to know where to go. People fleeing inside the country could either use the map on their phone or ask locals. The same government agency also asked people to set up barricades on certain roads, which people did with sandbags, tires, and homemade anti-vehicle and anti-tank obstacles. In the United States, there are often heated debates about whether property damage (such as taking down confederate statues or breaking pipeline mechanisms) counts as nonviolent action or not, but in this context, this seems quite clearly to be obstructive nonviolent action for peace.

A larger scale form of sabotage has also played an important role in resisting the invasion. The country of Belarus borders both Ukraine and Russia, and is ruled by President Alexander Lukashenko, an autocrat close to Russia's President Putin. It seemed likely that Belarusian troops would also invade, but early in the invasion Belarusian railroad workers switched and sabotaged the rails such that neither Russian troops nor supplies for them could be moved in by rail. This may have convinced Lukashenko that invading would create enough domestic difficulties for him, in a context of years of ongoing protests for democratic and social freedom in Belarus, that he has so far refrained from supporting Putin with troops (Daza Sierra 2022b). Like protective accompaniment, this points again to how international dynamics shape very local ones and vice versa – but also how nonviolent activists leverage these.

Street signs also became a way of showing international solidarity. Activists in Ukraine asked people around the world to change the names of the street signs that Russian embassies were on to "free Ukraine street," as a way of pressuring Russia. This turned into an official request by the Ukrainian ministry of transportation and some cities, including Stockholm and Oslo, officially changed street names ("List of Streets Renamed Due to the 2022 Russian Invasion of Ukraine" 2022). In 2023, the campaign was ongoing, and petitions were being signed to pressure city officials. If the war is still ongoing when you are reading this, you could petition to rename a street near you at uastreet. world. It seems the new tactic of huge word murals painted on the actual street (in sizes visible from the sky) that was popularized by the Black Lives Matter movement has not yet been used to paint "free Ukraine" in huge letters on the streets in front of embassies, but you could be the one to start a trend.

Another much more direct way of getting through to Russians was launched by a group of activists in Lithuania, which set up a web site (callrussia.org) to coordinate phone calls by volunteer Russian speakers around the world to random ordinary people in Russia. Many Russians believe their state propaganda that Ukrainians want Russians to come and "save" them from neo-Nazis. Rather than engage in arguments, the volunteers were trained to do deep listening first and build real personal connections that made it easier for Russians to then hear stories of the impacts of the war. By December of 2022, over 49,000 volunteers in 149 countries made more than 170,000 calls, according to the site. With voice over internet protocols, these calls were free; again, ICTs allowed people to build connections across distance and difference as a way to build peace.

Ukrainian President Volodymyr Zelensky has repeatedly recorded videos to the Russian people (in Russian) asking them to resist and desert and offering money, amnesty, citizenship, or help reaching another country to any that do (Hunter 2022). Though these pleas have been blocked in Russia, they can be forwarded on messaging services like Telegram and WhatsApp. Most Ukrainians have friends and family in Russia they can send messages to. Many ordinary Ukrainians have also simply walked up to Russian troops and urged them to desert. Like the phone calls and messages, this undermined the Russians' sense of legitimacy, of being wanted, and instead appealed to their common humanity (Van Hook 2022). For example, a video was widely circulated on social media of a deserter who, upon deserting, was given tea, a pastry, and a chance to video call his mother (Christopher Miller [@ChristopherJM] 2022). As Maria Stephan noted, what was notable was the repeated efforts not to dehumanize the Russian troops (Van Hook 2022). The Ukrainian government said by mid-December 2022 they had received over 1.2 million calls to the "I want to live hotline" where troops could surrender (news desk 2022).

As the invasion increasingly turned into trench warfare, it was harder for ordinary Ukrainians to reach Russians in person to pressure them to desert, but also for Russians to surrender across mined areas. Technology became more important not just for the pressure (via phone calls and text messages) but also for the mechanics of the process of surrender. Ukraine offered drones to lead Russians who had called the hotline across the mined area on a safer route (Santora 2022). Though initially Ukraine was offering payment and help moving to another country if desired, as of December 2022 Russian deserters were being held as prisoners of war (for exchange with Ukrainian prisoners), though promised good treatment on the "I want to live" hotline website (hochuzit.com). This strategy would be even more effective if people fleeing Russian military conscription were being widely welcomed in Europe or Canada, as Canada once welcomed U.S. conscientious objectors of the Vietnam War. As of September 2022, only Germany was accommodating deserters (Studer 2022) perhaps because of concern that they could be spies (Daza Sierra 2022b). At the same period of the war, Russia seemed concerned enough about this strategy that in July 2022, they increased the penalty for desertion to a 20-year prison term (Daza Sierra 2022a, 27).

Some incredibly brave Ukrainian civilians appealed to the humanity of Russian troops simply by putting their unarmed bodies in front of the tanks rolling into their towns (Christo Grozev [@christogrozev] 2022; УНIАН 2022;

Hunter 2022). A careful study of nonviolent actions across Ukraine in the first months of the invasion found at least 14 such actions of interposition (Daza Sierra 2022a, 22). In at least two cases (Enerhodar and Slavutich) when Russian troops kidnapped the town's mayor, the people of the town came out en masse and surrounded them until the mayor was released (Daza Sierra 2022a; Van Hook 2022). This tactic has also been used in Colombia, primarily by indigenous groups who have en masse surrounded guerrillas who kidnapped people to gain their release (Ulloa 2012). But as Daza Sierra (2022a) points out, this sort of resistance was not possible in other cities where the repression and violence by the troops was much greater. These sorts of protest actions, and demonstrations and rallies generally, were also more common early in the invasion. Later it became more dangerous under occupation as activists were increasingly tortured, massacred, disappeared, had their children taken away, and more (Daza Sierra 2022a).

Yet even under occupation, Daza Sierra and his team documented widespread noncooperation, a more subtle, but important, form of nonviolent action. Russian flags were removed. Ukrainian flags and ribbons were secretly put up. Teachers refused to teach in Russian. City officials and civil servants refused to legitimize Russian control, making it more difficult for Russia to collect taxes and build defensive infrastructure. But when even noncooperation became too dangerous, nonviolent action for peace continued in the form of protective action, smuggling food and supplies in through back roads, and people out. Horizontal mutual aid networks strengthened self-governance and autonomy even in the midst of war (Daza Sierra 2022a). Even in the depths of brutal occupation, in these small ways forms of positive peace are nurtured.

Nonviolent action has been used in Ukraine to flee, to stay, and to stop the invading forces. It has been used to get through to Russians at embassies, on the phone, and via video forwards. Ukrainians have used various ways to make it clear to Russian recruits that they are not wanted or seen as saviors, as they were led to believe they would be, and to encourage desertions. Nonviolent action has continued even in areas that have been occupied, primarily through various forms of noncooperation. Nonviolent action in Ukraine has been both obstructive and constructive. Nonviolent tactics have been used to obstruct, or stop, what those using them do not want – in this case the invasion. This has happened most dramatically by putting bodies in front of tanks, but also by acts as common as making calls to Russian homes and as creative as changing street names. Nonviolent tactics have also been used to construct, or build, what those using them do want. In particular in the examples here that has been a safe place to sleep, whether while fleeing the bombing, or while staying in a town after it has been bombed and one's home destroyed. From train stations to embassy streets to WhatsApp networks, a range of different spatialities have both made possible and shaped different nonviolent actions for peace in Ukraine.

Conclusion

Peace and war are a false binary. This chapter has been full of stories of making spaces of peace even in the midst of war. War and peace are entangled, as are other overly simplified binaries like the personal/political and global/intimate (Christian, Dowler, and Cuomo 2016). When peace is portrayed as a mythical singular, an end point after war, it can seem impossible to reach. But peace is not one thing. Even in one place, at one time, peace can have different emphases for different people. Peace really is not even a thing; it is a doing, an ongoing process that we all engage in together in different contexts, in different ways. Peace is made through spatial practices (Macaspac and Moore 2022). Peace is a socio-spatial relation that is always made and made again (Agnew 2009).

One of the powerful ways peace is made is through nonviolent action. Nonviolent action too is not one thing. It is not just a street protest, and not just obstructive action, as often imagined, but also constructive. It is the construction of homes, but also the construction of networks that help people flee bombing of their homes. Nonviolent action is not just action taken in peacetime – but is also widely done in the midst of war. Unarmed conflict exists within and alongside armed conflict.

Socio-spatial dynamics shape how we can and do work for peace. How we work for peace also can and does change socio-spatial dynamics. This is perhaps easiest to see with networks. The networks that helped women flee the bombing in Ukraine were strengthened as they were used, and became both broader and denser through the first year of the invasion. Meanwhile other social networks, the fabric between neighbors and local businesses for example, were ripped to shreds. So too socio-ecological networks were destroyed as trees that mushrooms rely on were cut for bunkers, or surrounded by land mines. Geographers have long looked at how social movements and different spatialities (notably territory, place, scale, and network) shape each other.[6]

Peace and peace activisms are shaped in and through the spaces and times in which they are made, as they too shape those spaces and spatialities. The international comes in close on a street sign pointing to The Hague, which makes it harder for troops to find their way through a local intersection. Online access to the phone numbers of ordinary Russians and voice over internet calling makes it possible to organize thousands of one-on-one conversations across borders about the invasion, bringing the impact of the war into Russian kitchens. A regional connection between a Slovakian pastor and a Ukrainian pastor opens the way for U.S. churches to fund tiny hope homes that allow grandmothers to stay next to the rubble of their bombed homes in Ukraine, in the process bringing churches closer, but also family appealed to for funding. Church basements in the United States host dinners where accompaniers working in Colombia tell stories of hikes through jungle areas controlled by paramilitaries, inspiring pressure calls to the U.S. State Department.

Nonviolent action works against violence either by being obstructive, working to stop what we do not want, or by being constructive, working to build what we do want. Sometimes it is both. Protective accompaniment is obstructive nonviolence in that it is working to stop violent attacks on the ground (and usually also to stop U.S. military aid). But it is often working to support constructive nonviolence by the activists who are accompanied, who are building alternative social structures, like peace communities that help each other live well together as they resist displacement. Standing in front of a tank invading your country is dramatic obstructive action, but building Telegram app networks that share connections for rides for those fleeing, and homes for them to stay in, is equally important protective constructive action. Perhaps not all of these actions would count as doing alter-geopolitics (Koopman 2011b), as some of them might be seen as making "us" safer by making "them" less safe – but it could also be argued that even the railway sabotage ultimately kept Russian troops safer by keeping them away from combat.

Nonviolent action takes many forms, and geographers too work for peace in various ways. Yes, we can raise funds for tiny homes and heaters and other aid. We can buy Ukrainian digital music and art directly from producers.[7] We can fly the flag, and put stickers on our computers. This is not an empty gesture, but a symbolic action that raises awareness and echoes those secretly putting the flag and stickers in occupied areas. But also, through our academic work that questions imaginaries of "there" and "them" we can help to build a bigger "us" – and help to build connection and care across distance and difference. This might look like a close read of a speech by Putin alongside a review of ways contested territories have historically been shared to look for mediation options (see Megoran 2022). Or it might look like sharing stories of Ukrainian refugees and connecting them to other migrant struggles and organizing. Or like this chapter, drawing lines between the ways different nonviolent actions in different places use and shape space for peace. Geographic insights can help us recognize these different socio-spatial strategies, recognize their power, and both draw and foster connections across movements.

Many geographers are doing work on aspects of social justice and sustainability without using the term "peace." Understanding peacemaking more broadly allows us to also connect these pieces of peace (Koopman 2011a). Decolonial geographies and work on racial state violence are absolutely work for peace. Our discipline has long served and been shaped by war and conquest (Driver 2001). Audre Lorde (2007) famously argued that the master's tools will never dismantle the master's house, but I disagree. Geography may have been developed as a master's tool – but it can, if used carefully, not only dismantle that house of war and coloniality, but also be used to build a new house of peace and freedom.

Notes

1 Nova Ukraine has been organizing volunteer metalworkers to make rocket and pot-belly stoves. https://novaukraine.org/2022/12/08/volunteer-metalworkers-turn-to-heating-homes/
2 Often defined as actions limited to outside of normal institutional channels (e.g., voting and lobbying). But edges blur when voting is restricted or lobbying by the grassroots.
3 7.6 million people in a country of 51 million (https://reporting.unhcr.org/colombiasituation).
4 Though some areas have been mined by the Russian army, making foraging more dangerous and less of a haven for peace (Varenikova and Hoffman 2022).
5 To sign up for emails and support the home construction: https://choosingslovakia.com/ukraine
6 Recent work by geographers on social movements includes the assemblage approach, which also brings in the role of things, or more-than-human elements like mushrooms (Koopman 2015).
7 Artists https://thegeekiary.com/support-ukrainian-artists-and-other-creatives/105924 and musicians https://tinyurl.com/supportukrainianmusicians. A review of some musicians: tinyurl.com/ukrainianmusic. I am particularly enjoying Dakhabrakha. Many Ukrainian movies are also available for streaming online.

References

Addams, J. 1907/2007. *Newer Ideals of Peace*. University of Illinois Press.
Agnew, J. 2009. Killing for Cause? Geographies of War and Peace. *Annals of the Association of American Geographers* 99: 1054–1059.
Bella, T. 2022. Ukrainian Agency, Urging Removal of Road Signs, Posts Fake Photo with a Colorful Message for Russia. *Washington Post*, February 26. https://www.washingtonpost.com/world/2022/02/26/ukraine-russia-roads-signs-facebook/
Bregazzi, H., and M. Jackson. 2016. Agonism, Critical Political Geography, and the New Geographies of Peace. *Progress in Human Geography* 42: 72–91.
Brickell, K. 2015. Towards Intimate Geographies of Peace? Local Reconciliation of Domestic Violence in Cambodia. *Transactions of the Institute of British Geographers* 40: 321–333.
Carroll, B.A., and C.F. Fink. 2007. Introduction. In *Newer Ideals of Peace*, by Jane Addams. University of Illinois Press.
Christian, J., L. Dowler, and D. Cuomo. 2016. Fear, Feminist Geopolitics and the Hot and Banal. *Political Geography*, Special Issue: Banal Nationalism 20 years on, 54: 64–72.
Christo, G. [@christogrozev]. 2022. "Video of Ukrainians in the North-Eastern Town of Bakhmach." *Twitter*. https://twitter.com/christogrozev/status/1497562255084474368
Christopher, M. [@ChristopherJM]. 2022. "Remarkable Video Circulating on Telegram. Ukrainians Gave a Captured Russian Soldier Food." https://T.Co/KtbHad8XLm. *Twitter*. https://twitter.com/ChristopherJM/status/1499060828817043474
Daza Sierra, F. 2022a. Ukranian Nonviolent Resistance in the Face of War. *ICIP International Catalan Institute for Peace*. https://novact.org/wp-content/uploads/2022/10/ENG_VF.pdf
Daza Sierra, F. 2022b. Civil Resistance in Ukraine and the Region. *Kroc Institute*, March 22. https://www.youtube.com/watch?v=CcttVAA-_-0

Driver, F. 2001. *Geography Militant: Cultures of Exploration and Empire*. Wiley-Blackwell.

Enloe, C. 2000. *Maneuvers: The International Politics of Militarizing Women's Lives*. University of California Press.

Galtung, J. 1969. Violence, Peace, and Peace Research. *Journal of Peace Research* 6: 167–191.

Gandhi, M. 1948. *Constructive Programme (Its Meaning and Place)*. Navajivan.

Haraway, D. 1992. The Promises of Monsters. In *Cultural Studies*, eds. L. Grossberg, C. Nelson, and P. Treichler, 295–337. Routledge.

Hunter, D. 2022. Ukraine's Secret Weapon May Prove to Be Civilian Resistance. *Waging Nonviolence*. February 27, 2022. https://wagingnonviolence.org/2022/02/ukraine-secret-weapon-civilian-resistance/

Isacson, A. 2022. U.S. Aid to Colombia. *Colombia Peace*. April 14. https://colombiapeace.org/u-s-aid-to-colombia/

King Jr, M. L. 1963. Letter from Birmingham Jail. April 16, 125–126.

Koopman, S. 2011a. Let's Take Peace to Pieces. *Political Geography* 30:193–194.

Koopman, S. 2011b. Alter-Geopolitics: Other Securities Are Happening. *Geoforum* 42: 274–284.

Koopman, S. 2015. Social Movements. In *The Wiley Blackwell Companion to Political Geography*, eds. John Agnew, Virginie Mamadouh, Anna J. Secor, Joanne Sharp: 339–351. Wiley-Blackwell.

Koopman, S. 2017. Peace. In *International Encyclopedia of Geography*, eds. D. Richardson, N. Castree, M. Goodchild, A. Kobayashi, W. Liu, and R. Marston, People, the Earth, Environment and Technology. Wiley-Blackwell.

Kropotkin, P.A. 1902. *Mutual Aid: A Factor of Evolution*. Kessinger Publishing.

Lakey, G. 2022. "Ukraine Doesn't Need to Match Russia's Military Might to Defend against Invasion." *Waging Nonviolence* (blog). February 25. https://wagingnonviolence.org/2022/02/ukraine-doesnt-need-to-match-russias-military-might-to-defend-against-invasion/

Lefebvre, H. 1992. *The Production of Space*. Wiley-Blackwell.

"List of Streets Renamed Due to the 2022 Russian Invasion of Ukraine." 2022. *Wikipedia*. https://en.wikipedia.org/w/index.php?title=List_of_streets_renamed_due_to_the_2022_Russian_invasion_of_Ukraine&oldid=1127642994

Lorde, A. 2007. *Sister Outsider: Essays and Speeches*. Crossing Press.

Macaspac, N.V., and A. Moore. 2022. Peace Geographies and the Spatial Turn in Peace and Conflict Studies. *Geography Compass* 16: e12614.

Mahony, L., and E. Eguren. 1997. *Unarmed Bodyguards: International Accompaniment for the Protection of Human Rights*. Kumarian Press.

McConnell, F., N. Megoran, and P. Williams, eds. 2014. *The Geographies of Peace: New Approaches to Boundaries, Diplomacy and Conflict Resolution*. I.B. Tauris.

Megoran, N. 2022. Geographical Contributions to Understanding and Resolving the Russo-Ukraine Conflict. *Geopolitica(s). Revista de Estudios Sobre Espacio y Poder* 13: 285–309.

news desk. 2022. Ukraine's Intelligence: Over 1.2 Million Calls Made to Surrender Hotline Service. *The Kyiv Independent*, December 17. https://kyivindependent.com/news-feed/ukraines-intelligence-over-1-2-million-calls-made-to-surrender-hotline-service

Pain, R. 2015. Intimate War. *Political Geography* 44: 64–73.

Pain, R., and S. Smith, eds. 2008. *Fear: Critical Geopolitics and Everyday Life.* Ashgate.

Petrasyuk, V. 2023. Last Holdouts: The Basement Lives Of Ukrainians Who Refuse To Flee Frontline Towns. *Worldcrunch*, March 8. https://worldcrunch.com/focus/ukraine-civilians-living-frontline

Santora, M. 2022. Surrender to a Drone? Ukraine Is Urging Russian Soldiers to Do Just That. *New York Times*, December 20, World. https://www.nytimes.com/2022/12/20/world/europe/russian-soldier-drone-surrenders.html

Sheehan, J. 2014. What Is Nonviolence, and Why Use It? *War Register's Handbook for Nonviolent Campaigns.* https://www.wri-irg.org/en/story/2014/what-nonviolence-and-why-use-it

Studer, P. 2022. Are Russian Deserters Entitled to Asylum? *Swissinfo*, September 29. https://www.swissinfo.ch/eng/business/are-russian-deserters-entitled-to-asylum-/47940296

Ulloa, A. 2012. Los territorios indígenas en Colombia: de escenarios de apropriación transnacional a territorialidades alternativas. *GeoCritica* 16, 23.

UNHCR. 2022. Ukraine Refugee Situation. https://data.unhcr.org/en/situations/ukraine

Van Hook, S, dir. 2022. Amplifying Stories of Nonviolent Struggle from Ukraine, Russia and the World at Large. *Nonviolence Radio.* https://wagingnonviolence.org/metta/podcast/amplifying-stories-nonviolent-struggle-ukraine-russia-maria-stephan/

Varenikova, M., and B. Hoffman. 2022. In Forests Full of Mines, Ukrainians Find Mushrooms and Resilience. *New York Times*, December 4, World. https://www.nytimes.com/2022/12/04/world/europe/ukraine-mines-mushroom-hunters.html

Warren, S. 2019. Borders and the Freedom to Move. *Dialogues in Human Geography* 9: 223–225.

Williams, P., and F. McConnell. 2011. Critical Geographies of Peace. *Antipode* 43: 927–931.

УНІАН, dir. 2022. *Безоружные Жители в Запорожской Области Остановили Русские Танки.* https://www.youtube.com/watch?v=KOZ4VMJxHsY

11

PEACEWORK

Everyday negative peace across South Asian borderscapes

Md Azmeary Ferdoush

I have lived long enough; it is almost time for me to go now. But I have never lived in peace in my life. I always lived in fear. I have seen violence so frequently in my life! There has not been a single night I could go to bed with peace [in my mind]. I have lived in fear.... Now, I will finally sleep in peace at night.

(fieldnote 2015)

Rahman Ali (pseudonym) shared this statement during our June 2015 interview. A farmer in his early seventies, Rahman reflected on his experience of being an Indian enclave resident for the last seven decades in the northern district of Panchagarh inside Bangladesh. He also hoped finally to be in "peace" once the enclaves were exchanged and he became a regular Bangladeshi citizen. Enclaves are territories of one sovereign state completely surrounded by another (van Schendel 2002). Almost 200 enclaves existed as extraterritorial spaces between Bangladesh and India for the last 70 years until July 2015 when they were finally exchanged and merged as regular spaces of the hosting state (Ferdoush 2019a). Those who lived in these small pockets of land were commonly known as enclave residents, not Bangladeshis or Indians. They lacked a formal citizenship recognition from the host state since they were living inside another sovereign state's territory. However, at the same time, they were unable to access citizenship rights from the "home" state as it did not maintain a regular connection with its enclave populations. As such, they lived as de facto stateless people with a lack of citizenship status. Thus, a lack of usual protection provided by different state mechanisms for example, the police and law, underpinned a quotidian feeling of being unsafe—an absence of peace.

DOI: 10.4324/9781003345794-11

But Rahman was looking forward to "peace" as the enclaves were to be exchanged, making him a Bangladeshi citizen with rights soon. For Rahman, this was what "peace" meant—a feeling of safety, that no one can easily harm him because he can go to the police and the court for justice. A state of affairs most of us take for granted.

In a different part of Bangladesh, at a larger scale and bigger population, "peace" takes a different shape. The three south-western hill districts in Bangladesh—Khagrachari, Rangamati, and Bandarban—together known as the Chittagong Hill Tracts (CHT) that border Myanmar and India, host the majority of the country's indigenous populations with distinct linguistic, cultural, and land use practices. *Shanti Bahini* (Peace Force), the armed wing of Jono Shonghoti Shmoiti (JSS) representing the CHT *jumma* people, fought the Bangladesh army for 22 years (1975–1997) in demand for self-determination and special status (van Schendel 2009). While the Bangladesh government and the JSS signed a treaty in 1997 calling for an end to the armed conflict with the state's numerous promises to resolve the crisis in the region including special administrative arrangements, land use, and taxation; none have been fully implemented to this day. As a result, frequent conflicts between the Bengali settlers (backed by state mechanisms) and the *jumma* people of CHT remain common. As Van Schendel (2009, 213) described, "peace did not return to the region." Even 12 years later, in an opinion piece marking the 24th anniversary of the 1997 Peace Accord, Mangal Chakma (2021), the Information and Publicity Secretary of the JSS, wrote, "peace is still missing in the CHT".

"Peace," therefore, may carry dramatically different meanings for different people varying in nature, scale, temporality, perspective, and context. As such, peace remains a notoriously slippery concept and defining it comes with its own challenges. Ironically though, a wide range of groups including academics, journalists, politicians, and activists often tend to assume that "everyone knows what 'peace' is" and leave the term undefined (Megoran 2011, 178). Serious engagement with peace in geography has been virtually nonexistent until the turn of the millennium when it started to gain attention during the last decade or so (Mamadouh 2005; Koopman 2011). By contrast, geographers have always been interested and "better in studying war" (Megoran 2011, 178). Perhaps, because we generally tend to consider war as an "active" state of affair while peace is "vegetating" thus, not worth the attention. But it is not. Peace and war are intimately intertwined phenomena that give meaning to each other; not necessarily as binary oppositions but in mutually constructive ways (Flint 2011). Paying less attention to one of them thus leaves us with a lopsided understanding of both. Similarly, peace is not only related to war but equally connected to conflict and violence as demonstrated in the excerpts above. For Rahman, peace means protection from violence. For the *jumma* people in the CHT, for whom peace was supposed to be attained with an end to the war, it was never achieved because conflict and violence continue to influence their daily lives.

What is "peace" then? Is it simply an absence of war? A process? A never-ending endeavor? A lack of insecurity? A resolution (temporary or permanent) to conflict and violence? A day-to-day arrangement on the ground? A given condition from the top? All of these, none of these, or a combination of these?

Johan Galtung, widely considered the trailblazer of peace studies in geography, categorized peace in two ways: positive and negative. Positive peace, for Galtung is, "the integration of human society" while negative peace is "the absence of violence" (Galtung 1964, 2). Accordingly, Galtung argues that negative and positive peace should be conceived as two different dimensions that can exist without one another. However, just five years after such conceptualization, Galtung's understanding of peace changed. In a 1969 article, Galtung defined positive peace as the "absence of structural violence" and negative peace as an "absence of personal violence" (Galtung 1969, 183). According to Galtung, the general idea of negative peace remained the same (i.e., an absence of violence), but more precisely personal violence. However, the significant change in the definition of positive peace came from the realization that "'positive peace' is constantly changing" and viewing it through an integration lens not only puts too much of a symmetric view of conflict but also perpetuates the "East-West conflict" (Galtung 1969, 190). Drawing on Galtung's conceptualization, Colin Flint defined peace as "not only the absence of war, but also the possibility of maximizing human potential" (Flint 2005, 7). He later defined peace through more of a negative lens in contrast with war as a "complex of social relations that produces winners and losers, and hence domination and conflict" (Flint 2011, 31).

As geographers brought their "tools" into the study of peace, geography of peace took a "spatial turn" with the questions of time, space, scale, and place positioned firmly into the discussion (Björkdahl and Buckley-Zistel 2016; Macaspac and Moore 2022). For example, Nick Megoran (2011) argues that peace is essentially a question of agency because we need to be critical of whose peace, for what, and by whom? Sara Koopman claims that peace is always plural because it assumes different meanings at "different scales, as well as to different groups, and at different times and places" (2011, 194). Such conceptualizations were complimented by Williams and McConnell (2011) as they called for peace to be viewed as "situated knowledges" immersed in various cultural, geographical, and historical settings. As a result, within a very short span of time, geographies of peace took three distinct strands, as shown by Jake Hodder (2017). First, rejecting the approach of viewing peace through a dualistic framework, geographers suggested peace to be analyzed as a continuous process of becoming. Second, the question of temporality enabled an understanding of peace not as a universally given representation but something that is experienced variedly across time and space. Thus, the focus has shifted into a deeper appreciation of how peace "takes place." Finally, peace has been decoupled from politics to now be positioned "beyond, after, above or without politics" (Hodder 2017, 32). In so doing, the question of power is foregrounded in the analysis of peace.

How, then, are we supposed to define peace? In answering this question, I draw from Anssi Paasi (2011) as he undertook a similar venture in relation to the question of a universal theory of borders. Paasi's answer was a straightforward one to this question—a general border theory is not only unattainable, but perhaps undesirable for two primary reasons (Paasi 2011, 27). First, every border is deeply characterized by its local context and power relations as their meaning changes with time. Second, since borders are firmly rooted in social, cultural, political, and economic practices and discourses, they cannot be universalized in the form of a theory.

Defining peace is similarly problematic. It changes across time (situated), takes different shape in different places (positive and negative), is forged from above (signing of a peace treaty), and is also made from below (everyday arrangement by the locals). The range of themes that animate this volume and bind the chapters together further demonstrates that "peace" is perceived from various perspectives which may not always "talk" to each other—individual agencies, scales, geographies, empathy, othering, and resistance—yet remains equally relevant. As I analyze peace in this chapter, I specifically engage with the themes of individual agency, scale, othering, and mutual understanding. Therefore, I do not attempt to offer a universal definition of peace. Rather, I define it in a context-based manner that would enable an appreciation of peace in the settings of postcolonial South Asia, especially regarding its frequent boundary making and bordering practices. In so doing, I contend that postcolonial South Asian states are characterized by "negative peace" as frequent (re)drawing of borders not only results in recurrent violence but also keep providing the logic for new boundaries and categories that simultaneously divide people between and within national borders. In other words, it functions against the possibility of maximizing human potential and thus against an integration of society.

The remaining discussion in this chapter supports this argument in three sections. The first section traces a brief history of frequent border making in postcolonial South Asia to show how borders not only result in massive violence but also provide grounds for creating difference based on new boundaries and categories. The second section demonstrates that the use of those boundaries and categories are not limited to a single border but are transferred across numerous borderscapes. The concluding section focuses on the strategies and actions of "ordinary people" in dismantling those boundaries and categories—a process I call peacework. In so doing, I contend that ordinary people engage in peacework to avoid or defer the possibilities of violence on a daily basis across South Asian borderscapes. As such, the nature of every-day peace attained across South Asian borderscapes is best understood as "negative peace."

A violent production of postcolonial South Asia: Borders, boundaries, and categories

In August 1947, the British colony of India was carved into two national territories to become "homes" for two distinct "nations" that previously had yet to exist. Religion was the primary criterion utilized for defining these nations. Thus, one territory was created that contained non-Muslims (India), while the other (Pakistan) contained Muslims. The idea itself—that non-Muslims and Muslims would form two distinct nations and need different territories—was not discussed until the 1930s. It gained traction in the 1940s when the Muslim League formally adopted the concept as a resolution known as the "two-nation theory" (van Schendel 2009).

The premise behind such a territorial division was that non-Muslims and Muslims were distinct people incapable of forming one single nation. At the same time, it was asserted that Muslims were a community within the nation of India and unique in various ways that justified the production of a "home-land." However, such thinking was ahistorical as it completely ignored the fact that these people lived side-by-side for hundreds of years with closely shared cultural and linguistic ties. Furthermore, in places like Bengal, the categories of "Muslims" and "Hindus" (as we conceive them today) did not exist until the twentieth century (Ahmed 1981; Jones 2014). It is also evident from a number of studies that suggest, to this day, a vast majority of the population lack a sense of belonging to a particular nation; not just because they refuse to view religion as the "glue" tying them to a nation but also because postcolonial India is yet to become a nation. Instead, it remains suspended between a "former colony" and "not-yet-nation" (Krishna 1994, 508; Feldman 2003; Chatterji 2007; Zamindar 2007).

The irony, as we reflect on postcolonial South Asia scholarship, is that the creation of these two national territories achieved anything but peace. Instead, the territorial division generated grounds for unforeseen difference, conflict, and violence. While the cartographic scissoring of the Indian subcontinent can be productively analyzed from a multitude of angles, I specifically focus on how the newly drawn border created, and continues to create, boundaries and categories that work against achieving "positive peace" in present-day South Asia.

Drawing on Reece Jones' conceptualization (2009a, 2014), in this chapter, a border is viewed a line on the ground that separates the territorial limit of one nation-state from another. A boundary is conceived in a broader sense to refer to any type of division; for example, between nations, people, ethnicity, and class.

In this sense, a border is a boundary too, but one that is specifically reserved for describing the division between nation-states. Along these lines, Lakoff and Johnson's (1999) container schema suggests that categories are the "containers" of division in a socio-cognitive sense—they have an inside, an outside, and a boundary. Such cognitive effects, Jones contends, drive us to imagine each of them with fixed boundaries that are inherently closed and exclusive to each other. Consequently, they have a strong effect on how we perceive our everyday lives as we tend to order sociopolitical affairs based on the "fixity" of categories. Concurrently, we use those categories to provide logic for the existing order, although, intellectually, we are aware that the boundaries between those categories are "almost always fluid and permeable" (Jones 2009a, 179).

Cyril Radcliffe headed the 1947 Boundary Commission responsible for drawing the border between Pakistan and India. Radcliffe was a lawyer who had never been to India and was unaware of the site-specific complexities of the subcontinent. The instruction for the Boundary Commission was to "demarcate the boundaries of the two parts of [the province] on the basis of ascertaining contiguous majority areas of Muslims and non-Muslims" but, at the same time, to take "other factors" into consideration (Chatterji 1999, 196). Besides not being familiar with the subcontinent, Radcliffe faced a multitude of other challenges. Specifically, he was tasked with making sense of the vagueness of "area" and "other factors" in only six weeks—an extremely short period of time (van Schendel 2005). The result was chaotic, confusing, and incomplete for both those who were given the task of materializing the border on the ground and for those who were moving across them. For instance, many Muslim-majority areas were allocated to India and non-Muslim-majority areas were made part of Pakistan. Several princely states were given the choice to decide whether to join India or Pakistan, leaving the people within in limbo. Natural boundaries such as the rivers and mountains were not clearly marked; there was little attention to detail, especially in the case of border enclaves that remained 70 years after the partition (Chatterji 1999; van Schendel 2005). People rushed to move to the "right" side of the border amid mass violence, and the chaos and confusion of partition. Although estimates vary, it is widely believed that almost 14 million people moved across the border and a million were killed in the immediate aftermath of the partition (Jones 2014) —an utter demonstration that "peace" was not soon to follow. In this context, I contend peace is a condition that would allow the different religious groups to move across borders and settle in their newfound homes without facing discrimination, conflict, and violence—both structural and personal.

Another immediate effect of drawing new borders was the creation of new political categories that had not previously existed. For the first time in the Indian subcontinent, the category "refugee" came into use with the establishment of the Ministry of Relief and Rehabilitation by India. Those who moved between the two new states were categorized differently based on their religious identities. In India, "displaced persons" was used to refer to only non-Muslims, while the category reserved for the Muslims was "evacuees." Such categories were intentional creation of the state as they implied that the non-Muslims moving to India had to be cared for, and one way of doing so was to allot them the former properties of the Muslims who departed (i.e., the evacuees). Different categories were reserved for Muslims and non-Muslims in Pakistan too. Although the Urdu term for refugee is *panaghir*, the official term used by the Pakistani state was *mujahir*. However, this was only reserved for the Muslims who moved to Pakistan from India. For the non-Muslims, it was *sharanatis*. The categorization of Muslim refugees as *mujahir* is telling because it allowed the state to ideologically reinforce their status in Pakistan by invoking the migration of Prophet Muhammad and his followers from Mecca to Medina in AD 622 (see, for details, Zamindar 2007). Such categories reasserted the "Muslimness" of the Pakistani state and offered legitimation for excluding non-Muslims based on their imagined boundaries. Soon after, Pakistan and India introduced passport and visa in 1952 to control cross-border movements of people (van Schendel 2005). This established further categories—"citizens," "noncitizens," "legal," and "illegal"—upon whom the newly created borders and boundaries of the nation came to be performed and gave meaning to their very existence.

However, social categories are not neutral; they are rife with uneven power relations—discursive, symbolic, and otherwise. As Jones stated, "categories do not simply mimetically represent the world but instead simultaneously *create it* and *limit it*" (2009a, 177). Explaining power relations in society, Pierre Bourdieu (1991) pointed out that substantial power comes from the ability to define and maintain boundaries between categories. Such power is crucial because it allows to make and unmake groups who are included or excluded from the existing resources. This is further elaborated by Rogers Brubaker's (2015) thought experiment with horizontal and vertical categories. Brubaker demonstrates that difference between categories does not necessarily imply unequal if they are horizontally imagined. But whenever categories are made contingent upon unequal access to resources (which they are in the real world), they end up being vertical (i.e., one is placed above another in terms of prestige, power, resource, access, and such.) Hence, socially imagined categories, placed vertically above one another, almost in all cases work against what Galtung called an "integration of society" or in other words, against positive peace (Galtung 1964, 2). Indeed, postcolonial South Asian borders continue to provide the logic for creating categories like those of noncitizens, *mujahirs*, evacuees, and refugees that justifies the vertical placing of one group over another and hence, ruptures the possibility of achieving positive peace. Consequently, I contend that South Asian nations are characterized by everyday negative peace where it is primarily achieved on a day-to-day basis in an effort to avoid conflict and minimize partisanship; not to realize human potential. In the following section, I delve into this exact discussion in detail and the concept of borderscapes.

South Asian borderscapes and everyday (negative) peace

Borderscapes are not just landscapes immediate to a physical border; they are spaces of power where struggles for inclusion and exclusion are constantly negotiated (Rajaram and Grundy-Warr 2007). Rajaram and Grundy-Warr's 2007 edited collection first brought the idea to the fore, in which they borrowed the term from one of the contributors in the volume, Suvendrini Perera. Although Perera focused specifically on Australian borders in coining the term *borderscapes*, Rajaram and Grundy-Warr utilized it more broadly. It gained further traction with a nuanced theorization by others (e.g., Brambilla 2015; Brambilla and Jones 2020).

In laying out the idea of borderscapes, Rajaram and Grundy-Warr (2007) began with a view of the border not as a line but as a zone that contains actors, processes, power, and resistance. In so doing, they knotted the concepts of landscape, power, and border together. Landscapes are not fixed, rather always in the process of becoming as they are made and unmade by human relations. They are socially constructed and produced through numerous spatial relations. Some landscapes, for example national landscapes, are imbued with hierarchy of land use and power relations (e.g., some are excluded and some are included to attribute a sense of belonging and nonbelonging to a particular nation). Thus, Rajaram and Grundy-Warr (2007) argue that national landscapes and their construction concurrently work as instruments of governmentality and coercing behavior. Accordingly, landscape functions as a strong foundation for the practice of power specifically, hegemonic power that enables certain ideologies, ideas, and practices to be more tangible, natural, meaningful, and sustainable. Consequently, practices, meanings, and ideologies that are deemed less relevant or "unnatural" are considered to be out of that landscape or "out of place" (Rajaram and Grundy-Warr 2007, xxvii). This is precisely where the border becomes relevant as the landscape of "state-citizen-nation nexus" brings the border into play by naturalizing who belongs, who does not, who has the right to move, and whose movements are labeled illegal (ibid., xxvii). Therefore, borderscapes—the landscape of power in relation to borders—assert the "inherent contestability of the meaning of the border between belonging and nonbelonging" (ibid., xxviii). Consequently, borderscapes are not contained in a specific space but perpetuate the power relations that legitimize the boundaries between categories across numerous spaces.

Brambilla (2015) offers three different axes from which borders can be conceptualized beyond a given geo-politico-territorial phenomenon by tapping into the potential of the borderscapes concept. Borderscapes enables us to "read" a border from epistemological, ontological, and methodological viewpoints. Epistemologically, borderscapes allows us to move beyond a static view of borders as a given line in finding alternative spatiotemporal topologies that does not privilege the Western hegemonic thought of viewing it as a clear line separating binary oppositions. Instead, it offers analytical space for a multi-sited and multi-layered organization of society. In this sense, borderscapes works as a kaleidoscopic lens that can grasp the variations of borders in time and space across a range of sociocultural settings which are constantly negotiated by a multitude of actors, not just by the state. Ontologically, the borderscapes concept allows us to appreciate the lived and embodied conditions of individuals that constitute their reality arising from existing social orders. What constitutes reality depends on individual "understanding and praxis" (Brambilla 2015, 26). Hence, an ontology of borders recognizes that reality is in a constant process of becoming which evolves in time and space, such that the meaning of a border is in a constant process of *being* and *becoming*. In such a fashion, an ontological reflection of borders remains a powerful tool in the face of the "big stories"[1] of nation-states that often silence individual stories (Brambilla 2015, 26). Methodologically too, borderscapes pushes us to bring forward the phenomenological (i.e., experiences and representations of, at, and from borders). Therefore, it privileges a dynamic understanding through those who experience, live, and interpret borders on an everyday basis (e.g., border guards, borderlanders, border crossers, smugglers, and migrants). In the process, borderscapes generates an appreciation of the dynamic "processual, de-territorialised and dispersed nature of borders," their ensuing regimes, and the ensembles of practices (Brambilla 2015, 22).

The categories and boundaries that form with the creation of new borders are perpetuated, complicated, and internalized within numerous borderscapes in postcolonial South Asia. Thus, the concept of borderscapes remains a productive steppingstone for an unearthing of peace in postcolonial South Asian settings as it not only allows an analysis of peace along its borders. It also broadens the horizon to foreground the categories and boundaries that continue to puncture the idea of "peace" envisioned during the drawing of those borders. This is primarily due to the fact that an abstract empty space on the map was turned into a place of national belonging for a certain group of people (Kaviraj 2003; Ferdoush 2019a). Such geopolitical forces imply that the same space was turned also into a place of nonbelonging for other groups who did not fit the criteria set by these nation-states. However, since those criteria are not fixed, often overlap, and are not always clearly distinguishable, their creation and constant maintenance remain a primary task of the postcolonial South Asian states. The larger number of borders, the more created categories. As a result, the entire national landscape is effectively turned into borderscapes upon which the categories of "Us" are vertically placed over "Them," perpetually rupturing the possibility of attaining positive peace.

For example, the present-day Indian state of West Bengal and the sovereign state of Bangladesh was split in 1947 between India and Pakistan. The Bengal Province that contained this entire territory before 1947 was the "home" of the Bengali-speaking populations—both Muslims and non-Muslims—for hundreds of years. People shared similar cultural ties based on a common language and history. However, as the province of Bengal was split to create new "homes" for Muslims and non-Muslims, the old category of Bengali was revised into new categories of "Bengali Muslims" and "Bengali non-Muslims" to make sense of those abstract spaces on the map. The implication was that the Bengali Muslims will make Pakistan their "home" while Bengali non-Muslims will find "home" inside the territorial limit of India. This idea was further sharpened a few years after the birth of Bangladesh as the country revised its constitution to change secular "Bengali" nationalism to "Bangladeshi" nationalism in an effort to distinguish its population from the Bengalis across the border based on territorial and religious logics (Huda 2004; Jones 2007). As the national landscape of Bangladesh was converted into a place of belonging for the Muslims, Bengali-speaking Muslims came to be dominantly identified as "Bangladeshis," while "Bengali" remained primarily reserved to denote non-Muslim Bengali speakers in India. The old category of Bengali is now sharply divided into "Bangladeshis" and "Bengalis" that provide meaning for the existence of the nation by turning its entire territory into borderscapes. "Bangladeshis" and "Bengalis" are now vertically put against each other that creates a strong sense of "Us" versus "Them" in inverted ways across the border. Bengali-speaking Muslims are frequently marginalized as "Bangladeshi infiltrators" especially in the Indian state of Assam, although it has been a home for a large Bengali-speaking Muslim population since before the creation of the Indian state (Sur 2021). These divisions have ontological effects too. Even the non-Muslims who moved from the present-day Bangladesh to the present-day Indian state of West Bengal at different periods after the partition are not necessarily categorized as "Us"; they are often categorized as "Bangal" (i.e., the "Them"). A phrase that carries a connotation of a refugee or migrant from across the border with distinct territorial ties and accent. Such epistemological and ontological divides between categories are rubbed against each other on a daily basis to limit or broaden their access to certain resources. In other words, those categories are frequently manifested in the day-to-day governance to fashion structural violence. Consequently, I suggest, structural violence based on the logic of categorical boundaries dominates the terrain of peace across South Asia.

However, I do not intend to draw an over encompassing picture of South Asian borderscapes marked exclusively with structural violence and negative peace. There are frequent efforts by the states to "weave" peace within and between nations (best seen as postscripts to frequent and chaotic border making). For example, in 2015, former enclave residents of India and Bangladesh, like that of Rahman Ali, were recategorized with the redrawing of the border. Before the exchange, the enclave residents lived in a state of de facto statelessness transitioning between overlapping identity categories such as citizens, proxy citizens, noncitizens, and enclave people (van Schendel 2002; Ferdoush 2019b). All these categories contributed to the creation and maintenance of boundaries between the host and the home nations. Since the exchange of the enclaves revised the border, the boundaries between those categories were also revisited. They officially became citizens either of the host or of the home countries—no longer categorized as enclave residents. Hence, the newly categorized "citizens" of Bangladesh (like Rahman Ali) hoped to have "peace" as the old category of "enclave residents" no longer existed to exclude them from basic state services.

The nature of everyday peace on the ground: Peacework

As I demonstrated, peace is not always produced from above nor is it always "noteworthy" stories of "big peace" attained by political leaders and state actors. It is also the everyday peace which is "far removed" from the "big peace" but constitutes the daily life of millions across the world (Mac Ginty 2021, 2). Everyday peace takes place across multiple geographies, scales, and time. As such, everyday peace is not universal but highly context based (Mac Ginty 2021). This is precisely why Koopman (2011) calls for taking "peace into pieces" in rethinking geographic approach to the study of peace. Therefore, I specifically focus on the idea of everyday peace in this section to further unlock the nature of peace across South Asian borderscapes.

Everyday peace is achieved by the so-called "ordinary people" (Mac Ginty 2021). Ordinary people, in this regard, refers to both citizens and noncitizens who do not necessarily possess the power to cause structural change in the existing social settings but use their agency to rupture conflict and stop violence so that peace is sustained as an ongoing process (Rumford 2012; Williams 2014; Mac Ginty 2021; Dempsey 2022). However, it must be acknowledged that the nature of everyday peace is a negative one that is primarily driven toward "preventing conflict" but lacks the capacity to bring structural change (Dalby 2014). It is concerned with small acts that may range from simply avoiding a tense situation to mutual understanding to defer conflict. The strategies and actions that are used by the ordinary people in achieving everyday peace is therefore what I call peacework. In so doing, I build on the idea of border scholar Chris Rumford's "borderwork" to contextualize peace, while at the same time, juxtapose it with Mac Ginty's "everyday peace" to unearth the strategies that ordinary people employ in attending peace on a day-to-day fashion (Rumford 2006, 2012, 2013; Mac Ginty 2014, 2021).

Peace is asymmetric. It does not affect everyone equally even within a similarly given context. Therefore, working toward attending and maintaining peace remains a continuous process, "peace is always being built, never done" (Koopman 2019, 209). In mundane lives ordinary people make peace "on-the-go" as they navigate the complexities of social and political world. They do so by "reading" social situations that may involve avoiding a certain street corner, making judgments of whether to engage in a particular activity, by exchanging a smile, avoiding eye contact, or coming to terms to rationalize a situation that might "involve violence or the threat of violence" (Mac Ginty 2021, 9). Such efforts of ordinary people that weave peace across numerous sociopolitical settings are precisely what characterize peacework. Indeed, it is not necessarily about dismantling the existing social structure but being able to create a state where violence (both symbolic and physical) is dodged or pacified. Peacework is the quotidian rupturing, undoing, or deferring of conflict on the ground. As such it enables us to appreciate peace in a state of constant production instead of a given situation from above. To elaborate further, I describe three broader contexts of peacework across South Asian borderscapes—in situ peace negotiations, dismantling of boundaries and categories, and expression of people power.

Peace has increasingly been intertwined with security, as peace *and* security, not just peace. Given the practice of "big actors" such as the United Nations (UN) it suggests peace without security is not important (Dalby 2014). In the name of achieving peace therefore, numerous acts of violence under the guise of security are justified. This rhetoric is raucously pronounced at the borders of the nation-states, especially those designed to prevent human mobilities such as the US-Mexico, India–Bangladesh, and the external EU borders (Jones 2012, 2016). The India–Bangladesh border is one of the most "secured" and lethal borders in the world, and claims an average of 60 Bangladeshi lives every year. To be exact, from 2000 to 2020, a total of 1,236 Bangladeshi people were killed by the Indian Border Security Force (BSF) along this border (Shahriar 2021). In almost all cases, such killings are justified by India in the name of "securing" the "edge" of the nation-state to maintain peace within (Jones 2012). Simultaneously it creates a "permanent state of exception" and fear on the Bangladesh side of the border as mundane activities could end up claiming their life (Jones 2009b). As a result, borderlanders, border guards, and other (non)state actors often resort to in situ negotiation of peace to appease conflict and deescalate possibilities of violence along this "killer border" (van Schendel 2005). One of the most familiar and effective tools is a "flag meeting" held between the border guards to settle various cross-border issues that do not necessarily involve top-level policymakers (Prothom Alo 2022). Local border guards arrange these encounters, but in many cases, local people initiate flag meetings to resolve local issues. For example, meetings are regularly arranged to return bodies of persons killed by the guards for crossing the border illegally, to deescalate a tense face-off between the border guards, or to settle disputes among borderlanders. Flag meetings are frequently used as an in-situ tool for weaving peace involving ordinary citizens and state actors and therefore remain one of the most visible examples of peacework across the Bangladesh-India border.

As demonstrated above, numerous categories and boundaries are frequently created and practiced by the states in South Asia to legitimize the presence of borders. In the process, these categories and boundaries also play a significant role in fostering a sense of "Us" and "Them" between those who reside across borders and those within. Consequently, these categories rupture the "big peace" between nation-states and the "everyday peace" among ordinary people. However, through peacework, ordinary people regularly engage in dismantling these boundaries and categories across different borderscapes. As Rahman's experience in the introduction demonstrates, when the border enclaves existed between Bangladesh and India, these residents were commonly categorized as noncitizens in the host country and lacked basic state services such as education, health care, and protection by law. Such exclusionary categorization created a sense of nonbelonging and lack of peace among the enclave residents. Nevertheless, the enclave residents (the noncitizens), their neighbors, and state actors (the citizens) often made informal arrangements that dismantled such categorization so residents were included within the state system as a "humanitarian" gesture. For example, during my fieldwork in 2017–18 in the former enclaves inside Bangladesh, one of the elected public representatives shared how he often made arrangements that during natural disasters government aid reached the poor enclave residents too. In justifying his acts, he said, "They are human first, then enclave residents. I could not just turn away from them.... It is a humanitarian issue" (fieldnote 2017). While state-imposed categories continue to create boundaries, those categories are often dismantled by citizens and noncitizens to avoid violence and conflict as well as to address humanitarian crises.

The third type of peacework commonly found across South Asian borderscapes is best understood as what Rumford called the "expression of people power" (2008, 7). Peace is attained in numerous occasions by ordinary people as an expression of shared power when a number of people come together to attain peace by going against the grain. This may include challenging established rules of society (both formal and informal), a disregard for conventional wisdom, or in response to an unprecedented event. As such, peacework—as an expression of people power—comes across as temporary arrangements for conflict resolution before a relatively permanent arrangement is reached. One of the most important features of this peacework is that it is often turned into "endorsed" practices if they continue to be expressions of what people desire in a given circumstance. The early phase of Rohingya refugees arriving in Bangladesh fleeing the state-sponsored massacre in Myanmar offers a powerful example.

The first groups of Rohingya people began crossing into Bangladesh in August and early September in 2017. Initially, the Bangladesh government was unsure how to respond. As a result, large groups of Rohingyas remained stranded in the "no man's land" between the border of Bangladesh and Myanmar at the hands of the Border Guard Bangladesh (BGB) (Roy and Jinnat 2017). Thousands of others crossed the border and entered the town of *Ukhiya*. There were no government-provided shelters, infrastructure, or arrangements at this time to help the asylum seekers, but the locals welcomed them by allowing them to shelter in their yards and homes, and providing them food. A few civil society organizations also quickly stepped in. Although more than a million Rohingya people were later allowed to enter the country, camps were established, and the international community eventually contributed— the earliest phase of peacework was established by the local people. The first steps of attaining peace were taken by the locals, even though it meant going against the convention at that point in time—a clear expression of people power in peacework.

Conclusion

Although one of the most powerful discourses supporting the redrawing of South Asian borders was that the Muslims and the non-Muslims needed separate "homes" to live in "peace," in this chapter, I demonstrate that "peace" is never achieved as a result of those borders. The frequent drawing of borders instead produces new opposing categories and boundaries that underpin unequal access to resources, power, and prestige. In so doing, I demonstrate that as peace operates across multiple geographies and scales, it also takes different meaning and shape for different populations. Thus, it is crucial that we remain attentive to how ordinary people use their agency to achieve peace on a daily basis. As new categories are put into practice, they galvanize the sense of "Us" and "Them" along their boundaries. Such a sense of sharp division functions against positive peace.

Applying this to South Asian borders, I demonstrate that a powerful sense of division between numerous categories is not just contained along its borders but extend across numerous borderscapes where those categories and boundaries are drawn on a daily basis to create "winners and losers" (Flint 201, 31). However, it is equally true that the meaning of those boundaries and categories are not fixed. Instead, they are negotiated and disregarded on the ground by ordinary people as they draw on their agency and empathy. They do so to avoid conflict or violence, and to deescalate tense situations. In other words, to make peace. In this sense, "peace" remains in the making across South Asian borderscape predominantly in a "negative" fashion. It is more about creating avenues for avoiding violence and conflict instead an "integration" of society, a prospect that is perpetually challenged by the presence of its borders.

Note

1 Big stories of the nation-state, in this sense, refers to events which are worthy enough to gain attention at a national or international level concerning a larger population as opposed to everyday events that affect a small group of people at a local scale.

References

Ahmed, R. 1981. *The Bengal Muslims, 1871–1906: A quest for identity*. Oxford University Press.

Prothom Alo. 2022. BGB, BSF hold flag meeting [online]. *Prothomalo*. Available from: https://en.prothomalo.com/bangladesh/bgb-bsf-hold-flag-meeting [Accessed 8 Dec 2022].

Björkdahl, A. and S. Buckley-Zistel. 2016. Spatializing peace and conflict: An introduction. In *Spatializing peace and conflict: Mapping the production of places, sites and scales of violence*, eds. A. Björkdahl and S. Buckley-Zistel, 1–24. Palgrave Macmillan.

Bourdieu, P. 1991. *Language and symbolic power*. Harvard University Press.

Brambilla, C. 2015. Exploring the critical potential of the borderscapes concept. *Geopolitics* 20: 14–34.

Brambilla, C. and R. Jones. 2020. Rethinking borders, violence, and conflict: From sovereign power to borderscapes as sites of struggles. *Environment and Planning D: Society and Space* 38: 287–305.

Brubaker, R. 2015. *Grounds for difference*. Harvard University Press.

Chakma, M.K. 2021. Why is peace still missing in the CHT? *The Daily Star*, 2 Dec.

Chatterji, J. 1999. The fashioning of a frontier: The Radcliffe Line and Bengal's border landscape, 1947–52. *Modern Asian Studies* 33: 185–242.

Chatterji, J. 2007. *The spoils of partition: Bengal and India, 1947–1967*. Cambridge University Press.

Dalby, S. 2014. Peace and critical geopolitics. In *Geographies of peace*, eds. F. McConnell, N. Megoran, and P. Williams, 29–46. I.B. Tauris.

Dempsey, K.E. 2022. *Introduction to the geopolitics of conflict, nationalism, and reconciliation in Ireland*. Routledge.

Feldman, S. 2003. Bengali state and nation making: partition and displacement revisited. *International Social Science Journal* 55: 111–121.

Ferdoush, M.A. 2019a. Symbolic spaces: Nationalism and compromise in the former border enclaves of Bangladesh and India. *Area* 51: 763–770.

Ferdoush, M.A. 2019b. Acts of belonging: The choice of citizenship in the former border enclaves of Bangladesh and India. *Political Geography* 70: 83–91.

Flint, C. 2005. Geography of war and peace. In *The geography of war and peace: From death camps to diplomats*, ed. C. Flint, 3–18. Oxford University Press.

Flint, C. 2011. Intertwined spaces of peace and war: The perpetual dynamism of geopolitical landscapes. In *Reconstructing conflict: Integrating war and post-war geographies*, eds. S. Kirsch and C. Flint, 31–48. Ashgate.

Galtung, J. 1964. What is peace research? *Journal of Peace Research* 1: 1–4.

Galtung, J. 1969. Violence, peace, and peace research. *Journal of Peace Research* 6: 167–191.

Hodder, J. 2017. Waging peace: Militarising pacifism in Central Africa and the problem of geography, 1962. *Transactions of the Institute of British Geographers* 42: 29–43.

Huda, Z. 2004. Problem of national identity of the middle class in Bangladesh and state-satellite television. unpublished doctoral dissertation. University of Warwick, UK.

Jones, R. 2007. Sacred cows and thumping drums: Claiming territory as 'zones of tradition' in British India. *Area* 39: 55–65.

Jones, R. 2009a. Categories, borders and boundaries. *Progress in Human Geography* 33: 174–189.

Jones, R. 2009b. Agents of exception: Border security and the marginalization of Muslims in India. *Environment and Planning D: Society and Space* 27: 879–897.

Jones, R. 2012. *Border walls: Security and the war on terror in the United States, India, and Israel*. Zed Books.

Jones, R. 2014. The false premise of partition. *Space and Polity* 18: 285–300.

Jones, R. 2016. *Violent borders: Refugees and the right to move*. Verso.

Kaviraj, S. 2003. Crisis of the nation-state in India. *Political Studies* XLII: 115–129.

Koopman, S. 2011. Let's take peace to pieces. *Political Geography* 30: 193–194.

Koopman, S. 2019. Peace. In *Keywords in radical geography: Antipode at 50*, ed. Antipode Editorial Collective. Wiley-Blackwell, 207–211.

Krishna, S. 1994. Cartographic anxiety: Mapping the body politic in India. *Alternatives* 19: 507–521.

Lakoff, G. and M. Johnson. 1999. *Philosophy in the flesh: The embodied mind and its challenge to western thought*. Basic Books.

Mac Ginty, R. 2014. Everyday peace: Bottom-up and local agency in conflict-affected societies. *Security Dialogue* 45: 548–564.

Mac Ginty, R. 2021. *Everyday peace: How so-called ordinary people can disrupt violent conflict*. Oxford University Press.

Macaspac, N.V. and A. Moore. 2022. Peace geographies and the spatial turn in peace and conflict studies: Integrating parallel conversations through spatial practices. *Geography Compass* 16: e12614.

Mamadouh, V. 2005. Geography and war, geographers and peace. In *The geography of war and peace: From death camps to diplomats*, ed. C. Flint Oxford, 26–60. Oxford University Press.

Megoran, N. 2011. War and peace? An agenda for peace research and practice in geography. *Political Geography* 30: 178–189.

Paasi, A. 2011. A border theory: An unattainable dream or a realistic aim for border scholars. In *The Ashgate research companion to border studies*. ed. D. Wastle-Walter, 11–31. Routledge.

Rajaram, P.K. and C. Grundy-Warr. eds., 2007. *Borderscapes: Hidden geographies and politics at territory's edge*. University of Minnesota Press.

Roy, P. and M.A. Jinnat. 2017. Stranded in no man's land: Several thousand Rohingyas from Myanmar build makeshift shelters there. *The Daily Star*. 30 Aug 1 & 2.

Rumford, C. 2006. Theorizing borders. *European Journal of Social Theory* 9: 155–169.

Rumford, C. 2008. Introduction: Citizens and Borderwork in Europe. *Space and Polity* 12: 1–12.

Rumford, C. 2012. Towards a multiperspectival study of borders. *Geopolitics* 17: 887–902.

Rumford, C. 2013. Towards a vernacularized border studies: The case of citizen borderwork. *Journal of Borderlands Studies* 28: 169–180.

Shahriar, S. 2021. Bangladesh-India border issues: A critical review. *Geoforum* 124: 257–260.

Sur, M. 2021. *Jungle passports: Fences, mobility, and citizenship at the northeast India-Bangladesh border*. University of Pennsylvania Press.

van Schendel, W. 2002. Stateless in South Asia: The making of the India-Bangladesh enclaves. *The Journal of Asian Studies* 61: 115–147.

van Schendel, W. 2005. *The Bengal borderland: Beyond state and nation in South Asia*. Anthem Press.

van Schendel, W. 2009. *A history of Bangladesh*. Cambridge University Press.

Williams, P. 2014. Everyday peace, agency and legitimacy in north India. In *Geographies of peace*, eds. F. McConnell, N. Megoran, and P. Williams, 194–211. I.B. Tauris.

Williams, P. and F. McConnell. 2011. Critical geographies of peace. *Antipode* 43: 927–931.

Zamindar, V.F.-Y. 2007. *The long partition and the making of modern South Asia: Refugees, boundaries, histories*. Columbia University Press.

12

HYBRID NETWORKS

Technology, geopolitics and ontology in digital warfare

Ian Slesinger

Introduction

This chapter will evaluate how digital technologies modulate and mediate the contemporary geopolitics of war and peace, with a particular focus on internet-based warfare. It will expand upon several unifying themes of the book by examining ways in which technology dynamically reconfigures how agency circulates in geopolitical conflicts in relation to the ever-changing spatial topology of digitised networks and infrastructures. In doing so, it will address how contemporary electronic warfare confounds the notion of there being discrete scales, localities and relations of proximity and distance. This complexity and ambiguity is expressed through the chapter's discussion of how digital networks elide, or blur, the boundaries between the thresholds of war and peace, and what constitutes a military or civilian tool or target.

The discussion here will also tie into key ideas in the literature on technology in contemporary warfare in geography and related disciplines, and will illustrate that in many ways internet warfare shares some of the same features and theoretical implications as other recent trends in technological warfare. This includes the time-space compression and risk transference endemic to the use of unmanned aerial vehicles (UAVs), as well as the elision between usage of commercial and consumer UAVs for civilian and military purposes, as well as the opposite trend in the use of satellite imagery from military intelligence and targeting to open-source intelligence (OSINT) investigations by journalists, human rights NGOs and non-state insurgent due to the increased commercialisation and availability of this technology.

In order to examine the relationship between contemporary digital technologies and war, this chapter will employ an ontological approach, meaning a

DOI: 10.4324/9781003345794-12

philosophical position on the nature of being, based on materialist ontology and assemblage theory to explain geopolitical phenomena. A materialist ontology understands the world – and sociopolitical phenomena within it – as "more-than-human" (Whatmore 2006, 603). This conceptualisation affirms that power is relational between humans, technology and the spatial environment and that agency (the power to act in the world) is not a solely human resource. This requires an epistemology (framework for obtaining knowledge) that can grapple with how technological objects in geopolitical conflicts are depicted, "held accountable and interrogated" (Squire 2016, 9). Such an understanding of technological agency must relate technologies' capacities to act in the world with their existence within their social, temporal and spatial environment. Assemblage theory accounts for the heterogeneous assemblage of a "socio-material network" of human and non-human things through which agency is "distributed" (Müller 2012, 382). Agency does not come from an assemblage's individual components, but rather through the connections and relations between them (De Landa 2006). Furthermore, assemblages are not static and fixed, but can dynamically reorganise, rearrange and incorporate new elements gives them limitless possibility to enact new forms of social action (Dittmer 2014).

The increased pervasiveness of digital technology in nearly all aspects of daily life – both in the most economically developed countries and increasingly in lower- and middle-income countries – demonstrates the elided and problematic nature of the distinction between civilian and military targets, and even the very notion of war as an exceptional spatiotemporal state in contrast to peace. Multiple geopolitical events over approximately the last fifteen years highlight that networked digital infrastructures are (and have always been) material assemblages, rather than abstracted virtual spaces. These assemblages are increasingly enmeshed in the ontological fabric of states, societies, economies and infrastructures, and in turn make infrastructures themselves objects of security protection (Aradau 2010). Part of this "technologisation" of the world the networked relations between human and non-human agencies that iteratively produce and shape contemporary existence are increasingly becoming both an attack surface for violent conflict and a weaponised medium through which such conflict is perpetuated. However, as will be evidenced in the literature review that follows, existing International Relations approaches to internet warfare have tended to either focus on the elite construction of cyber threat, or state-level normative approaches to mitigating cyber war through establishing cyber norms and engaging in diplomacy around the governance of cyberspace. Such approaches, however, fail to engage substantively with the material and everyday impacts of internet-based conflict and the underlying challenges to conventional statist approaches to security based on geopolitical sovereignty networked technologies present.

The following section will outline some of the main themes in the literature on technology and war from international relations, geography, security studies

and critical theory. It will then engage more specifically with the current literature on cyber war, which forms the starting point for this piece's theoretical and
empirical interventions and critiques. I will continue by providing a theoretical
outline of the significance of the internet as a geopolitical vector for conflict, as
well as a historical overview of key incidents of internet-based warfare that are
geopolitically significant. The discussion will then theoretically engage with and
exemplify the spatial and temporal dimensions of internet warfare, particularly
in terms of how internet warfare elides statist conceptions of sovereignty and
security. The final section will evaluate the challenges and implications of internet warfare for peacebuilding, particularly in conventional arenas for international diplomacy, before relating internet warfare back to other technologically
mediated forms of geopolitical violence in the conclusion.

Literature review

Geopolitics and war

Place is both an object of violence, and the "medium and a means and a momentum" through which geopolitical violence is directed (Thrift 2007, 274).
The spatial complexities of geopolitical conflicts that emerge through the
unique conditions of place is evidenced by the highly precarious distinction
between "civilian" and "military" that comes about through the convergence
of practices, infrastructures, topology, and technology within the state (Perugini
and Gordon 2017). These conditions and practices are also implicated in actions against the state. This ambivalence is implicit in the military origins and
continuities of the logistical techniques that enable the "'pipelines' of flow"
through which goods and bodies can elide the boundaries of the nation-state
(Cowen 2014, 4). The proliferation of weaponry beyond state actors and the
growing elisions between military and civilian technological domains mean
that the parameters of violent conflict extend beyond the eventful intensity of
war, and violence becomes an ever-present danger (Coaffee, Wood, and Rogers
2009).

This blurring of dyadic distinctions disrupts the premise of geopolitical stability as a default state. Instead, it becomes clear that "war and peace should
not be seen as dichotomous periods and spaces" and that "places, state, regions, and landscapes are continually being destroyed and re-built by organized violence, or war" (Kirsch and Flint 2011, 19). Rather, as Barkawi (2015,
56) argues, it is through the "organisation and facilitation of coercive capacities" by political actors including the state that the ostensible condition of
"peace" is produced. In geopolitics the temporality of peace, conflict and (in)
security is modulated through the shifting intensities of violence and power.
Spinks (2001, 24) notes that "political decision is itself produced by a series of
inhuman or pre-subjective forces and intensities." Such intensities can be

understood as the dynamic temporal shifts in the concentration of violent power that occur through the interplay between human intentionality, technological capacities and space.

The relationship between spatiality, technology and international politics can be better addressed by supplementing formal geopolitical analysis with an engagement with the "day-to-day" experiences and perspectives of "ordinary people" (Jeffrey 2006, 474). Such understanding is complemented by considering the dynamic practices of hybridity and exchange in technological practices between social and cultural groupings. Whilst a clear historical precedent exists for the trans-global transfer of military technologies and knowledges, the recent acceleration of globalisation has had an acute impact on how both tactics and technologies shift between theatres of war and social groups. Rather than a particular style of fighting or a type of weapon being bounded to a specific geographical location, global telecommunications, media and economic networks allow combatants to borrow tactics and techniques from fighters in other geographically disparate conflicts, and to share training and expertise. Likewise, small-scale insurgents now have greater access to weaponisable technologies that had previously been only available to major powers (Kaldor 2012). The increased global flow of tactics and technologies mean that experts must be careful to attend to the interdependent flows between place and the regional and global geopolitical dynamics of a conflict in which a particular technology or tactic is applied.

Technology, space and warfare

Several historians and military theorists trace histories of warfare through what they identify as key technological Revolutions in Military Affairs (RMAs) that have reshaped the strategy and conduct of war, several of which presage the current information technology RMA by centuries (Boot 2006; van Creveld 1989). The current network-centric warfare RMA doctrine was codified to account for the changes in warfare brought about by the coordination of networked telecommunications with precision-guided munitions in the 1990s and 2000s (Arquilla and Ronfeldt 1997). Hirst (2005, 143) critiques the naivety of proponents of the network centric-warfare concept that suggests that omniscient command through information networks combined with surgical strike capability can create a "cleaner" form of war that will "eliminate the 'fog of war'." Heng (2006, 13) argues that contemporary warfare by Western states has shifted from a linear application of coercive force to a policy-driven process of "risk management" predicated on "active anticipation" and "reflexive" consideration of possible adverse consequences that have yet to occur. In one of the few articles in the geography literature to examine the present RMA from a more-than-human perspective, Ek (2000, 866) suggests that "the interaction between humans and nonhumans, and among nonhumans creates … geopolitical risks."

In addition to the risks faced in the field, political actors are also faced with the volatility of public opinion and the affective ties between the military and the population. Martin Shaw's (2005, 71) *The New Western Way of War* argues that democratic states work to navigate the tensions between domestic political and foreign relations. Military policy imperatives seek to manage this tension through "risk-transfer war." This approach to governing conflict displaces the risks of war from the military to a target population perceived as aligned to an enemy state or militant organisation. This is achieved, largely, through a reliance on air power and advanced technologies such as drones and robots, to minimise risk to Western troops as high casualty rates could cause a "risk rebound," leading to dissent, anger and political fallout at home.

The concept of risk-transfer war is also relevant to the literature on "military urbanism" that engages in-depth with the material and technological dimensions of geopolitics and urban spaces (Graham 2011, 60). This is because the complexity of urban terrain undermines the efficacy of conventional warfighting tactics used by large standing armies and lends itself to strategies of guerrilla warfare (Hills 2004). Powerful states have sought to overcome the challenges of urban warfare through reliance on increasingly advanced weaponry and reconnaissance technologies. Key themes within this literature include the vertical control of urban space (Graham and Hewitt 2013; Adey 2010; Weizman 2007), and the fixation of military forces on "urbicide" (Coward 2010, 35) - a neologism for the destruction of urban infrastructure and the built environment in pursuit of political goals.

Visuality is a privileged and significant sensory mode in the spatial exercise and circulation of power in geopolitics (MacDonald, Hughes, and Dodds 2010). Within the critical geopolitics literature, emphasis has been placed on how visual power is incompletely accumulated and projected through the "eye of God" (Chamayou 2015, 37) or "God trick" (Gregory 2011, 204; Wilcox 2017, 3; Shaw and Akhter 2012, 1495), borrowing from Haraway's feminist critique of scientific objectivity presenting a totalising "vision from everywhere and nowhere" (Haraway 1988, 584). This expression of power is especially relevant to the military use of UAVs, satellite imagery and aerial photography. This alludes to the sense of omnipotence imbued by UAV and other scopic technologies' powers of vertical perception and surveillance, and their perceived ability to project targeted force over large distances whilst mitigating risk to the force projecting said technological power (Kindervater 2017; Shaw 2017; Neocleous 2013; Sauer and Schornig 2012). In recent conflicts images have been gathered and utilised through drones and the computer screens of command and control centres, in which the horizontal view of terrain from above produces the illusion of domination over space from which it is impossible to escape or hide secrets (Graham and Hewitt 2013; Williams 2013). Considering the scopic interface of many RMA technologies as a system of ordering and orienting the battlespace (Bousquet 2018) – for example satellite imagery, night vision, thermal imaging, aerial photography – it is imperative to pay attention to the risks and limitations of the visual domain in war, in addition to its power.

Work on aerial geopolitics deals explicitly with the assemblage of military airpower as both a socio-technical and a performative political agent in relation to its privileging of the visual domain. Williams' (2011, 386) research on UAVs shows that the more-than-human collective of machine and operator produces an ambivalence in which "the humans-in-the-loop" both benefit from the enhanced visualisation capabilities of the UAV's sensors and cameras that extent beyond the capability of human vision, and limit the potential capacities of the aircraft due to their mortal propensities for fatigue, inattention, less than instantaneous decision-making and error. The significance of the interface between technology and the ocular in how dominant institutions, especially the state, attempt to govern space is not limited to the panoptic gaze. It also informs the diffuse ways in which geopolitical knowledges are assembled, disseminated, internalised and reproduced, the totalising vision of reality from above, or the mesmeric "eye of God," is present in the contemporary usage of satellite imagery and geographical information systems to produce political truths (Kurgan 2013).

The literature on UAVs in particular, along with a concomitant emphasis of visual modes of enacting power (MacDonald, Hughes, and Dodds 2010), has dominated discussions in the critical geopolitics' literature, as these phenomena have dominated the representation of war from the First Gulf War in 1991 to the present. However, examining other forms of military technologies alongside UAVs, and to a lesser extent satellite imagery, in scholarly analyses of war can allow for different perspectives to emerge on the nature and location of technological agency in geopolitical conflict. For this reason, I will focus here on internet-based conflict, or the debated phenomenon of cyber war, as a means to consider differently the ontological stakes of the imbrication between technology, geopolitics and the conditions of daily life.

The internet as battlespace

In one of the earliest works to discuss internet-based warfare International Relations scholar Thomas Rid (2013) sought to deflate hyperbolic claims of impending cyber war by definitively claiming that "cyber war will not take place." This contention is based on a theoretical interpretation derived from constructivist security studies that suggests cyberattacks have and will occur below the threshold of lethality and are more means of espionage or subversion rather than acts of explicit violence, and do not meet the test of direct political attributability of an act which the author defines as a precondition for war. However, the nature of digital networks and their ontological relations to other infrastructures have changed significantly in the ten years since the book was written. Therefore the argument does not adequately account for the growing integration of "smart" digital control with the infrastructures that sustain daily life and the complexity of relational effects between them. Furthermore, as this chapter will argue, it is precisely the murky and eliding properties of digital warfare that subvert accountability, and clear thresholds of violence and distinction in targets, that make it both powerful and emblematic of how technology mediates the nature of violence in contemporary warfare.

Betz and Stevens (2013) argue that imprecise analogical reasoning around digital technologies has led to a false distinction between a realm of imagined virtual space and the tangible space of the "real world," which in turn colours how digital security is conceptualised and enacted. Rather, there is a "mutable 'geography'" in relation to the geopolitics of sovereignty (Betz and Stevens 2013, 152). They point out that in actuality what is imagined as cyberspace materially exists as an interlinked infrastructure of digital hardware that comprises physical nodes and linkages. This chapter will extend this materialist understanding of what constitutes digital "space" to underscore the inextricable linkages between what Bratton (2016, 4–5) conceptualises as the "interdependent" and "planetary-scale" computational "megastructure" of "the Stack" that is imbricated with the infrastructures and systems that sustain contemporary life, and the implications this has for the conduct and effects of war. For this reason, I prefer to use in this piece the terms "digital" and "internet-based" (instead of "cyber") to describe malicious and violent activities that work through or target networked telecommunications infrastructures. In a similar mode to Betz and Stevens, Ashraf (2021) argues that there is a lack of conceptual clarity and agreement in the literature about the definition of "cyber war," and instead proposes a malleable framework of actions, actors, effects, geography and targets for thinking through what he labels as the alarmist, sceptic and realist positions in the literature.

There is also a growing discussion in the literature of cyber war from the perspectives of critical international relations and political geography. Tim Stevens (2016) examines the "chronopolitics" of cyber security to argue that political decision-making about cyber security operates along a temporal continuum from the decelerated process of the routinisation of insecurity to realtime decision-making to the accelerated temporality of emergent disaster. Dunn Cavelty (2013) argues that a Cold War–inflected language of threat has turned cyber security into a militarised domain, with the attached security logics and strategies this invites, to the exclusion of other less binary and combative approaches to the means and objectives of security. Similarly, Kaiser (2015, 19) uses the case of the 2007 attacks on Estonian government, economic and media organisations in the geopolitical context of a dispute between Estonia and Russia over the relocation of a Soviet-era war memorial in Tallinn to argue that security elites were led to "performatively enact" and institutionalise a discursive threat of cyber war.

This chapter shares these authors' calls for alternative framings of digital security in geopolitics that emphasise ontological security rather than martial securitisation. However, the pitfall of such a constructivist critique of the linguistics of threat is that it can minimise or ignore the very real material dangers and harms that both violent and ostensibly non-violent digital attacks can create. As a corrective, Collier (2018) recommends applying the concept of assemblages to better account for the complex material configurations of human and non-human actors in cyber security. Clare Stevens (2020, 129) uses assemblage in a slightly different way to emphasise the process, more so than the physically networked arrangement of actors, to highlight how geopolitical knowledges become part of the "social, technical and material alliances" of forensic malware analysis to influence how cyber security practices are "made," including how cyber security threats are attributed to particular state and non-state actors. Dwyer (2021) also takes a materialist-discursive approach that calls for greater attention to the more-than-human ecology of cyber security by theorising the relationship between human practices and language and the computational grammars of code that enable a computer cognition that differs substantially from its human counterpart. The next section will give a brief historical outline of events that have led to the current trajectory, and probable futures, of digital warfare in order to contextualise the theoretical discussion that follows.

The origins of digital war

The internet as a geopolitically relevant attack surface can be traced back to Cold War espionage operations in the 1980s, notably a 1986 attack by a West German hacker on the Lawrence Berkeley National Laboratory in order to sell US state technological secrets to the KGB (Stoll 2005). However, until relatively recently hacking attacks were primarily targeted at gaining access to data or for financial gain, rather than as a direct means of achieving geopolitical ends. One of the earliest cases of a geopolitically motivated cyberattack being coordinated with other forms of warfare occurred during Russian military action against Georgia in 2008 (Hollis 2011). These attacks consisted mainly of Distributed Denial of Service (DDoS) attacks to shut down access to Georgian government and media websites to limit the Georgian state's internal and external communications in coordination with localised bomb strikes, as well as for psychological or propaganda value. In both the Georgia incident and 2007 Estonia incident mentioned in the previous section, it is difficult to definitively attribute Russian state sponsorship for these attacks. However, there was intelligence gathered in advance of the Estonia incident providing warning of impending DDoS attacks which indicated at least some degree of political coordination behind the attacks (Mansfield-Devine 2012). This murkiness exemplifies one of the key debates in academic discourse on cyber war called "the attribution problem," which will be discussed in the following section. That said, one of the most famous incidents, and likely the first case of a state-sponsored cyber-physical attack[1] on another country, is the Stuxnet virus, a joint US-Israeli covert intelligence operation that in 2010 destroyed uranium enrichment centrifuges in Iran's Nantez nuclear facility to disrupt Iranian nuclear weapons research (Dunn Cavelty 2013; Rid 2013).

Examples of digital war that have emerged since the mid-2000s provide timely examples for understanding the deep ontological enrolment of networked digital technologies and infrastructures in war. They can be seen both as an attack surface and the medium through which patterns of daily life are disrupted and fractured, as well as a spatial strategy for shaping the boundaries of territory. Doing so will demonstrate that the assemblage of technologies that comprise so-called "cyber space" is not virtual or an abstraction, and that the harm is not immaterial, or grossly exaggerated. Rather, digital networks have become the basis of increased interdependency between social, political and economic domains, as well as foundational to the existence and governance of the state. Therefore attacks on the internet threaten state security inasmuch as they affect the economic, societal and infrastructural bases of both the state and the daily lives it governs.

So far, these attacks have been defined and limited in their aims and objectives. In the case of the 2008 Georgia cyberattacks, Hollis (2011, 4) argues that the targets for attacks were strategically calibrated to "trigger comparative 'inconvenience'" rather than cause profound "chaos or injury." However, over time both the targets and consequences of attacks have become more pervasive and have caused more significant and widespread impacts. In 2017, the WannaCry virus was used to perpetuate ransomware attacks in a number of countries on computers using outdated versions of the Windows operating system. It caused significant disruption to the UK's National Health Service (NHS) and cost the organisation approximately £92m (Moore 2018). The attack was attributed by the US and the UK governments to the North Korean state-sponsored Lazarus hacking group. However, according to a BBC investigative journalism podcast this attack was likely a proof-of-concept test for a future economically motivated attack to obtain hard currency for the North Korean state that went awry, rather than a direct geopolitical act of aggression (BBC World Service 2022).

Also in 2017, the NotPetya cyberattack targeted key points of Ukraine's national infrastructure and financial system, including energy suppliers, hospitals, airports, banks and payment systems as well as computer systems used by scientists to monitor the Chernobyl nuclear disaster site (Greenberg 2019). A significant knock-on effect of NotPetya was that it spread across the Danish multinational shipping company Maersk's global logistics network, causing significant disruption to its network, which is responsible for approximately one-fifth of all global shipping. The US, UK and Dutch governments have publicly attributed this attack to the Russian state (Bannelier et al. 2019). However, the clandestine nature of cyberattacks and often convoluted chains of affiliation and causality make it difficult to definitively attribute and evidence state sponsorship to cyberattacks. An earlier example was the BlackEnergy attack in 2015 that temporarily disabled power infrastructure in the Ukraine, affecting approximately 225,000 customers (Anon, 2016). Actions within the 2022 Ukraine war, which was ongoing on at the time of writing, and the Not-Petya and BlackEnergy examples, demonstrate that regardless of linear command or causality cyberattacks can and have weaponised the infrastructures that underpin the social, economic and material conditions of contemporary existence to cause meaningful and real damage to the very same infrastructures and the human lives enrolled and sustained by them. The next section will evaluate how digital warfare relates to grand geopolitical conceptualisations of sovereignty and territory, and how it complicates and exceeds these normative frameworks.

Sovereignty, (in)security and cyber war

In conventional grand geopolitics, the internet is both a vector of attack and an attack surface that targets the sovereign territory of state space. However, in reality the internet complicates conventional logics of territorial control and regimes of sovereignty. This is a result of the seemingly seamlessly interconnected telecommunications infrastructures and date flows across borders and jurisdictions of governance. However, in actuality these networks are often not seamless or non-hierarchal. This is evidenced by the attempts of governments to bound the free access of data, often with unintended consequences. For example, in 2008 the Pakistani government attempted to block access to the popular YouTube video-sharing platform within Pakistan; however, their "clumsy" attempt to control the routing of internet traffic from servers located outside the country's borders shut down the YouTube website globally (Broeders 2016, 73–74). Likewise, there is an increasing tendency by state actors to attempt to territorially contain and border the infrastructure of the internet. The most famous example of this is China's so-called "Great Firewall" (Deibert et al. 2010). However, another example which clearly illustrates the geopolitical logic of territorialising internet infrastructure occurred in 2022 as part of Russia's annexation of the Kherson region in Eastern Ukraine when Russian authorities re-routed internet and mobile data networks from the Ukrainian telecommunications network to a Russian Telecommunications network connected via Russian-occupied Crimea. This allowed the Russian occupation authorities to cut off civilians' access to social media, independent news and communication with family members elsewhere in Ukraine (Satariano and Reinhard 2022).

Furthermore, internet-based attacks can cause significant economic damage with relative ease and low cost for an attacking party. This allows for the incorporation of the broader national economy as a target for attack in geopolitical conflict as a means to cause damage to the state. The likelihood and potential to cause harm through such means is exacerbated by the financialization of contemporary economies, which incorporates short-term risk calculation and just-in-time decision-making in investment and other market-based decisions. The most noteworthy example of an internet-based attack directed at a national economy is NotPetya, which deliberately targeted the Ukrainian economy by means of its digital infrastructure by introducing malware via a backdoor in M.E. Doc, an accounting software used widely and exclusively in Ukraine (Dwyer 2021).

Remote digital warfare undermines one of the core tenets of realist geopolitical thought, which is the assumption of the nation-state as the primary geopolitical actor. There exists an attribution problem with malicious internet activity, whereby it is difficult to find an actor responsible or link an identified actor to the direction of a nation-state to a high enough evidentiary standard for an attribution of responsibility to be plausible. This can lead to false attributions, the inclusion of red herring forensic evidence to influence false attributions and a potential for escalation of tension or conflict between nation-states over attributions made.

Compounding this is the spatiality of distance, and the dynamic of depersonalisation an attacker's perception of distance can invite. Similar to arguments made about UAV warfare, one of the key advantages of internet-based attacks for its belligerents is that it can project force over distance with little risk to the attacker (Chamayou 2015; Gregory 2011). The affective nature of hacking, combined with the dynamics afforded by spatial distance, negates the possibility of empathic engagement with those affected by an attack since the attackers are dealing with the medium of code, and the intellectual challenges of technical problem-solving. The section that follows will engage further with the temporal dimension of digital warfare and how this temporality has geopolitical implications across scales.

The spatiality and temporality of networked violence

The scale of digital-based warfare compresses the relationship between distance and time. In temporal terms the raison d'etre of the internet is speed. The networked digital telecommunications infrastructure(s) that comprise the internet accelerate the transmission of communication data in a way that defies a linear relationship in which greater distance equals greater time for an object to move from point A to point B, as well as an ability to permeate communication and data between and beyond geopolitical borders. However, these very features are precisely what enables digital violence to become effective.

Digital attacks can be understood as an insurgent practice (Reid 2015) that manipulates and undermines the structural relations of power, space and time that conventional state-centric theorisations of warfare and strategy are based upon. Compared to conventional weapons there is a low barrier to access in terms of financial cost and resources, as well as a high reward-to-cost ratio in terms of the low existential and financial risk to the attacker compared to the extensive disruption or damage they can cause. Furthermore, the networked connectivity of so many of the infrastructures that sustain contemporary life to enable better centralised control or greater convenience also has the unintended capacity allow for a wide range of potential available targets for consequential attacks.

Israel and Iran have already begun mutually attacking each other's critical national infrastructures in a tit-for-tat manner. In 2020, there were attacks on Israeli regional water distribution networks, which Israel attributed to Iran. Several weeks later, the Iranian port of Shahid Rajaee located in the Strait of Hormuz was severely disrupted by an internet-based attack. *The Washington Post* reported (based on intelligence sources) that the attack was carried out by Israeli state-sponsored hackers in retaliation for the earlier attack (Bergman and Halbfinger 2020; Warrick and Nakashima 2020). In 2021, there was an internet based-attack on Iran's national fuel distribution network in which petrol pumps "suddenly stopped working and a digital message directed customers to complain to Iran's supreme leader, Ayatollah Ali Khamenei, displaying the phone number of his office" (Bergman and Halbfinger 2020). Shortly thereafter followed an attack on an Israeli hospital and mass data breaches of a popular Israeli LGBTQ dating site and group of private health clinics, which resulted in the private data of both organisations members being leaked publicly. Although it has been suggested in *The New York Times* that the Israeli and Iranian governments are likely behind these attacks, neither Israel nor Iran has publicly claimed responsibility for them. Based on these events security analysts believe that "cyber conflict" could focus on attacking private companies as softer targets than government agencies (Fassihi and Bergman 2021).

As indicated by the previous example, much of the malign activity from digital attacks occurs below the bona fide threshold of war. However, these acts are intended to enact and perpetuate geopolitical conflicts with modulating degrees of violence in a way that elides the ostensible dichotomy between temporal states of war and peace. Furthermore, there is a developing trend in digital conflict that indicates the potential for future scenarios in which there will be greater frequency and severity (including lethality) of internet-enabled attacks. In other words, the present definition of what constitutes "cyber war" is not yet clearly defined, and the contours of digital warfare are rather murky for now. But this does not mean that the scale and pervasiveness of digital warfare won't become more expansive in the near future, and indeed recent events suggest that it will be. The next section will critically interrogate how a state-centric management of digital (in)security is developing at the international level to respond to this trajectory, and briefly identify some challenges and blind spots of these approaches.

Digital peacebuilding?

An emerging doctrine in the International Relations sphere for the security governance of digital networks is the development of cyber diplomacy that will codify a body of cyber norms, and ensure adherence to these norms. Cyber norms are voluntary "non-binding principles that shape state conduct in cyberspace," which would deter cyberattacks by making them "politically costly to the point of being disadvantageous for the state actors who launch them" (Taddeo 2017, 389). According to proponents of this approach, norm entrepreneurs – which can be either states (Crandall and Allan 2015 cast small states as particularly oriented to this role) or private entities (Hurel and Lobato 2018) – are able to take leadership roles in shaping and prompting the creation of such norms. However, the cyber norms approach has been criticised from a Realist position as toothless and ineffectual in deterring the violation of norms (Hampson and Sulmeyer 2017). Compounding this is the problem of consensus building between both state actors in elite diplomatic venues, and between scales and entities outside this milieu, around both defining norms and the consequences for their violation. Furthermore, the policing of cyber norms can recapitulate and reify unequal power relations between nation states, whereby more conventionally powerful state actors dominate both the creation of norms and the enforcement of norms in a way that furthers these state's interests at the expense of other states and citizenries.

From a critical geopolitics perspective, the emphasis of the cyber norms discourse reifies the state as the primary locus of sovereign power and security, and with it the ostensible primacy of the elite diplomatic level in shaping and constituting geopolitics. Such an approach ignores the ontological depth of geopolitics and the multiplicity of human and non-human agents that mediate it. This short-sightedness becomes particularly acute in conceptualising the geopolitics of digital warfare precisely because of how the hybrid and entangled agencies in digital networks as physical entities, dynamic processes and amalgamations of practices create the attribution problem in cyber international relations. It is extremely difficult to ascribe – let alone prove – the causal agency of a given actor in perpetrating an attack or intending a particular set of consequences from that attack, especially when this is made at the state level since it is highly debatable whether the state can be considered a coherent or unitary actor.

Eschewing the cyber norms approach to digital diplomacy requires other venues, practices and vehicles for mitigating the proliferation and harms of digital warfare in its stead. This must be sensitive to both the vulnerability and the ontological stakes of computer-inflicted violence in a world where digital networks are becoming ever-closer entwined with the infrastructures that sustain life. This requires moving the locus of protection for the digital away from a totalising paradigm of military power towards a paradigm in which state and non-state actors emphasise a shared policy response that both takes seriously the material consequences of digital warfare, and mitigates against its harms. This would (and, to some extent, already does in an ad hoc way) take the shape of a hybrid approach that combines scenario forecasting, emergency preparedness and cooperation between experts and affected individuals across scales. Actors involved in these responses include independent hacker groups, cyber security response teams in both industry and government, emergency services, civil society organisation and powerful economic entities including central banks, financial regulators and markets. The extent to which cyber warfare can be mitigated against once "Pandora's box" has been opened, and what digital peacebuilding might look like, are questions for further investigation raised by the present discussion.

Conclusion: Entangled relations of ontology, technology and war

This chapter has argued that the digital networks and infrastructures that comprise so-called "cyber space," rather than being an artificially militarised and securitised arena, are in fact significant vectors and attack surfaces of geopolitical insecurity in war. This new terrain is materially salient because of how digital infrastructures are now pervasively enmeshed with the ontological conditions of space and time in daily life. However, state-centric attempts to securitise and militarise the digital is not smooth, unproblematic or uncontested. Networked digital infrastructures become both the means of attack and the object of attack in targeting the conditions of the everyday. Furthermore, because of the non-linear and nodal spatial relations of digital networks, the supposedly clear categorical distinctions between military and civilian actors and targets become profoundly elided and murky. This complexity has been evidenced in several ways, including the national economy as target and vector of attack, and the attribution problem in determining both the initiator and motive for a given digital attack, and the unclear threshold for when an internet-based attack constitutes an act of war, and how such hostilities play out.

These aspects of internet warfare are not entirely unique in terms of theorising the contemporary nexus of technology and war, and have clear parallels to other forms of recent technological warfare. There is a concomitant elision of the distinction between civilian and military means and objects of war. Previously commercial technologies are being used by a range of state and asymmetrical non-state actors for military purposes, from laptops and smartphones used for real-time battlefield intelligence-gathering and command networks (C4ISR), to the modification of commercially available UAVs for surveillance, combat and police use of force. Conversely previously militarised technologies are now pervasive in society-at-large and have been appropriated by civil society to monitor and countermand political violence, such as the use of now easily accessible satellite and aerial imagery by investigative journalists and human rights organisations to conduct OSINT investigations. Likewise, internet warfare shares with other technologically advanced warfare an affordance to enable an acutely urbicidal form of warfare (cf. Coward 2010) in which attackers deliberately focus violence on destroying the built environment and infrastructures that sustain the lives of cities. This is due to a confluence of the intensive pattern of urbanisation in the contemporary world with the capacities of technologies to concentrate violent force on the urban terrain as a strategy to mitigate the spatially challenging topography and terrain characteristic of urban warfare (Hills 2004). In the case of internet warfare, the increasing connectivity of smart cities and their integration into the urban environment afford the ability for attackers to target the infrastructures that sustain the conditions of daily life in cities to maximise damage at a relatively low cost in terms of resources and political risk. Whilst the means of attack may be relatively novel, internet-derived attacks on infrastructure are a continuation of historical and contemporary phenomena and trends in the relationship between technology and warfare.

The material nature and consequences of internet-based digital warfare require us to take seriously the very real stakes of digital violence for the ontological security of daily life. This leads to a question of what means and approaches can best secure vulnerable technologically dependent infrastructures. Because of the threat to the conditions of everyday life and in certain cases to life itself, alongside the elision of any distinction between military and civilian – or military and economic – target in digital warfare there is a need for an approach to digital security that attends to the everyday vulnerabilities and consequences of such violence, rather than abstracting or gamifying it. it is also important that vulnerability is addressed in a way that mitigates the reification of an iterative feedback loop in which an increased militarisation of digital infrastructures leads to increased targeting of these infrastructures, and a detrimental shift towards their fortification in a way that limits their capacity to co-create the very ontological security of daily life they are meant to enable.

This chapter has contributed to the wider themes of this volume by illustrating the more-than-human agency of technology in co-producing with human actors the political geographies of war and peace. It has done so in particular by demonstrating how internet-based warfare complicates the dynamics of power between human political actors in profoundly asymmetrical and non-linear ways that disrupt conventional notions of militarism and state power. Internet-based warfare also extends the reach and potential means of political violence at multiple geographical settings and scales by affecting particular places and territorial arrangements through the physical and spatialised topology of digital communications networks enmeshed with other material infrastructures. In doing so, such digital violence simultaneously reshapes these spatial configurations in complex and unanticipated ways.

Note

1 Cyber-physical attack means an attack on physical objects or infrastructures executed by digital infiltration and manipulation of a computer or network.

References

Adey, P. 2010. Vertical security in the megacity: Legibility, mobility and aerial politics. *Theory, Culture & Society* 27: 51–67.

Anon. 2016. Cyber-attack against Ukrainian critical infrastructure | CISA. February 25. Accessed December 19 2022. https://www.cisa.gov/uscert/ics/alerts/IR-ALERT-H-16-056-01

Aradau, C. 2010. Security that matters: Critical infrastructure and objects of protection. *Security Dialogue* 41: 491–514.

Arquilla, J., and D. Ronfeldt. 1997. *In Athena's Camp: Preparing for Conflict in the Information Age*. Rand Corporation.

Ashraf, C. 2021. Defining cyberwar: Towards a definitional framework. *Defense & Security Analysis* 37: 274–294.

Bannelier, K., N. Bozhkov, F. Delerue, F. Giumelli, E. Moret, and M. Van Horenbeeck. 2019. Cosmic dust: Attribution and evidentiary standards. In *Guardian of the Galaxy: EU Cyber Sanctions and Norms in Cyberspace*. European Union Institute for Security Studies (EUISS). https://www.jstor.org/stable/resrep21136.9

Barkawi, T. 2015. Diplomacy, war and world politics. In *Diplomacy and the Making of World Politics*, eds. O.J. Sending, V. Pouliot, and I.B. Neumann, 55–79. Cambridge University Press.

BBC World Service. 2022. Kill switch. *The Lazarus heist*. June 20. Accessed December 19 2022. https://www.bbc.co.uk/programmes/w13xtvg9/episodes/downloads

Bergman, R., and D.M. Halbfinger. 2020. Israel hack of Iran port is latest salvo in exchange of cyberattacks. *The New York Times*, May 20. Accessed December 29, 2022. https://www.nytimes.com/2020/05/19/world/middleeast/israel-iran-cyberattacks.html

Betz, D.J., and T. Stevens. 2013. Analogical reasoning and cyber security. *Security Dialogue* 44: 147–164.

Boot, M. 2006. *War Made New: Technology, Warfare and the Course of History, 1500 to Today*. Gotham Books.

Bousquet, A. 2018. *The Eye of War: Military Perception from the Telescope to the Drone*. University of Minnesota Press.

Bratton, B. 2016. *The Stack: On Software and Sovereignty*. MIT Press.

Broeders, D. 2016. *The Public Core of the Internet: An International Agenda for Internet Governance*. Amsterdam University Press.

Chamayou, G. 2015. *A Theory of the Drone*. New Press.

Coaffee, J., D.M. Wood, and P. Rogers. 2009. *The Everyday Resilience of the City*. Palgrave Macmillan.

Collier, J. 2018. Cyber security assemblages: A framework for understanding the dynamic and contested nature of security provision. *Politics and Governance* 6: 13–21.

Coward, M. 2010. *Urbicide: The Politics of Urban Destruction*. Routledge.

Cowen, D. 2014. *The Deadly Life of Logistics: Mapping Violence in Global Trade*. University of Minnesota Press.

Crandall, M., and C. Allan. 2015. Small states and big ideas: Estonia's battle for cybersecurity norms. *Contemporary Security Policy* 36: 346–368.

De Landa, M. 2006. *A New Philosophy of Society: Assemblage Theory and Social Complexity*. Continuum.

Deibert, R., J. Palfrey, R. Rohozinski, and J. Zittrain. 2010. China. In *Access Controlled: The Shaping of Power, Rights, and Rule in Cyberspace*, eds. R. Deibert, J. Palfrey, R. Rohozinski, and J. Zittrain. The MIT Press. https://doi.org/10.7551/mitpress/8551.003.0043

Dittmer, J. 2014. Geopolitical assemblages and complexity. *Progress in Human Geography* 38: 385–401.

Dunn Cavelty, M. 2013. From cyber-bombs to political fallout: Threat representations with an impact in the cyber-security discourse. *International Studies Review* 15: 105–122.

Dwyer, A.C. 2021. Cybersecurity's grammars: A more-than-human geopolitics of computation. *Area* 55: 10–17.

Ek, R. 2000. A revolution in military geopolitics? *Political Geography* 19: 841–874.

Fassihi, F., and R. Bergman. 2021. Israel and Iran broaden cyberwar to attack civilian targets. *The New York Times*, November 27, sec. World.

Graham, S. 2011. *Cities Under Siege: The New Military Urbanism*. Verso.

Graham, S., and L. Hewitt. 2013. Getting off the ground: On the politics of urban verticality. *Progress in Human Geography* 37: 72–92.

Greenberg, A. 2019. *Sandworm: A New Era of Cyberwar and the Hunt for the Kremlin's Most Dangerous Hackers*. Doubleday.

Gregory, D. 2011. From a view to a kill: Drones and late modern war. *Theory Culture & Society* 28: 188–215.

Hampson, F.O., and M. Sulmeyer. 2017. *Getting beyond Norms: New Approaches to International Cyber Security Challenges*. Centre for International Governance Innovation. https://www.jstor.org/stable/resrep05241

Haraway, D. 1988. Situated knowledges: The science question in feminism and the privilege of partial perspective. *Feminist Studies* 14: 575–599.

Heng, Y.-K. 2006. *War As Risk Management: Strategy and Conflict in an Age of Globalised Risks*. Routledge.

Hills, A. 2004. *Future War in Cities: Rethinking a Liberal Dilemma*. F. Cass.

Hirst, P. 2005. *Space and Power: Politics, War and Architecture*. Polity.

Hollis, D. 2011. Cyberwar case study: Georgia 2008. *Small Wars Journal*. January 6. Accessed December 19, 2022. https://smallwarsjournal.com/blog/journal/docs-temp/639-hollis.pdf

Hurel, L.M., and L.C. Lobato. 2018. Unpacking Cyber Norms: Private Companies as Norm Entrepreneurs. In *GigaNet: Global Internet Governance Academic Network, Annual Symposium 2017* January 22. Accessed December 19, 2022. http://dx.doi.org/10.2139/ssrn.3107237

Jeffrey, A. 2006. Book review: "The Colonial Present" by Derek Gregory. *Area* 38: 472–474.

Kaiser, R. 2015. The birth of cyberwar. *Political Geography* 46: 11–20.

Kaldor, M. 2012. *New and Old Wars: Organized Violence in a Global Era*. 3rd ed. Stanford University Press.

Kindervater, K. 2017. Drone strikes, ephemeral sovereignty, and changing conceptions of territory. *Territory, Politics, Governance* 5: 207–221.

Kirsch, S., and C. Flint. 2011. Introduction: Reconstruction and the worlds that war makes. In *Reconstructing Conflict: Integrating War and Post-War Geographies*, eds. S. Kirsch and C. Flint, 3–28. Ashgate.

Kurgan, L. 2013. *Close up at a Distance: Mapping, Technology and Politics*. Zone Books.

MacDonald, F., R. Hughes, and K. Dodds. 2010. *Observant States Geopolitics and Visual Culture*. I.B. Tauris.

Mansfield-Devine, S. 2012. Estonia: What doesn't kill you makes you stronger. *Network Security* 2012: 12–20.

Moore, M. 2018. WannaCry attack cost the NHS £92m. *TechRadar* October 12. Accessed December 19, 2022. https://www.techradar.com/news/wannacry-attack-cost-the-nhs-pound92m

Müller, M. 2012. Opening the black box of the organization: Socio-material practices of geopolitical ordering. *Political Geography* 31: 379–388.

Neocleous, M. 2013. Air power as police power. *Environment and Planning D: Society and Space* 31: 578–593.

Perugini, N., and N. Gordon. 2017. Distinction and the ethics of violence: On the legal construction of liminal subjects and spaces. *Antipode* 49: 1385–1405.

Reid, J. 2015. Deleuze's War Machine: Nomadism Against the State. *Millennium: Journal of International Studies* 32: 57–85.

Rid, T. 2013. *Cyber War Will Not Take Place*. Oxford University Press.

Satariano, A., and S. Reinhard. 2022. How Russia took over Ukraine's Internet in occupied territories. *The New York Times*, August 9, sec. Technology.

Sauer, F., and N. Schornig. 2012. Killer drones: The "silver bullet" of democratic warfare? *Security Dialogue* 43: 363–380.

Shaw, I. 2017. Robot wars: US empire and geopolitics in the robotic age. *Security Dialogue* 48: 1–20.

Shaw, I., and M. Akhter. 2012. The unbearable humanness of drone warfare in FATA, Pakistan. *Antipode* 44: 1490–1509.

Shaw, M. 2005. *The New Western Way of War: Risk-Transfer and Its Crisis in Iraq*. Polity.

Spinks, L. 2001. Thinking the post-human: Literature, affect and the politics of style. *Textual Practice* 15: 23–46.

Squire, R. 2016. Rock, water, air and fire: Foregrounding the elements in the Gibraltar-Spain dispute. *Environment and Planning D: Society and Space* 34: 545–563.

Stevens, C. 2020. Assembling cybersecurity: The politics and materiality of technical malware reports and the case of Stuxnet. *Contemporary Security Policy* 41: 129–152.

Stevens, T. 2016. *Cyber Security and the Politics of Time*. Cambridge University Press.

Stoll, C. 2005. *The Cuckoo's Egg: Tracking a Spy Through the Maze of Computer Espionage*. Simon and Schuster.

Taddeo, M. 2017. Deterrence by norms to stop interstate cyber attacks. *Minds and Machines* 27: 387–392.

Thrift, N. 2007. Immaculate warfare? The spatial politics of extreme violence. In *Violent Geographies: Fear, Terror, and Political Violence*, eds. D. Gregory and A. Pred, 273–294. Routledge.

van Creveld, M. 1989. *Technology and War: From 2000 B.C. to the Present*. Free Press.

Warrick, J., and E. Nakashima. 2020. Officials: Israel linked to a disruptive cyberattack on Iranian port facility. *The Washington Post*, May 18. Accessed December 29, 2022. https://link.gale.com/apps/doc/A624285998/AONE?u=rho_ttda&sid=bookmark-AONE&xid=b6022f0f

Weizman, E. 2007. *Hollow Land: Israel's Architecture of Occupation*. Verso.

Whatmore, S. 2006. Materialist returns: Practising cultural geography in and for a more-than-human world. *Cultural Geographies* 13: 600–609.

Wilcox, L. 2017. Embodying algorithmic war: Gender, race, and the posthuman in drone warfare. *Security Dialogue* 48: 11–28.

Williams, A. 2011. Enabling persistent presence? Performing the embodied geopolitics of the unmanned aerial vehicle assemblage. *Political Geography* 30: 381–390.

Williams, A. 2013. Re-orientating vertical geopolitics. *Geopolitics* 18: 225–246.

13

GEOGRAPHIES OF ENVIRONMENTAL PEACE AND CONFLICT

Shannon O'Lear

Oil wars. Conflict diamonds. Climate change exacerbated conflict. Environmental features are frequently associated with violence and conflict, but how well do we understand these connections? A geographic perspective is valuable for critiquing assumptions about how environmental features are related to conflict. This chapter reviews key themes in scholarship on resource-related conflict. It also considers how different ways of managing and valuing environmental resources can create multiple geographies of haves and have-nots and of environmental benefits and harms. Even seemingly peaceful relationships involving natural resources warrant close attention through a geographic perspective: What may appear to be peaceful at one spatial scale or through the eyes of particular actors may not be as peaceful from another vantage point.

This chapter connects with the overarching themes of this book in several ways. Human interactions with environmental systems involve various forms of agency: Who gets to decide how environmental resources are valued or devalued, who gets to benefit from their use, and who faces negative impacts of these decisions without any agency. Importantly, a critical aspect of human agency in regard to environmental features has to do with our understanding of environmental features and resources and how that understanding informs behavior and practice. This chapter includes examples of how certain ideas about environmental features were formed and how those ideas have continued to shape how we utilize (or disregard) environmental resources. How we value or disregard particular aspects of "the environment" contributes to another theme of this book, namely, multiple geographies. Humans have altered the surface of the planet in many ways with a multitude of results and impacts and

DOI: 10.4324/9781003345794-13

at a variety of spatial scales. Yet there are also different approaches to understanding environmental features and how human systems interact with, draw from, or alter environmental systems. Much of the Western world's relationship to environmental features is grounded in othering and partitioning, which is deeply rooted in the notion that humans and environmental features are separate and distinct. Other ways of understanding the world view humans and environmental features as integrated and dependent on each other for survival. Spaces that are constructed through human interactions with their environmental surroundings reflect variations in political power: power to shape how environmental resources and systems are valued, power to impose a particular set of values on a landscape or waterway, and power to voice alternative ways of valuing environmental systems and our relationship to them. These themes of agency, multiple geographies and spatial scales, partitioning or inclusive views of environmental features, and political aspects of how humans value and (de)value environmental features all come into play in this chapter's consideration of geographies of environmental peace and conflict.

As a starting point, it is helpful to recognize resource determinism. Resource determinism is the assumption that the resource context, or the environmental context more broadly, dictates the human response. The idea that water scarcity will lead to conflict is resource determinism. Similarly, the idea that a supply of oil (or diamonds, lithium, cobalt, avocadoes, etc.) will lead to conflict over access to those resources is also resource determinism. How can both scarcity of resources and abundance of resources lead to the same outcome? There seems to be more to the story than just the absence or presence of a desirable resource, and there is. Resource determinism is anti-geographical thinking because it disregards the complexities of human systems and human-environment interactions, as well as the nuance and place-specific impact and connections among places. Instead, a geographical approach is useful for understanding how resources and other environmental features are intertwined both with situations of dispute and collaboration. A geographical approach pays attention to the interaction of multiple geographies, the construction of policies and spaces, dynamics of othering and inclusivity, and processes of resistance and militarism.

Environmental resources, conflict, and violence

The 1990s saw the end of the Cold War and its focus on imminent, international conflict over political ideologies. Many scholars and policymakers' attention shifted to how environmental resources could become the next focal points for conflict (Gleick 1991; Deudney 1991). They anticipated how international grabs for specific resources could lead to tension and conflict. "Water wars" emerged as a particularly popular idea with transboundary water resources believed to be the key issue that would mobilize international conflict.

In the face of that trend, Aaron Wolf (1999) led a research project that focused, instead, on how shared water resources contributed to peaceful relations. At the time, the trend in political science was to use quantitative conflict databases to generate statistical studies with a large number of cases of states that share a water body (large-N studies). Wolf and his colleagues, however, took a qualitative, empirical approach in which they examined water treaties signed between states and analyzed the specific features of those negotiations and treaties that made them successful at establishing reliable, ongoing relations over shared water resources. Through fieldwork and in their conversations with people in places with transboundary water resources, Wolf and his team found that conflict or peace is not directly related to the presence or absence of a transboundary water resource. Instead, the institutional capacity to manage water and any emerging disputes shapes relationships over water in each context (Yoffe, Wolf, and Giordano 2003). This example illustrates the importance of how policies over shared water resources may be intentionally constructed in collaborative ways and how those policies emerge from and map onto spaces of water use.

The importance of developing a geographic understanding of transboundary waters was expanded by Kathryn Furlong (2006) in her work on transboundary waters in Southern Africa. In her work, she examines how transboundary waters and their potential for conflict had been studied using IR (International Relations) approaches and assumptions. These IR approaches and assumptions have been taken up by policymakers working to develop and manage transboundary water. Furlong argues that these approaches too were problematic. A central problem with taking an IR approach to understanding transboundary waters, Furlong argues, is that it falls into the "territorial trap" (Agnew 1994) of prioritizing the state level and state-to-state relations. This type of scholarship tends to assume that states are fixed containers of both uniform activity and identity and that domestic and foreign policy are distinct and separate. Focusing on the state and interstate interactions overlooks activities and tensions in other spaces, such as among groups within a state or antagonism between a state government and its own people.

An IR approach also tends to be ahistorical and does not take into account ongoing legacies of colonialism and imperialism that have shaped current political, economic, social, and environmental systems. Instead, Furlong makes a compelling case for a geographic approach to understanding transboundary water systems which recognizes that water is a multifaceted resource that means different things to different groups of people (e.g., basic survival, public provision, economic development). Her work also highlights the importance of seeing how colonial legacies have contributed to disparities in justice and how development projects can further compound inequities in access and benefits. These are some of the vital, context-dependent characteristics of water resources – or other environmental features – that can be helpfully brought to light and examined through a geographic perspective. Specifically, this work illustrates how resource management can involve dynamics of othering as some groups of people are disenfranchised from access to resources. Furthermore, it shows how those practices of marginalization may be perpetuated by research that does not pay attention to unequal power dynamics. This work also illustrates the importance of recognizing multiple, simultaneous geographies of human interactions with environmental features, such as water, and how those environmental features are valued, utilized, and managed differently for different purposes by different groups of people.

Shared or managed resources can foster peaceful relations, but it is important to consider what is happening at multiple spatial scales. A helpful example comes from work of Kimberley Thomas (2017, 2021) who has studied local impacts of water diversion projects near international boundaries. In her work, she observes that water diversion projects in India and Thailand serve to keep relations between the neighboring states peaceful, and conform to agreements with their downstream neighbors, Bangladesh and Laos, respectively. These infrastructure projects allow neighboring states to avoid direct conflict over water. However, she finds that these same infrastructure projects contribute to mundane violence that communities in border areas experience. For example, these large-scale projects alter access to reliable water supplies and other resources such as fish, thus making seasonal and sometimes day-to-day existence for these groups of people more difficult, if not in impossible in some cases. Ultimately, what may appear to be a peaceful arrangement for shared resources at one spatial scale may contribute to violence or harm unfolding at other spatial scales.

Other environmental resources can certainly be connected to conflict, both visible and less visible, particularly in cases where a resource commodity may be used to finance an ongoing dispute. Yet, resources differ in terms of how they become entangled within a conflict as well as the spatial strategies and form(s) of harm that they motivate. These are some of the insights offered in Philippe Le Billon's work on the political ecology of war (2001). Similar to other work that critiques an oversimplified resource determinism in which the presence or absence of an environmental resource dictates conflict, Lebillon demonstrates that resource dependence is an important aspect for consideration. That is, how is a particular group of people relying on a particular resource for economic benefit or to stay (or get) in power? Le Billon considers the complex social, economic, and political relationships built around particular resources. He argues that it is the human relationship to the environmental resource – not the resource itself – that shapes different spatial strategies of power and different forms of violence. For instance, both diamonds and timber may be sold on international markets for funds to purchase weapons or other tools in waging a battle over territory and influence. However, to control a supply of diamonds, it may be necessary to control certain locations where diamond mines are located. To control a supply of poached timber, it may be necessary to control a large, forested area. Not only are the spatial strategies of control different for these different resource commodities, each one has a different requirement for getting to market: diamonds are small and may be easily smuggled but moving timber would require different logistics altogether. These kinds of features are part of the political ecology of how resources are tied to economic agendas of belligerent groups as well as how those resources are valued or made valuable. The process of securing a supply of a particular resource and gaining benefit from it also has implications for populations of people caught up in the violence of these agendas. Additionally, there are implications for economic globalization that makes this kind of resource capture profitable despite the localized violence it causes. This example highlights why it matters how particular, environmental resources are economically valued and why the logistics of those market values can matter when they are entangled in ongoing processes of resistance.

Stuck on scarcity (and abundance)

These empirical examples from geographic literature provide nuance that the overarching notion of resource conflict just does not deliver. Why, then, does the notion of resource conflict persist in academic work, popular media, and public discourse? The general acceptance of the idea that scarce resources lead to conflict goes back to the work of Reverend Robert Thomas Malthus (1766–1834). Malthus famously speculated that since food production could only increase mathematically (1, 2, 3, 4, 5, etc.), and since irresponsible humans were likely to reproduce geometrically (1, 2, 4, 8, 16, 32, etc.), it was only a matter of time before the human population surpassed the food supply and brought about horrific suffering and the downfall of society (Appleman 1976). Malthus was writing as the Industrial Revolution was gaining momentum in Britain. On the ground, one feature of the Industrial Revolution was enclosure, or the forced removal of people from land that sustained them and closing public access to that land. With no other option for survival, people had to sell their labor. Malthus's writings disparaged the system of place-specific economic support for many people living in rural, agricultural areas and argued that these less well-off were destined to make poor choices, have large families, and be a drain on society (Malthus [1798] 1993). Therefore, he argued, they should be uprooted from the countryside and detached from parish handouts to become mobile labor for the expanding, industrial economy. In short, rather than point to the consumption patterns of wealthy groups in society and their consolidation of economic wealth, Malthus helped to construct the notion of disempowered people as a less wealthy, less intelligent "them" who could not be trusted to make socially responsible decisions. Malthus effectively touted victim-blaming and the notion that wealthy people were endowed with the good sense to make responsible decisions for everyone. Despite these claims providing no actual data or numbers to support or prove them, these ideas prevailed, especially among wealthier groups of people.

Over a century later, Garret Hardin (1968) rewarmed the same Malthusian scarcity narrative. In his widely read piece, The Tragedy of the Commons, he makes the case for the privatization of public lands so they can be managed by an elite and supposedly responsible people. He acknowledges that this process of transferring power and the control of resources to an elite group of people in society will certainly be violent but necessary for the greater good. Again, this idea is not grounded in real-world empirics. The idea that an elite group of wealthy people knows the best way to manage environmental resources helped to construct that very arrangement in Western society. Yet rather than focus on this skewed distribution of power, the focus tends to be on the "other," poorer sectors of society – as the ones to watch for disruption around environmental resources that have supposedly been rendered scarce. That focus serves to distract from acts of enclosure that enable established powerful groups to gain control over more land and resources.

Notions about scarcity and enclosure are not just academic ideas in books; they are visible on the landscape in the United States and around the world. The practice of establishing national parks demonstrates enclosure:

> many national parks and other lands set aside for wilderness conservation are also the ancestral homelands of Native peoples. These communities were forced off their lands during European colonization of North America.
>
> *(Kashwan 2020)*

Well-known figures in the U.S. conversation movement, John Muir, John James Audubon, Theodore Roosevelt, and Aldo Leopold, have been recognized as having racist opinions against Indigenous groups in Africa and the United States. Their interests in conserving wild areas were informed by their views that diminished the self-determination, integrity, and well-being of Indigenous people and people of color both at home and abroad. Their values shaped not only the National Parks system and conservation practices in the United States; they also shaped conservation practices in other countries:

> Similar injustices continued to unfold even after independence in other parts of the world … the largest areas of national parks were set aside in countries with high levels of economic inequality and poor or nonexistent democratic institutions. The poorest countries – including the Republic of the Congo, Namibia, Tanzania and Zambia – had each set aside more than 30% of national territories exclusively for wildlife and biodiversity conservation.
>
> *(Kashwan 2020)*

In stark contrast to this approach to conservation, there are other models for stewarding places and resources that are not based on exclusion. Such socially just conservation practices are

> possible under two main conditions: Indigenous and rural communities have concrete stakes in protecting those resources and can participate in policy decisions.
>
> *(Kashwan 2020)*

This kind of approach rejects scarcity as the starting point for shaping human interactions with the environment. Instead, it perceives humans and environmental features and processes as inevitably interdependent.

Resource scarcity narratives can be seen to play out in other areas of environmental stewardship and management. For instance, the impacts of enclosure and exclusion are also disregarded on a much larger scale, namely, climate change policy. International efforts to address climate change impacts often recognize that some groups of people in particular places are more vulnerable to the changing climate, but the historical causes for those vulnerabilities are often overlooked:

> We know the history of extractive relations that created the marginality and poverty that makes people vulnerable in the face of climate events. Historians and social scientists find this self-evident. But many technocrats and climate scientists do not recognize that vulnerabilities have historical causes, an omission that can be seen in many "climate impact" models (as if damages that follow climate events can be assumed to be "impacts" of climate).
>
> *(Kashwan and Ribot 2021, 327)*

By disregarding systems of injustice and exploitation that contributed to current, inequitable conditions, mainstream narratives about climate change can perpetuate these injustices. Again, this is an example where an elite group of people and countries diverts attention in ways that alleviate their responsibility. For instance:

> Although they acknowledge distributional inequalities, the IPCC [Intergovernmental Panel on Climate Change] reports frame their analysis of vulnerability by identifying, in very general terms, who is vulnerable – paying inadequate attention to why different groups are vulnerable.
>
> *(Kashwan and Ribot 2021, 330)*

Leaders who rely on policies that overlook historical legacies of colonialism or socioeconomic roots of inequalities within particular countries point to simplistic explanations of climate change impacts as a way to avoid more difficult questions of social causes of vulnerability.

These examples demonstrate why focusing on resource scarcity is usually misleading. A focus on scarcity tends to overlook how the scarcity was created, for whom and where, and it almost certainly avoids discussion of who and which places benefited from the creation of that scarcity. Peter Haas has critiqued the idea that resource scarcity is a key contributing factor to conflict. His research demonstrates that there is scant empirical research to show resource scarcity leads to conflict. Additionally, there is little, if any, research that examines cases where resource scarcity exists but does not lead to conflict (2002). He has argues that narratives or discourses about resource scarcity and conflict are largely unfounded and that:

> These discourses persist not because they are accurate, but because they are politically embedded discourses … they acquire a taken-for-granted quality as bureaucrats are socialized into accepting the worldview within the administration through standard operating procedures, and because the arguments are often expedient for achieving other goals. In addition, lazy thinkers acquire the ideas in graduate school.
>
> *(9)*

Perhaps it is time, after all, to question if Malthusian assumptions are still viable (indeed, if they ever were).

As for the argument that resource abundance can lead to conflict, there has been a great deal of work examining what happens when a country is largely dependent on the export and income from a single, key commodity resource. There are indeed examples of countries in which resource dependence contributes to skewed governmental policies, poor economic diversification, and internal destabilization. These features are associated with an abundance of a particular resource or a "resource curse." Yet other examples show that some countries manage their resource wealth in ways that are productive and contributive to society. Furthermore, it is less clear that countries come into military conflict predominantly over abundant resources. Even in the case of oil, with the common assumption that it leads to wars, there is not a clear case that resource abundance leads to military conflict. In her book *The Oil Wars Myth: Petroleum and the Causes of International Conflict*, Emily Meierding (2020) looks at several real-world examples to demonstrate that foreign invasion and retaliation, territorial occupation, and damage to relations with oil companies carry too high a price for countries to be inclined to fight for oil. Similar to assumptions about resource scarcity causing conflict, assumptions about resource abundance driving conflict does not seem reliable in the real-world.

Both types of resource determinism overlook the fact that complex relationships and systems emerge around resources as well as how they are valued by different groups of people in different places. There are other ways to think about how societies interact with resources (see O'Lear 2018, chapter 3). Michael Watts and Nancy Peluso (2014) suggest the notion of a "resource complex":

> The resource complex ... examines both how resources are made regulable objects, how they are governed as parts of particular systems of rule, and what are the political and power relations by which the complex is, or is not, stabilized and rendered self-producing.
>
> *(196)*

Instead of seeing a particular resource as a stand-alone variable in a conflict scenario, this resource complex perspective positions a given resource within systems of production, transportation, and consumption. It also understands environmental resources as being associated with and shaped by particular institutions and actors, policies, and practices, with unique spatial traits and characteristics within these interactions. A resource complex approach focuses on how a particular resource commodity – such as diamonds, oil, or grain, or how an environmental feature, such as a transboundary water resource – is integrated into human interaction and systems of value. Here again, beyond resource scarcity or abundance, geographic approaches such as the resource complex bring greater nuance to understanding how environmental resources come to be the focus of peaceful negotiations as well as entangled in complex political ecologies of violence.

Environmental slow violence

Another way to reconsider the notion of resource conflict is to look for ways that environmental decisions and practices that benefit some people can lead to environmental harm or diminish well-being for other groups of people. It can be difficult to be attentive to harm that unfolds slowly and irreparably over time, yet this kind of harm is widespread and ongoing. This is slow violence. Rob Nixon's book *Slow Violence and the Environmentalism of the Poor* (2011) draws attention to ways in which people and ecosystems in the Global South are enduring long term and persistent harm from exploitative processes set in motion by colonialism, capitalism, and extractivism. Nixon argues that it is important to find ways to tell the stories of these experiences so that they are visible. Once visible, they may be recognized, addressed, and, ideally, alleviated. This slow violence approach is rooted in earlier work by sociologist Johan Galtung (1969) in which he argued that looking at peace and conflict as opposites is not as helpful as understanding violence. Instead, he framed it as having multiple dimensions: direct or indirect, immediate or latent, intentional or unintentional. Later, Galtung (1990) further argued that harmful practices are embedded in the very structures of society: culture, religion, art, language, and even the way that science is conducted. These normalized practices often involve or enact harm on particular environments or groups of people who are already marginalized, vulnerable, or otherwise unable to alter or escape these forms of harm. When we think of conflict, we tend to think of something that is immediately visible, measurable, and urgent. Taking slow violence into consideration, however, encourages us to see forms of violence and harm that are less visible. In his work on slow violence in Louisiana's Cancer Alley, Thom Davies (2018, 2021) encourages us to see and understand slow violence through "slow observation" that is attentive to negative or harmful changes in a place over time through the eyes of people living the experience.

This chapter has already provided examples of slow violence in the effects of enclosure and exclusion in environmental conservation and climate change policies. Slow violence is also evident in the example of negative, local impacts of infrastructure projects that keep the peace at the level of state-to-state interactions. Paying attention to slow violence is useful for a geographic approach to peace and conflict, because it considers how multiple, simultaneous geographies can be connected, how policies and spaces are constructed, the impulse and impact of practices of othering, and of inclusivity. The following example highlights how the militarization of conservation can contribute to slow violence.

Rosaleen Duffy (2014) and others (Duffy et al. 2019) have examined the international concern to save charismatic megafauna such as tigers, gorillas, elephants and rhinos in sub-Saharan Africa. International wildlife preservation campaigns often center their messaging against organized, criminal poachers and urge support for militarized campaigns to protect wildlife. These military groups may be state agents, but they are often private companies hired to protect specific wildlife from illegal poaching and from perceived pressures from a growing human population. The very premise of conservation and militarized conservation efforts are a form of colonial enclosure (as discussed previously in this chapter), and they can employ various, military technologies including drones, camera traps, and microchipping. This research enquires into the very concept of poaching, how it is used, and what it hides. These studies find that poaching is often the label assigned to local actors defying state-level, militarized colonization efforts. The category labeled as "poaching" is tied up with global trade markets and international interest in trophy hunting and photography tourism. "Poaching" is also related to how traditional hunting methods were criminalized by colonizing government administrations. Duffy cautions us to think about international messaging on conservation: What is actually being conserved or protected, for whom and how?

This example of militarized wildlife conservation demonstrates multiple geographies related to resources and conflict from international campaigns to local ecosystems, actions, and impacts. It also depicts policies that are rooted in colonial valuation of environmental features as exportable commodities and how those policies for militarized conservation create spaces for certain kinds of use (tourism and trophy hunting versus subsistence), while also creating dynamics of exclusion and slowly unfolding harm for people who live in those places and the ecosystems in those places.

Technological solutions: Peace *and* conflict?

The previous example mentions the use of surveillance technologies in militarized conservation efforts. These technologies make it possible for some groups of people to gain advantage by tracking, monitoring, and managing environmental features through an expanded view of processes on the ground. It would seem that opening up technological ways of seeing and expanding participation to more groups of people would be a good way to reduce conflict and tension over environmental features and minimize forms of environment-related slow violence. Here again, a geographic perspective allows an investigation into how technologies can be used to shape multiple geographies of environmental use and advantage while simultaneously excluding voices or generating harm that is difficult to see. Even what appear to be "global" views of environmental features can quietly reinforce dynamics of uneven benefits from environmental resources. An increasing number of digital, global, environmental monitoring systems present a synoptic, "all at once," view of specific environmental systems and features. The final product appears to be an all-seeing view that suggests an all-knowing understanding of the feature in question. This synoptic, global view, referred to as the "god trick" of seemingly objective science (Harraway 1988, 581), can be intertwined with geographies of unequal environmental and strategic benefit (Adey et al. 2013).

The case of Global Forest Watch, hosted by the World Resources Institute, is an illustrative example of a global scale technology contributing to uneven geographies (Schneider and Olman 2020). The project started as a static atlas of the state of forests in a few African locations where unregulated deforestation was rampant in the late 1990s. By 2014, the project evolved into an interactive platform on Google Maps where anyone could assess the status of forest cover with optional map layers such as roads or political boundaries. It is promoted as a digital service that makes open-source data on the state of forest available to governments, citizens, NGOs and other stakeholders. The platform offers a global view down to a 30 m × 30 m resolution.

On this platform, loss and gain in forest are underrepresented depending on land cover type and age, and the accuracy varies by biome. Temporally, the mapping platform is problematic in its focus on more recent trends of harmful deforestation practices and its silence on colonial-era destructive practices. Since the platform "defines forests based on algorithms applied globally to optical landsat data" (Schneider and Olman 2020, 46), there is a lack of sensitivity to local interpretations of what a forest is: what kinds of land cover, of what age, exotic or native species, etc. As with other forms of cartography, maps generated on this platform produce places through the resolution of data and spatial scale selected to portray or obscure certain features and places.

An objective of the Global Forest Watch project is to support the work of local activists by providing a venue through which they may upload stories or photos of illegal forest harvesting for view by a wider audience. Local activists may also draw on the platform's downscaled data to support a claim or fight a legal battle regarding forest management. However,

> the fact they grounded their actions on global-synoptic views of their environment transfers significant agency back to the transnational agencies that generated those views. This agency tends to manifest in sweeping geoengineering solutions to local environmental problems.
>
> *(Schneider and Olman 2020, 50)*

The technologies integrated in and enabling these transactions may indeed be useful to local activists, but they reinforce the dominant scientific and political hegemony, structures that arguably contribute to conditions of slow violence for people whose livelihoods are threatened by timber poaching and by externally generated "solutions." The actors who build the technology to construct the global view tend to hold the power both to define and to manage the problem at multiple spatial scales.

There has been growing interest in recent years in the concept of food-energy-water networks, and there are good reasons to understand how these life-sustaining processes are intertwined (Andrews-Speed et al. 2014). Indeed, it seems obvious and even compelling that food supplies are connected to water management, and both of those systems are also tied to energy supply and demand. Calls for cross-sector, interdisciplinary collaboration on food-energy-water networks would seem to be a promising shift toward big picture thinking and problem-solving. In their book *Divided Environments: An International Political Ecology of Climate Change, Water and Security* (2022), Jan Selby and his colleagues take a careful and critical look at how these arguments tend to be constructed and by whom. In their research, they find that the "water nexus" narrative and "nexus talk" more broadly emerged from the World Economic Forum which generates ideas and policy strategies. This work is done with the support of powerful corporations such as Coca-Cola, Nestle, and Dow Chemicals who have established track records of environmental depletion and degradation as well inflicting various forms of harm on local populations in places where these companies operate. Their interest in promoting nexus thinking reflects their corporate concerns about maintaining, if not expanding, their operations. Ultimately, nexus thinking tends to promote technology-and market-based approaches to water "problems":

> throughout nexus discourse there is an assumed but uninterrogated water-centrism, according to which it is imagined, for instance, that nations with "a legacy of difficult hydrology have remained poor," that water is "the gossamer that links together the web of food, energy, climate, economic growth and human security." Linked to this, contemporary water challenges are assumed to arise, at root, from scarcity, for instance from the fact that "[t] hirst is now global" and from the "world's vulnerability to the shock of diminishing resources."
>
> *(Selby et al. 2022, 272)*

This type of narrative encourages policies that favor free-market capitalism and government deregulation that enable corporate freedom of activity. Selby et al. call out this "eco-determinist crisis rhetoric" as not only rooted in resource determinism and scarcity thinking, as discussed earlier in this chapter, but as harmfully narrow. This narrative from the policy world, for instance, does not acknowledge that corporate activity is a significant *contributor* to water problems in the places where these corporations operate. Nor does nexus thinking clarify why some factors such as water, energy, and food are essential to these interactions while other factors such as "land, labour, soil, politics, or gender, to name but a few" (Selby et al. 2022, 272) are not part of the discussion. Nexus thinking tends to favor advances and investment in irrigation technologies without taking into account how these technologies may serve exploitative corporate interests that heighten water insecurity for some populations. Expanding the potential for the production, transportation, and consumption of ever more commodities, as this perspective promotes, would seem to promise an economic tide that could lift all boats, yet this enhanced activity tends to reinforce established patterns of structural and hierarchical inequities.

In their extensive empirical research, Selby and his colleagues find, time and again, that water is rarely, if ever, the reason or causal mechanism where depletion, degradation, and vulnerability are advancing. Stepping out of the eco-determinist crisis rhetoric and the water-centrism of nexus thinking, Selby and his colleagues argue that in order to understand water (in)security, it is important to look at how water is related to trade, agricultural production, energy use and circulation, and capital. Similar to the idea of the resource complex discussed earlier in this chapter, the argument here is that water, like other resources, must be understood in terms of how it is not only an environmental feature with multiple uses, but deeply and dynamically connected to economic, political, and cultural systems, values, and practices.

Environmental peacebuilding and climate adaptation

Environmental peacebuilding and climate adaptation are two overlapping topics that warrant consideration in a chapter on geographies of peace and conflict as they relate to environmental features. From the direction of environmental peacebuilding, there is growing attention paid to how to rebuild relationships in conflict-prone areas to support sustainable and collaborative livelihoods in peaceful conditions. Instead of focusing on "environmental triggers of conflict," environmental peacebuilding is an approach to "pinpoint the cooperative triggers of peace that shared environmental problems might make available" (Conca and Dabelko 2002, 5).

Following the end of the Cold War and the preponderance of scholarly work focused on interstate war, attention shifted to other aspects of conflict such as civil war and post-conflict peacebuilding. A growing awareness of potential impacts of a changing climate merged this "broad field of environment, conflict and peace, weaving together a number of related threads that address both environmental risks of conflict and environmental opportunities for peace" (Ide et al. 2021, 2). There has been a surge in scholarly and applied work that reflects "the value of environmental peacebuilding both as an integrative research field and as a practice furthering peace, sustainability and development" (Ide et al. 2021, 16). Environmental peacebuilding, which tends to focus on environmental cooperation to prevent conflict (Carius 2006), is not a one-size-fits-all panacea to conflict. It can be a way to understand how space is socially constructed in ways that encourage groups of people to cooperate or to be confrontational, how resources, territory, and social identities are intertwined, and how peacebuilding processes may be exclusionary for some groups of people (Ide 2017). Environmental peacebuilding, depending on how it is implemented, may have positive effects in terms of environmental protection measures and conflict resolution, but it can also have one or more negative effects including: depoliticization, particularly of technical approaches to problems; displacement of people without their consent; discrimination related to who benefits from environmental peacebuilding efforts; deterioration into conflict rather than conflict de-escalation; delegitimization of the state if local or international actions are perceived as more successful than state efforts at providing public goods; and degradation of the environment if unsustainable political or economic systems persist (Ide et al. 2021).

Recently, as attention is shifting from peacebuilding to sustaining peace, impacts of climate change add new dimensions to this area of work (Hardt and Scheffran 2019). Indeed, Simon Dalby has observed that

If environmental change is a stressor, or threat multiplier, then thinking through how to prevent conflict, or perhaps better still make peace by using environmental cooperation as a tool, would seem to be a promising way to think about constructive policy.

(Dalby 2022, 157)

Yet Dalby cautions that projects touted as peacebuilding may in fact be greenwashed versions of economic development projects:

If what is being secured here is the extension of conventional development projects and the further colonization of vulnerable landscapes, then peacebuilding might be much less than environmentally oriented.... Where uneven access to land, water and commons is involved, complex issues of property, redistribution and equity also have to be considered.

(162)

Climate adaptation efforts are often envisioned at the international or even national scale and may be foisted upon local contexts in a similarly harmful way as economic development projects. There are concerns about how adaption projects, conceived as addressing biosphere-level problems and initiated from state-level interests, have an impact at the point of intervention (Swatuk and Wirkus 2018). The construction of windfarms, solar panel fields, carbon capture and storage facilities, or other efforts to facilitate a global scale reduction in carbon dioxide emissions, for instance, may be "good" for the climate but not necessarily what is wanted by or beneficial to local populations. Michael Mikulewicz's empirical work in the Global South examines how technical solutions in the name of climate adaptation can operate similarly to colonial expansion and economic development projects. Mikulewicz (2020) considers how the discursive violence on the part of agents of development compounds rather than alleviates vulnerability, and he argues for the democratization of local responses to climate change (2018). This is an example of what a focus on climate justice looks like in practice (see Jafry, Helwig and Mikulewicz 2019).

In light of these kinds of concerns, Dalby offers this insight about environmental security in his book *Rethinking Environmental Security* (2022): "It isn't about rival powers, but about vulnerabilities built into the landscapes and infrastructures that feed and supply most human needs" (164).

This reframing of security encourages a shift in focus away from poor and marginalized populations as the potentially destabilizing elements to focusing on elites as major climate change agents. This stance is another rejection of the "eco-determinist crisis rhetoric" that Selby et al. (2022) referred to which perpetuates an unhelpful and imbalanced focus on scarcities as a starting point for understanding. A geographic approach, with attention on multiple spatial scales of human activity and impacts, is a powerful antidote to this anti-geographical thinking.

This discussion on climate security returns us to overarching themes of this book. The very notion that the global climate can somehow be made predictable and secure glosses over a range of disparities and vulnerabilities both in the historical roots of industrialization and fossil fuel-powered capitalism. Referring to the global scale misses the many complex geographies of places and spaces, both human and nonhuman, that have been altered by human activity and the uneven distribution of benefits wrought by these changes. Multiple geographies, multiple scales, and the construction of uneven advantage and disadvantage through the agency of powerful actors are underlying storylines of the changing climate. Arguably, this predominant, current state of the planet is a result not only of the systemic practices of exerting control over territory, but it also reflects a world view that divides humans from our environmental surroundings and life support system.

Another theme of this book has to do with dynamics of empathy instead of othering and inclusivity instead of partitioning. Stepping back from the dominant, Western worldview that has shaped much of the human–environment interactions discussed in this chapter, we can look to other ontologies or world views for a different perspective. Robin Kimmerer (2014) offers an Indigenous perspective that takes a reciprocal view of our relationship to the Earth. Rather than looking at the environment in terms of categories of "natural resources" or "ecosystem services" as if the planet was merely a stockpile for human use, Kimmerer suggests asking the question, "What does the Earth ask of us?" and recognizing the Earth's animacy (more than being an "it" that is easily exploitable). She encourages us to think in terms of reciprocity. She writes:

> I don't believe that we are entering the Anthropocene but that we are living in a transient period of profoundly painful error and correction on our way to a humbler consideration of ourselves. In the geologic scope of things, the Industrial Revolution that fueled the expansion of the exploitative, mechanistic worldview was only an eye blink ago. For eons before that, there was a long time on this planet when humans lived well, in relative homeostasis with biotic processes, embodying a worldview of reciprocity....We are a species that can learn from the global mistakes we are making....We are a species who can change.
>
> *(23)*

This perspective captures another theme of this book: resistance. This resistance, however, does not merely replicate the system that it is challenging. Instead, it invites us to consider other geographies – places, spaces, and relationships – that are possible if we start from a different understanding of our relationship with our immediate and our planetary environments.

References

Adey, P., M. Whitehead and A. J. Williams, eds. 2013. *From above: War, violence and verticality*. Oxford University Press.

Agnew, J. 1994. The territorial trap: The geographical assumptions of international relations theory. *Review of international political economy* 1: 53–80.

Andrews-Speed, P., Bleischwitz, R., Boersma, T., Johnson, C., Kemp, G. and S.D. VanDeveer. 2014. *Want, waste or war?: The global resource nexus and the struggle for land, energy, food, water and minerals*. Routledge.

Appleman, P. 1976. Introduction. In *An essay on the principle of population: Text, sources, and background criticism*, ed. P. Appleman, xi–xxvii. W.W. Norton.

Carius, A. 2006. Environmental peacebuilding. *Environmental Cooperation as an Instrument for Crisis Prevention and Peacebuilding. Conditions for Success. Adelphi Report 3*, no. 7. Accessed November 21, 2022 https://www.adelphi.de/en/system/files/mediathek/bilder/us_503_-_carius_environmental_peacemaking_06-07-02_1.pdf

Conca, K. and Dabelko, G.D. eds. 2002. *Environmental peacemaking*. Woodrow Wilson Center Press.

Dalby, S. 2022. *Rethinking environmental security*. Edward Elgar Publishing.

Davies, T. 2018. Toxic space and time: Slow violence, necropolitics, and petrochemical pollution. *Annals of the American Association of Geographers* 108: 1537–1553.

Davies, T. 2021. Geography, time, and toxic pollution: Slow observation in Louisiana. In *A research agenda for geographies of slow violence*, ed. S. O'Lear: 21–40. Edward Elgar Publishing.

Deudney, D. 1991. Muddled thinking. *Bulletin of the Atomic Scientists* 47: 22–28.

Duffy, R. 2014. Waging a war to save biodiversity: The rise of militarized conservation. *International Affairs* 90: 819–834.

Duffy, R., Massé, F., Smidt, E., Marijnen, E., Büscher, B., Verweijen, J., Ramutsindela, M., Simlai, T., Joanny, L. and E. Lunstrum. 2019. Why we must question the militarisation of conservation. *Biological conservation* 232: 66–73.

Furlong, K. 2006. Hidden theories, troubled waters: International relations, the 'territorial trap', and the Southern African Development Community's transboundary waters. *Political Geography* 25: 438–458.

Galtung, J. 1969. Violence, peace, and peace research. *Journal of Peace Research* 6: 167–191.

Galtung, J. 1990. Cultural violence. *Journal of Peace Research* 27: 291–305.

Gleick, P.H. 1991. Environment and security: The clear connections. *Bulletin of the Atomic Scientists* 47: 16–21.

Haas, P.M. 2002. Constructing environmental conflicts from resource scarcity. *Global Environmental Politics* 2: 1–11.

Haraway, D. 1988. Situated knowledges: The science question in feminism and the privilege of partial perspective. *Feminist Studies* 14: 575–599.

Hardin, G. 1968. The tragedy of the commons: The population problem has no technical solution; it requires a fundamental extension in morality. *Science* 162: 1243–1248.

Hardt, J. N. and J. Scheffran. 2019. Environmental peacebuilding and climate change: Peace and conflict studies at the edge of transformation. *Policy Brief* 68: 1–20. Accessed November 21, 2022. https://toda.org/assets/files/resources/policy-briefs/t-pb-68_hardt-and-scheffran_environmental-peacebuilding-and-climate-change.pdf

Ide, T. 2017. Space, discourse and environmental peacebuilding. *Third World Quarterly* 38: 544–562.

Ide, T., Bruch, C., Carius, A., Conca, K., Dabelko, G.D., Matthew, R. and E. Weinthal. 2021. The past and future (s) of environmental peacebuilding. *International Affairs* 97: 1–16.

Jafry, T., Helwig, K. and M. Mikulewicz. eds. 2019. *Routledge handbook of climate justice*. Routledge.

Kashwan, P. 2020. American environmentalism's racist roots have shaped global thinking about conservation. Accessed November 20, 2022. https://theconversation.com/american-environmentalisms-racist-roots-have-shaped-global-thinking-about-conservation-143783

Kashwan, P. and J. Ribot. 2021. Violent silence: The erasure of history and justice in global climate policy. *Current History* 120: 326–331.

Kimmerer, R. W. 2014. Returning the gift. *Minding Nature* 7: 18–24. Accessed 6 February 2023 https://www.turtlelodge.org/wp-content/uploads/2017/08/Kimmerer-Returning-the-Gift.pdf

Le Billon, P. 2001. The political ecology of war: Natural resources and armed conflicts. *Political Geography* 20: 561–584.

Malthus, Thomas Robert. 1993. An essay on the principle of population (1798). In *An essay on the principle of population: Text, sources, and background criticism*, ed. P. Appleman, 15–129. W.W. Norton.

Meierding, E. 2020. *The oil wars myth: Petroleum and the causes of international conflict*. Cornell University Press.

Mikulewicz, M. 2018. Politicizing vulnerability and adaptation: On the need to democratize local responses to climate impacts in developing countries. *Climate and Development* 10: 18–34.

Mikulewicz, M. 2020. The discursive politics of adaptation to climate change. *Annals of the American Association of Geographers* 110: 1807–1830.

Nixon, R. 2011. *Slow violence and the environmentalism of the poor*. Harvard University Press.

O'Lear, S. 2018. *Environmental geopolitics*. Rowman & Littlefield.

Schneider, B. and L. Olman. 2020. The geopolitics of environmental global mapping services: An analysis of Global Forest watch. In *A research agenda for environmental geopolitics*, ed. S. O'Lear: 44–57. Edward Elgar.

Selby, J., Daoust, G. and C. Hoffmann. 2022. *Divided environments: An international political ecology of climate change, water and security*. Cambridge University Press.

Swatuk, L. and L. Wirkus. eds. 2018. *Water, climate change and the boomerang effect: Unintentional consequences for resource insecurity*. Routledge.

Thomas, K.A. 2017. The river-border complex: A border-integrated approach to transboundary river governance illustrated by the Ganges River and Indo-Bangladeshi border. *Water International* 42: 34–53.

Thomas, K.A. 2021. Enduring infrastructure. In *A research agenda for geographies of slow violence*, ed. S. O'Lear: 107–122. Edward Elgar Publishing.

Watts, M. and N. Peluso. 2014. Resource violence. In *Critical environmental politics*, ed. C. Death, 184–197. Routledge.

Wolf, A.T. 1999. The transboundary freshwater dispute database project. *Water International* 24: 160–163.

Yoffe, S., Wolf, A.T. and M. Giordano. 2003. Conflict and cooperation over international freshwater resources: Indicators of basins at RISR. *JAWRA Journal of the American Water Resources Association* 39: 1109–1126.

14

CONFLICT AND COOPERATION

The adverse effects of climate change

Andrew Linke and Clionadh Raleigh

Nature–society interactions, climate change and conflict

Among policymakers and academics, the term "climate security" has been used to describe a paradigm for understanding the implications of climate change for political instability, including violent conflict. Over eighty countries were designated above the "warning" level of state fragility in 2022, according to the Fragile States Index (more than thirty lie in the more alarming "alert" category) (Fragile States Index 2022). Some of these countries endure large-scale civil wars and intractable insurgencies, while others are sites of low-level armed conflict that claims thousands of lives annually. Engaging with concerns raised in the climate security, recent analysis has shown that of the 855 million households (roughly 1.8 billion people) faced with environmental hazards, "the number of households exposed to climate hazards is about six times greater in conflict-prone areas compared with more peaceful areas" (Läderach et al. 2021, e856). While we do so with a critical eye, discussing these human–environment conflict interactions holds an important place in any evaluation of the global landscape of conflict and peacebuilding.

We can begin by asking, what the term "climate security" means? How robust are these relationships between changing weather patterns – one of the adverse effects of climate change – and organized armed violence? The UN has defined climate security as "recognizing climate change as a 'threat multiplier,' and addressing and coordinating UN efforts on climate related security risks is becoming increasingly important" (United Nations 2020). Climate security perspectives strive to mitigate the social and political instability and social risks that climate change may create.

DOI: 10.4324/9781003345794-14

But, in practice, climate security is a nebulous concept that often means so many things that its dynamics are difficult to pinpoint with the level of detail geographers appreciate. The climate security paradigm also too often focuses on violent conflict as a national security risk without due consideration for the adverse effects of climate change *whether or not conflict erupts*. In a sense, the climate security school is one focused heavily on *in*security. Geographers and others (e.g., also in political and human ecology) have critiqued overly simplistic and often politicized climate security discourses for years. They often stress the influential role of regional political, economic, and other social circumstances as moderating and mediating dimensions of human–environment relationships (e.g., Turner 2004; Raleigh 2010; Linke et al. 2018a). These conditions define a heterogeneous pattern of experiences with the harmful effects of varying weather patterns in agrarian societies. Traditions are also found in Geography from Watt's *Silent Violence* (1983), through a "slow violence" (O'Lear 2021) perspective of how environmental forces affect communities at the interplay of politics, uneven power relationships, and political economy.

As often as the adverse effects of climate change cause hardship, however, these challenges have not transformed into violent conflict. Instead, these economic and social challenges are often met with effective adaptation responses and even forms of cooperation. Stated bluntly, many places experience moderate to severe climate change impacts, including unpredictable weather and extreme meteorological events, but ultimately little or no violence. In Nebraska, USA, a 2012 drought – "the drought of 2012 was a multi-billion dollar agricultural disaster" (Rippy 2015), according to specialists – ruined the regional economy for the season, but nobody formed a rebel army. In contrast, when violence spikes in tumultuous northern Nigeria, to choose another context, dry weather is sometimes blamed (Eichelberger 2014) without critically reflecting on the fundamental (and well-known) differences in social circumstances surrounding such events in two places. Less for journalistic interpretations of contemporary events than for social science theory about human–environment interactions, the distinctions are crucially important.

Conflict remains a rare and complex outcome when and where it occurs, and the weight of any climatological factors in shaping conflict is far from certain (Mach et al. 2019). Something like a "chicken-or-the-egg" problem faces common interpretations of the relationship between unpredictable weather and violence; the greatest predictor of conflict is previous conflict, and the fact that households in regions of a country with armed conflict are six times more likely to experience extreme climate hazards than others presents a dilemma to a traditional climate security framework (Läderach et al. 2021). If most climate security research prioritizes cases where conflict exists (*in*security manifest in violence), the set of observations misses the vast majority of status quo or cooperation outcomes.

Does a status quo or cooperation outcome mean that the adverse effects of climate change are *not concerning*? Absolutely not. Warming projections for Africa, for example, under optimistic scenarios forecast a 2°C increase towards the end of the 21st century, compared to the 20th-century mean annual temperature. Under the more extreme emission scenarios, temperatures could reach between 3°C and 6°C above the same timescale (Niang et al. 2014). Forecasts predict that rainfall will become more unpredictable and droughts more common (Haile et al. 2019), with heavy rainfall also more frequent and intense (Tramblay, Villarini, and Zhang 2020). Higher temperatures result in greater rainfall in some places and precipitation deficits elsewhere (Kendon et al. 2019). Timelines for the beginning and end of growing seasons will be less predictable. Across Least Developed Countries (LDCs) with predominantly agricultural economies, intense rainfall, shifting seasons, and drought negatively affect climate sensitive livelihoods (Niang et al. 2014). Climate change scenarios of >2.5°C temperature rise should reduce gross domestic product (GDP) across Africa and Asia considerably (Plambeck and Hope 1996; Tol 2002), as hazards interact with social vulnerability to compromise production and immediate physical security (Wisner et al. 2004). Negative feedbacks towards other development challenges arise, influencing poor health outcomes and mortality rates from heat stress and malaria (McMichael et al. 2008).

At the local level, communities contend with climate variability in the form of flooding (Lopez-Marrero 2010; Mustafa 1998), protracted drought (Eriksen and Lind 2009; Stringer et al. 2009), and more intensive storms (Karim and Mimura 2008). The focus of our research in these cases should not only be where socio-economic hardships emerge from the damaging impacts of climate change, *but where they do not*. We can build upon our examples above. There are several glaring differences between the examples of Nebraska, USA, and northern Nigeria. In a developed country context, farmers enjoy crop insurance. Generally speaking, there are also accountable, representative, and inclusive political institutions. Perhaps most importantly, contemporary Nebraska lacks large active armed militia organizations living at baseline levels of subsistence, among other conditions *already* ripe for armed raids when rains fail to arrive. Exposure and sensitivity to extreme weather events are driven by underlying and chronic vulnerability, such as poverty and underdevelopment, including poor education and healthcare infrastructure (Mearns and Norton 2010; Wisner et al. 2004). Alternatives to climate-sensitive agrarian livelihoods are becoming more common overall, such as retail, brokerage, transport, labour, and administration, but a growing population means that in absolute terms, many more people remain reliant upon farming and pastoralist livelihoods (Barrett, Ndegwa, and Maggio 2021).

In sum, unpredictable weather and failed harvests affect the political economy of a locality, the politics of land use and livelihood change, migration, local and national government capacity, and other common issues present across all LDC states. However, these shifts are also not unconditionally linked to violent conflict. There is little basis to assume that these changes will *in principle* generate conflict (e.g., violence in Nigeria) when they do not in the vast majority of places and communities (e.g., 2012 Nebraska). Priorities linked with *national security* – threats posed by disenfranchised and marginalized communities of the Global South – tend to disregard ultimately peaceful outcomes in the wake of extreme weather events.

However, an alternative view of climate security weighted towards *human security* – everyday stability for households – allows us greater flexibility for appreciating non-violent outcomes as well. Climate change is reorganizing the circumstances communities navigate and the resources at their disposal. These forces include population relocation, land politics, food security, and acute disaster management responses. Thinking of the prevailing literature as a collection of climate *in*security research, a different climate security perspective would include resilience to unstable social changes that are likely to emerge as environments cannot support agricultural livelihoods. A failure to adapt does not suggest violence is imminent, but it can create vulnerabilities in communities who may rely on ineffective demographic or environmental management and resource access (e.g., diminishing rates of agricultural landownership (Linke and Tollefsen 2021)). But negotiation and peace-making efforts can integrate adaptations and resilience-building during key periods of community transitions.

In the literature about conflict and cooperation adaptations to climate change impacts, we find considerable evidence of agency among those navigating new challenges to their livelihoods. Actors at local, regional, and national scales create rules for natural resource management, share solutions to address scarcity, and enter negotiations to reconcile competing claims among communities. It is admittedly a glass-half-full perspective to suggest that the geographies of climate change impacts have not resulted in catastrophe all the time. But, as we demonstrate, any optimism is not without empirical evidence. Across varied geographical settings, which are defined by different territorial regulations, social institutions, and community networks, violent conflicts are not the sole adaptation to the adverse effects of climate change.

The first section evaluates evidence for central climate security assumption, concluding that unusual and extreme weather resulting from climate change is not an especially robust determinant of conflict risk. The second section is a discussion of community experiences with climate change. Rather than conflict, there is extensive cooperation around climate change impacts, especially at the local level in contexts highly vulnerable to environmental stress. Cooperation is the dominant, constant, and widespread reaction to climate change impacts. Rather than focusing exclusively on violent outcomes, a more comprehensive research agenda on climate security nature and society relationships ought to place greater weights on understanding "secure" outcomes.

Climate insecurity: nature–society interactions and violence?

The social impacts of climate change are often discussed through a national security perspective that over-predicts overt violent conflict as the likely outcome, despite it being a somewhat uncommon occurrence. The classical "climate security" literature presents three areas of concern. First, there are geopolitical threats posed by rising seas, widespread heat waves, routine flooding, and other extreme events. These might disrupt global military capacities, international agreements, and alliances. Here, the distribution of strategic resources, territorial control (e.g., of a melting Arctic and its shipping routes, see Dittmer et al. 2011), and capacities of international law are in flux. International diplomacy and high-level strategic geopolitics in the International Relations traditions dominate in this realm. The second concerns mobilization of state security resources to respond to acute climate change impacts such as droughts and flooding ("slow onset" and "fast onset" hazards, respectively). The principal concern here is the cost of managing damage to infrastructure and agriculture and how this may weaken state capabilities and resources in a manner that reduces political instability.

We focus more specifically on a third concern; that climate change effects lead directly to the emergence and intensification of violence. Early literature in this area often relied on poor evidence related to environmental stress indicators (e.g., national-level annual average temperature in Burke et al. 2009) and modelling or research design decisions (see O'Loughlin et al. 2014). More indirect links consider how climate change can lead to either temporary or permanent migration, economic and development progress, and/or land use and access, which have secondary effects leading towards conflict. But many of these studies report a powerful role for mediating or moderating forces, including intra-community dialogue during migration (e.g., Linke et al. 2018b) or unofficial customary resource use rules (e.g., Linke et al. 2018a).

Highly critical reviews of early large-N studies emerged, including claims that the effect estimates of how temperature led to African Civil wars (see Burke et al. 2009) were fragile and faltered when updated or with variables coded differently (see Buhaug 2010). The lack of a strong relationship in nuanced case study research suggests that "conclusions regarding the climate–conflict relationship are heavily influenced by data and modelling choices" (Koubi 2019, 352). There are also "streetlight effect" issues (Adams et al. 2018), where scholars have focused on convenient cases (e.g., Kenya), rather than where environmental stress is significant and understudied (e.g., Haiti or Malawi).

Sometimes the Syrian War (e.g., see Busby 2018b), the Lake Chad basin, and the pastoral conflicts across Africa are presented as key cases linking climate conditions to conflict. Reflecting on the diversity of conclusions among individual in-depth case studies, such claims are suspicious. When drawing a causal sequence between these events and outcomes, little evidence suggests that the climate conditions that occurred in each case create scarcity or vulnerability that match or exceed the powerful effects of existing conflict conditions found in these locations. Detailed analysis has questioned any evidence of the popular assumptions linking Syrian drought to migration and ultimately war (Ide 2018). More recent work returns consistent evidence for this claim, however (Ash and Obradovich 2019). We stress that migration – which *could* be managed – is the proximate or acute driver of the conflict. Herder–farmer relations in West Africa demonstrate that historical social tensions and power asymmetries are associated with contemporary conflict, rather than short-term natural resource scarcities: "struggles over resources are often only superficially so – they in fact reflect not only broader tensions ... between social groups, but also ... within these groups" (Turner 2004, 866). Further, conflict in the region has emerged because of unequally distributed resource rents extracted by a small elite (e.g., Le Billon 2001; Carment 2003; Bogale and Korf 2007). The geographic perspective emphasizes the importance of ongoing politically fractured contexts. However, they are largely missing in analysis searching for, and ultimately seeking to confirm, conflict as a terminal outcome of climate change.

A direct, robust, and consistent relationship is not found between conflict and temperature, rainfall anomalies, and increased climatic disasters at the state level, in Africa, or globally (Koubi 2019; Van Weezel 2015; Koubi et al. 2012; Bergholt and Lujala 2012). Furthermore, the scarcity narrative is poorly supported and contradicted by *resource abundance*, which can lead to violent conflict frequently (Auty and Gelb 2004; Collier and Hoeffler 2002). Precipitation trends are only weakly correlated with conflict events (Sharifi et al. 2021; Ayana et al. 2016), and others report "inconclusive" relationships between temperature and intergroup conflict (Gleditsch 2021; Bernauer et al. 2012). There is limited, if any, evidence that extreme drought events and competition over water or land resources generate armed conflict (von Uexkuell et al. 2016; Ghimire, Ferreira, and Dorfman 2015; Ambalam 2015; Mertz et al. 2016; Ngaruiya and Scheffran 2016; Sultana et al. 2022). Even in existing conflict contexts – such as the North Kivu region of Congo – no general association is found between drought exposure and support for the use of violence among the population. Armed conflict is a rare phenomenon and taking up arms should not be the default, and not even a frequent, response to a climate-related hazard (von Uexkull et al. 2020).

Rather than direct links, many researchers have sought to determine the influence of climate on intermediate factors on conflict. Changes in income, land use, and migration are partially related to rising intercommunal violence across livelihoods (Sharifi et al. 2021; Schilling et al. 2015; Detges 2017; Okpara et al. 2017). However, these same studies concede that existing political and socio-economic factors mainly contribute to such violence. Multiple experts conclude that "the causal factors that are judged to be most sensitive to climate are ranked as much less influential to the risk of conflict overall," and note,

> for the four conflict drivers that were ranked to be the most influential over-all, experts estimate that the climatic sensitivity of these drivers is relatively low (low socioeconomic development, low capabilities of the state, inter-group inequality and recent history of conflict).
>
> *(Mach et al. 2019, 195)*

Rather than causing conflict, research suggests that widespread environmental disasters often build solidarity and cooperation (also Theisen et al. 2012; Nardulli et al. 2015). The actions of individuals and policymakers alter the magnitude and directionality of an association between disasters and conflict (Brzoska 2018; Busby 2018a), and, thus, political factors and institutions are responsible for the severity of a disaster (Kelman et al. 2016). In turn, disasters may encourage processes of peacebuilding (Brzoska 2018; Schlessnuer et al. 2016).

As the Syria example above illustrates, migration is a contested explanatory mechanism linking dry conditions to instability (Weinthal et al. 2015; Byravan and Rajan 2017). A number of studies find that migrants have not contributed to conflict when populations have dispute resolution resources and mecha-nisms available (Bhavnani and Lacina 2015; Nishimura 2015; Nine 2016; Cattaneo and Bosetti 2017; Sharifi et al. 2021; Linke et al. 2018b). Indeed, the most impoverished – imagine a farmer whose harvests have all recently failed – rarely have the financial resources to relocate and are effectively "trapped" in their plight (Black et al. 2011). Additional justification for a critical reflection on migration effects is that where income is less sensitive to rain shocks – due to irrigation – climate variation and riots still co-occur (Sarsons 2015). Making agricultural production less sensitive to rain shocks does not mitigate risks unrelated to the environment (Koubi 2019).

Most studies concede that spaces highly vulnerable to environmental stress and climate change impacts have many additional conflict-generating vulnera-bilities, including poverty and both physical and institutional isolation. Many criticisms of the natural resource scarcity literature point towards other factors like "poor governance, corruption, institutional instability, and other location-specific and structural conditions as important confounding factors in the re-lationship between resource scarcity and conflict" (Koubi 2019; also Barnett and Adger 2007). For example, a study of twenty-three African countries ar-gues that drought *did not* influence political violence *unless* it intersects with social and political marginalization (Detges 2017). Drought and conflict most often coexist in the presence of agricultural dependency, limited coping strate-gies, and pre-existing tensions (Ide et al. 2014). Arguments that link climate change impacts and conflict in a bivariate and simplified perspective are overly deterministic since it removes violent conflict from its local social and political contexts (Raleigh, Linke, and O'Loughlin 2014).

What additional questions remain about these nature–society relationships? Substantial challenges lie ahead for those determined to further understand this relationship, including investigations of why conflicts are so rare when climate change impacts are ubiquitous. Greater focus upon ongoing adaptation and cooperation engagement with risks would also be valuable (Tache and Oba 2009). We might also seek an understanding of how socio-economic shifts call into question the deterministic nature of any climate–conflict relationships. Examples of such shifts include continuing urbanization, high rates of economic growth (McMillan, Rodrik and Verduzco-Gallo 2014), and increasing productivity in agriculture due to mechanization and technological change (McCullough 2017; Gleditsch 2021).

Adaptation and or cooperation in response to environmental stress

The "climate security" discussion often occurs far from the communities that are on the frontline of climate change. How communities experience climate change is mediated through existing governance – both formal and informal – institutions. Existing governance structures, including resource management institutions, land access and use legislation, agricultural ministry policy, and disaster management responses, are the site where environmental stress is negotiated. Climate change adaptation is leading to new practices and policies, and, as those using a political ecology conceptual framework remind us, these often intensify the political agendas and power relationships already in place in a region. Adaptation measures, however, tend to be designed nationally and are susceptible to the whims of political will and financing priorities.

The United Nations Framework Convention on Climate Change defines adaptation as

> adjustments in ecological, social, or economic systems in response to actual or expected climatic stimuli and their effects or impacts. It refers to changes in processes, practices, and structures to moderate potential damages or to benefit from opportunities associated with climate change.
>
> *(Biagini et al. 2014, 98)*

Further, there are several categories of adaptation actions that are encouraged in response to different climate change, including capacity building, management, and planning within institutions, developing new practices and behaviour, policy and information developments, improvements in physical and green infrastructure, warning or observing systems, and creating financing and technology developments.

Historically, cooperation as a response to climate impact risks and natural resource scarcity has been a more frequent result than conflict. Several studies in sub-Saharan Africa, including Uganda (Hisali et al. 2011), Tanzania (Paavola 2008; Below et al. 2012), South Africa (Thomas et al. 2007), Ethiopia (Deressa et al. 2008), Ghana (Westerhoff and Smit 2009), Mozambique, and South Africa (Osbahr et al. 2010) find that, in response to climate-induced stresses and shocks, households and communities adapt and cooperate in different ways. They sell their assets, withdraw children from school, send children to live elsewhere, migrate, borrow formally and informally, change food consumption habits, take on wage employment, rely on outside help, diversify livelihoods, change farming practices, plant trees, reduce household expenditure, rely on social networks, turn to faith and church groups, and engage in petty trading. Whether these are ultimately "good" or "bad" is a less important consideration than the fact that these are not violent outcomes. There are several existing mechanisms through which people cooperate, including advanced coping mechanisms that underlie safety nets such as nascent crop insurance schemes available to some, training services, improved storage, land tenure, cash, and adaptation programmes for diversification, among others.

Cooperation is mediated through established governance institutions on national, regional, local, and community levels (Sharifi et al. 2021; Rai et al 2017). Resource scarcity stimulates cooperation for the fair distribution of resources before and beyond any conflict mechanism (Ostrom 1990) and "most previous empirical analyses have focused on contexts in which climate variability has led to conflict, rather than resilient, cooperative and peaceful outcomes that are evident in ethnographic studies" (Mach et al. 2019, 196).

The best practices for managing adverse effects are community participation, comprehensive impact analyses, and building conciliatory and trusting relationships among communities, as difficult as this can be where conflicts are ongoing. In short, official (governmental) unofficial (traditional or customary) institutions are forums for dispute management that determine whether climate change impacts produce threats to human security or effective adaptation (Dellmuch et al. 2018; Salifu 2020). Indeed, resource scarcity can offer opportunities for asset-poor households to benefit from local sharing agreements and builds incentives for cooperative solutions (Bogale and Korf 2007).

In Kenya, natural resource regulations and governance arrangements are crucial in solving resource conflicts (Ngaruiya and Scheffran 2016), and citizen relationships determine the vulnerability to extreme climate events (Sharifi et al. 2021). Water scarcity leads groups to adopt cooperative approaches to manage water resources and foster resilience to drought (Abrahams and Carr 2017). Cooperative insurance provides employment opportunities for farmers during periods of drought in India and Kenya (Fetzer 2020); in Somalia (Maystadt and Ecker 2014), livestock markets are the primary channel available to alleviate risk by avoiding sell-offs (e.g., through insurance) that depress prices throughout the sector, driving some towards militant activities.

Recent work has demonstrated that scarcity actually *creates and escalates cooperation and inter-group trust building opportunities* (De Juan and Hänze 2021). Exposure to droughts led to an increase in self-reported trust within and across ethnic groups, conditional on the severity of group hazard exposure, because they understood the common rather than unique (e.g., one ethnic community) risks. Environmental stress can, therefore, reduce the risk of violence as the need for reconciliation presents itself. This innovative research is supplemented by multiple examples of inter-group cooperation from environmental challenges (e.g., Aksoy and Palma 2019). In rural China (Yin et al. 2020) water scarcity led community members to contribute more to a "common pot." Generosity, rather than conflict, predominated in Guatemala during periods of scarcity (Aksoy and Palma 2019). Existing expectations of trust have additional benefits: communities will expect that the first relief will come from each other, and disasters can provide opportunities for de-intensification and termination of collective violence (Broska 2021; see also Eastin 2018; Walch 2018; Skidmore and Toya 2014).

Despite considerable evidence of dispute resolution forces, cooperation is no guarantee or unconditional panacea for human security concerns. While it is the best solution for community-wide issues, it can reproduce local-level and regional power inequities. The power dynamics that underlie local authority are the same which can foster exclusion and/or competition, as well as stifle resource management innovation, which is needed most. The more narrowly community benefits and power structures are, the more likely it will be that "cooperation" (broadly defined) benefits the economically and socially secure.

One of the most important considerations for the adaptation and risk management literature is scale, ranging from the international (e.g., transboundary river treaties) to the local (e.g., a single village). While cooperation at the macro-level is well established in response to scarcity, the institutions at the local level are often overlooked despite developing front-line means to address crises (Schreurs 2008; Linke et al. 2018a; Detges 2017). Formal institutional management and natural resource regulation moderate access, use, and competition for water across and within countries (Tir and Stinnet 2011; Wolf 2007; Tir and Ackerman 2009; D'Exelle, Lecoutere, and Van Campenhout 2010). In addition,

> local official and unofficial rules, as well as an increasing number of such rules during the last decade, eliminate the harmful effects.... Our broadest conclusion is that while environmental change and shortages of rainfall represent a stress for many households' livelihoods, certain regulations may ameliorate these difficulties.
>
> *(Linke et al. 2018a, 1573)*

In the hierarchy of claims over property rights and land access, African communities have a pivotal role in determining the endowment of resources (e.g., control or access to the resource), entitlements (e.g., ownership), and their benefits and utilities (Devereux 1996).

Recent work on indigenous communities found that extensive community knowledge is organized for the preservation and management of natural resources (Filho et al. 2021). The knowledge of communities is vital in the success of adaptative policies: these include biophysical monitoring and weather forecasting, resource governance, traditional social insurance, and safety systems (ILO 2017; Tilahun et al. 2017; Iticha and Husan 2019; Bashiru 2020). However, advancements and extensions to these traditional systems are hampered by the access to resources and technologies (Filho et al. 2021).

Across Kenya, unofficial (non-governmental) conflict resolution networks facilitate negotiation, cooperation, dialogue, and possible resolution of contentious politics (Ngaruiya and Scheffran 2016). Clearly, understanding how to adapt and mitigate climate change through governing collective access to natural resources by nonstate actors is important (e.g., as in Ostrom 1990, McCabe 1990, and more recently Salifu 2020). Across dryland West Africa, "the mediation and management of conflicts as they arise in communities is one area in which various members of the community play important roles as shaped by community norms and obligations" (Turner et al. 2012, 342). In Ethiopia, relations between resource users under conditions of environmental scarcity and political instability suggest the crucial role of local institutions in governing competing resource claims. "In the violence-prone Somali Region… agro-pastoralist communities develop sharing arrangements on pasture resources with intruding pastoralist communities in drought years, even though this places additional pressure on their grazing resource" (Bogale and Korf 2007, 743). In southern Ethiopia, pasture management is a key element in fostering cooperation and dealing with increased heterogeneity among community members (McCarthy et al. 2001).

Evidence of effectiveness is no assurance of availability, however, and failures of governance systems can engender insecurity once communities have come to rely upon them (Kassahun, Snyman, and Smit 2008; Benjaminson and Lund 2003; Tache and Oba 2009). Where these institutions are not supported or disappear, management and entitlements can become a zero-sum game, and benefits can accrue to those in pre-existing authority positions, or to those who usurp that authority in a political vacuum.

In charting how to best accommodate the communities facing uncertain and overwhelming livelihood risks as a result of climate change, it is important to recall some lessons from the "environmental peace-making" agenda (Conca and Beevers 2018). These challenge the assumption that environmental degradation and natural resource scarcities lead to, or are drivers of, violent conflict. Through a range of mechanisms, including the assessment and reduction of grievances, the identification of joint human security gains, and forming new institutions and practices, communities can navigate increased conflict risk co-operatively and the environment is the key ground for cooperation and coordination (Conca and Dabelko 2002).

Conclusions

The classical "climate security" (i.e., national security) perspectives for understanding climate change impacts for armed conflict and insecurity at a national scale have largely overlooked the activities and agency of local-scale customary or official institutions to facilitate cooperation and effective adaptation under certain circumstances. Such a view ignores multiple regional and local social, economic, and political geographies in favour of meta-narratives of disorder and a chaotic geography of inevitable violence. The dominant perspective of climate security treats the climate as the first and foremost stressor that leads to instability, rather than one of the many issues affected by long-standing legacies of conflict and unaccountable national political institutions that dominate in much of the Global South. Traditionally, the climate security narrative also contributes to a militarized framing of climate change impacts, rather than a humanitarian lens where accommodations and effective institutional responses may intervene to dampen the risk of violent conflict.

These oversights have a fatalistic effect. Such a perspective places little value on communities' experiences and agency; it obscures from view the many opportunities that we know exist for effective climate change adaptation. Assuming that intervening to break the link between environmental stress and community hardships is futile, because conflict is inevitable, means abandoning the disenfranchised and often marginalized households that must navigate the dire impacts of a changing environment. Climate change creates highly variable risks that are difficult to predict. However, in practice, climate change adaptation policies are also mediated through existing resource management institutions, distribution and access constraints, and entrenched socio-economic power relationships. These institutions are designed to contend with disaster management, access to markets for crops, seasonal shifts in grassland or water resources, and increasingly irregular seasons. It is the regional distribution and availability of these mechanisms for adaptation and dispute settlement that deserve greater attention in a literature turning away from national security priorities in the study of human–environment interactions and conflict geographies.

References

Abrahams, D., and E.R. Carr. 2017. Understanding the connections between climate change and conflict: Contributions from geography and political ecology. *Current Climate Change Reports* 3: 233–242.

Adams, C., T. Ide, J. Barnett, and A. Detges. 2018. Sampling bias in climate-conflict research. *Nature Climate Change* 8: 200–203.

Aksoy, B., and M. A. Palma. 2019. The effects of scarcity on cheating and in-group favoritism. *Journal of Economic Behavior & Organization* 165: 100–117.

Ambalam, K. 2015. Security governance and climate change: A non-military perspective in African context. *Journal of Climate Change* 1: 109–118.

Ash, K., and N. Obradovich. 2019. Climate stress, internal migration, and Syrian civil war onset. *Journal of Conflict Resolution* 64: 3–31.

Auty, R. M., and A. H. Gelb. 2004. Political economy of resource-abundant states. In *Resource Abundance and Economic Development*, eds. R. M. Auty and A. H. Gelb, 126–144. Oxford University Press.

Ayana, E.K., P. Ceccato, J.R.B. Fisher, and R. DeFries. 2016. Examining the relationship between environmental factors and conflict in pastoralists areas of East Africa. *Science of the Total Environment* 558: 601–611.

Barnett, J., and W. N. Adger. 2007. Climate change, human security and violent conflict. *Political Geography* 26: 639–655.

Barrett, S., W. Ndegwa, and G. Maggio. 2021. The value of local climate and weather information: An economic valuation of the decentralised meteorological provision in Kenya. *Climate and Development* 13: 173–188.

Bashiru, S. 2020. Climate change impact and traditional coping mechanisms of Borana Pastoralists in Southern Ethiopia. Building Adaptive Capacity and Resilience from an Indigenous People's Perspective.

Below, T. B., K. D. Mutabazi, D. Kirschke, C. Franke, S. Sieber, R. Siebert, and K. Tscherning. 2012. Can farmers' adaptation to climate change be explained by socio-economic household-level variables? *Global Environmental Change* 22: 223–235.

Benjaminson, T. A., and C. Lund. 2003. *Securing Land Rights in Africa*. Frank Cass.

Bergholt, D, and P. Lujala. 2012. Climate-related natural disasters, economic growth, and armed civil conflict. *Journal of Peace Research* 49: 147–162.

Bernauer, T, T. Böhmelt, and V. Koubi. 2012. Environmental changes and violent conflict. *Environmental Research Letters* 7: 015601.

Bhavnani, R. R., and B. Lacina. 2015. The effects of weather-induced migration on Sons of the Soil riots in India. *World Politics* 67: 760–794.

Biagini, B., R. Bierbaum, M. Stults, S. Dobardzic, and S. M. McNeeley. 2014. A typology of adaptation actions: A global look at climate adaptation actions financed through the global environment facility. *Global Environmental Change* 25: 97–108.

Black, R., S. R. G. Bennet, S. M. Thomas, and J. R. Beddington. 2011. Migration as adaptation. *Nature* 478: 447–449.

Bogale, A., and B. Korf. 2007. To share or not to share? (Non-) violence, scarcity and resource access in Somali Region, Ethiopia. *The Journal of Development Studies* 43: 743–765.

Broska, L. H., 2021. It's all about community: On the interplay of social capital, social needs, and environmental concern in sustainable community action. *Energy Research & Social Science* 79: 102165.

Brzoska, M., 2018. Weather extremes, disasters, and collective violence: Conditions, mechanisms, and disaster-related policies in recent research. *Current Climate Change Reports* 4: 320–329.

Buhaug, H. 2010. Climate not to blame for African civil wars. *Proceedings of the National Academy of Sciences* 107: 16477–16482.

Burke, M. B., E. Miguel, S. Satyanath, J. A. Dykema, and D. B. Lobell. 2009. Warming increases the risk of civil war in Africa. *Proceedings of the National Academy of Sciences* 106: 20670–20674.

Busby, J. 2018a. Warming world: Why climate change matters more than anything else. *Foreign Affairs* 97: 49.

Busby, J. 2018b. Taking stock: The field of climate and security. *Current Climate Change Reports* 4: 338–346.

Byravan, S., and S. C. Rajan. 2017. Taking lessons from refugees in Europe to prepare for climate migrants and exiles. *Environmental Justice* 10: 108–111.

Carment, D. 2003. Assessing state failure: Implications for theory and policy. *Third World Quarterly* 24: 407–427.

Cattaneo, C., and V. Bosetti. 2017. Climate-induced international migration and conflicts. *CESifo Economic Studies* 63: 500–528.

Collier, P., and A. Hoeffler. 2002. On the incidence of civil war in Africa. *Journal of Conflict Resolution* 46: 13–28.

Conca, K., and M. Beevers. 2018. "Environmental Pathways to Peace." In *Routledge Handbook of Environmental Conflict and Peacebuilding*, eds. A. Swain and J. Öjendal, 54–72. Routledge.

Conca, K., and G. Dabelko. 2002. *Environmental Peacemaking*. The Johns Hopkins University Press.

De Juan, A., and N. Hänze. 2021. Climate and cohesion: The effects of droughts on intra-ethnic and inter-ethnic trust. *Journal of Peace Research* 58: 151–167.

Dellmuth, L. M., M.-T. Gustafsson, N. Bremberg, and M. Mobjörk. 2018. Intergovernmental organizations and climate security: Advancing the research agenda. *Review of Climate Change* 9: 1–13.

Deressa, T., R. M. Hassan, and C. Ringler. 2008. Measuring Ethiopian farmers' vulnerability to climate change across regional states. *International Food Policy Research Institute*. Working paper No. 15-5. Available: https://ebrary.ifpri.org/utils/getfile/collection/p15738coll2/id/13894/filename/13895.pdf

Detges, A. 2017. Droughts, state-citizen relations and support for political violence in Sub-Saharan Africa: A micro-level analysis. *Political Geography* 61: 88–98.

Devereux, S. 1996. Fuzzy entitlements and common property resources: Struggles over rights to communal land in Namibia. IDS Working Paper No. 44, Institute of Development Studies.

D'Exelle, B., E. Lecoutere, and B. Van Campenhout. 2010. *Who Engages in Water Scarcity Conflicts? A Field Experiment with Irrigators in Semi-arid Africa*. Working Paper. Microcon. https://ueaeprints.uea.ac.uk/id/eprint/27374/

Dittmer, J., Moisio, S., Ingram, A., and K. Dodds. 2011. Have you heard the one about the disappearing ice? Recasting Arctic geopolitics. *Political Geography* 30: 202–214.

Eastin, J. 2018. Hell and high water: Precipitation shocks and conflict violence in the Philippines. *Political Geography* 63:116–134.

Eichelberger, E. 2014. Drought, population explosion, and poverty are aggravating conflict in Nigeria. Climate change will likely add fuel to the fire. *Mother Jones* 10: https://www.motherjones.com/environment/2014/06/nigeria-environment-climate-change-boko-haram/

Eriksen, S., and J. Lind. 2009. Adaptation as a political process: Adjusting to drought and conflict in Kenya's drylands. *Environmental Management* 43: 817–835.

Fetzer, T. 2020. Can workfare programs moderate conflict? Evidence from India. *Journal of the European Economic Association* 18: 3337–3375.

Filho, W. L., N. R. Matandirotya, J. M. Lutz, E. Abate Alemu, F. Q. Brearley, A. A. Baidoo, A. Kateka, G. M. Ogendi, G. B. Adane, N. Emiru, and R. A. Mbih. 2021. Impacts of climate change to African indigenous communities and examples of adaptation responses. *Nature Communications* 12:6224.

Fragile States Index. 2022. Measuring fragility: Risk and vulnerability in 179 countries. *The Fund for Peace.* https://fragilestatesindex.org/

Ghimire, R., S. Ferreira, and J. Dorfman. 2015. Flood-induced displacement and civil conflict. *World Development* 66: 614–628.

Gleditsch, N. P., 2021. This time is different! Or is it? NeoMalthusians and environmental optimists in the age of climate change. *Journal of Peace Research* 58: 177–185.

Haile, G. G., Q. Tang, S. Sun, Z. Huang, X. Zhang, and X. Liu. 2019. Droughts in East Africa: Causes, impacts and resilience. *Earth-Science Reviews* 19: 146–161.

Hisali, E., P. Birungi, and F. Buyinza. 2011. Adaptation to climate change in Uganda: Evidence from micro level data. *Global Environmental Change* 21: 1245–1261.

Ide, T. 2018. Climate war in the Middle East? Drought, the Syrian civil war and the state of climate-conflict research. *Current Climate Change Reports* 4:347–354.

Ide, T., J. Schilling, J. S. A. Link, J. Scheffran, G. Ngaruiya, and T. Weinzierl. 2014. On exposure, vulnerability and violence: Spatial distribution of risk factors for climate change and violent conflict across Kenya and Uganda. *Political Geography* 43:68–81.

ILO. 2017. *Indigenous Peoples and Climate Change: From Victims to Change Agents Through Decent Work.* International Labour Office, Gender, Equality and Diversity Branch.

Iticha, B., and A. Husan. 2019. Adaptation to climate change using indigenous weather forecasting systems in Borana pastoralists of southern Ethiopia. *Climate and Development* 11: 564–573.

Karim, M. F., and N. Mimura. 2008. Impacts of climate change and sea-level rise on cyclonic storm surge floods in Bangladesh. *Global Environmental Change* 18: 490–500.

Kassahun, A., H. A. Snyman, and G. N. Smit. 2008. Impact of rangeland degradation on the pastoral production systems, livelihoods and perceptions of the Somali pastoralists in Eastern Ethiopia. *Journal of Arid Environments* 72: 1265–1281.

Kelman, I., J. C. Gaillard, J. Lewis, and J. Mercer. 2016. Learning from the history of disaster vulnerability and resilience research and practice for climate change. *Natural Hazards* 82: 129–143.

Kendon, E. J., R. A. Stratton, S. Tucker, J. H. Marsham, S. Berthou, D. P. Rowell, and C. A. Senior. 2019. Enhanced future changes in wet and dry extremes over Africa at convection-permitting scale. *Nature Communications* 10: 1–14.

Koubi, V. 2019. Climate change and conflict. *Annual Review of Political Science* 22: 343–360.

Koubi, V., T. Bernauer, A. Kalbhenn, and G. Spilker. 2012. Climate variability, economic growth, and conflict. *Journal of Peace Research* 49: 113–127.

Läderach, P., J. Ramirez-Villegas, G. Caroli, C. Sadoff, and G. Pacillo. 2021. Climate finance and peace-tackling the climate and humanitarian crisis. *The Lancet* 5: E856–E858.

Le Billon, P. 2001. Angola's political economy of war: The role of oil and diamonds, 1975–2000. *African Affairs* 100: 55–80.

Linke, A. M., and A. F. Tollefsen. 2021. Environmental stress and agricultural land-ownership in Africa. *Global Environmental Change* 67: 102237.

Linke, A. M., F. D. W. Witmer, J. O'Loughlin, J. T. McCabe, and J. Tir. 2018a. Drought, local institutional contexts, and support for violence in Kenya. *Journal of Conflict Resolution* 62: 1544–1578.

Linke, A. M., F. D. W. Witmer, J. O'Loughlin, J. T. McCabe, and J. Tir. 2018b. The consequences of relocating in response to drought: Human mobility and conflict in contemporary Kenya. *Environmental Research Letters* 13: 094014.

Lopez-Marrero, T. 2010. An integrative approach to study and promote natural hazards adaptive capacity: A case study of two flood-prone communities in Puerto Rico. *Geographical Journal* 176: 150–163.

Mach, K. J., C. M. Kraan, W. N. Adger, H. Buhaug, M. Burke, J. D. Fearon, C. B. Field, C. S. Hendrix, J.-F. Maystadt, J. O'Loughlin, and P. Roessler. 2019. Climate as a risk factor for armed conflict. *Nature* 571: 193–197.

Maystadt, J.-F., and O. Ecker. 2014. Extreme weather and civil war: Does drought fuel conflict in Somalia through livestock price shocks? *American Journal of Agricultural Economics* 96: 1157–1182.

McCabe, J. T. 1990. Turkana pastoralism: A case against the tragedy of the commons. *Human Ecology* 18: 81–103.

McCarthy, J. J., O. F. Canziani, N. A. Leary, D. J. Dokken, and K. S. White, eds. 2001. *Climate change 2001: Impacts, adaptation, and vulnerability: Contribution of Working Group II to the third assessment report of the Intergovernmental Panel on Climate Change* (Vol. 2). Cambridge University Press.

McCullough, A. 2017. Environmental impact assessments in developing countries: We need to talk about politics. *The Extractive Industries and Society* 4: 448–452.

McMichael, A. J., S. Friel, A. Nyong, and C. Corvalan. 2008. Global environmental change and health: Impacts, inequalities, and the health sector. *British Medical Journal* 336: 191–194.

McMillan, M., D. Rodrik, and Í. Verduzco-Gallo. 2014. Globalization, structural change, and productivity growth, with an update on Africa. *World Development* 63: 11–32.

Mearns, R., and A. Norton. 2010. *Social Dimensions of Climate Change: Equity and Vulnerability in a Warming World*. World Bank.

Mertz, O., K. Rasmussen, and L. V. Rasmussen. 2016. Weather and resource information as tools for dealing with farmer–pastoralist conflicts in the Sahel. *Earth System Dynamics* 7: 969–976.

Mustafa, D. 1998. Structural causes of vulnerability to flood hazard in Pakistan. *Economic Geography* 74: 289–305.

Nardulli, P. F., B. Peyton, and J. Bajjalieh. 2015. Climate change and civil unrest: The impact of rapid-onset disasters. *Journal of Conflict Resolution* 59: 310–335.

Ngaruiya, G. W. and J. Scheffran. 2016. Actors and networks in resource conflict resolution under climate change in rural Kenya. *Earth System Dynamics* 7: 441–452.

Niang, I., O. C. Ruppel, M. A. Abdrabo, A. Essel, C. Lennard, J. Padgham, and P. Urquhart. 2014. Africa. In *Climate Change 2014: Impacts, Adaptation, and Vulnerability. Part B: Regional Aspects. Contribution of Working Group II to the Fifth Assessment Report of the Intergovernmental Panel on Climate Change*, eds. V. R. Barros, C. B. Field, D. J. Dokken, M. D. Mastrandrea, K. J. Mach, T. E. Bilir, M. Chatterjee, K. L. Ebi, Y. O. Estrada, R. C. Genova, B. Girma, E. S. Kissel, A. N. Levy, S.

MacCracken, P. R. Mastrandrea, and L. L. White, 1199–1265. Cambridge University Press.

Nine, C. 2016. Water crisis adaptation: Defending a strong right against displacement from the home. *Res Publica* 22: 37–52.

Nishimura, L. 2015. 'Climate change migrants': Impediments to a protection framework and the need to incorporate migration into climate change adaptation strategies. *International Journal of Refugee Law* 27: 107–134.

O'Lear, S., ed. 2021. *A Research Agenda for Geographies of Slow Violence*. Edward Elgar.

O'Loughlin, J., A. M. Linke, F. D. W. Witmer. 2014. Modeling and data choices sway conclusions about climate-conflict links. *Proceedings of the National Academy of Sciences* 111: 2054–2055.

Okpara, U.T., L.C. Stringer, and A.J. Dougill. 2017. Using a novel climate–water conflict vulnerability index to capture double exposures in Lake Chad. *Regional Environmental Change* 17: 351–366.

Osbahr, H., C. Twyman, W. N. Adger, and D. S. Thomas. 2010. Evaluating successful livelihood adaptation to climate variability and change in southern Africa. *Ecology and Society* 15: 27.

Ostrom, E. 1990. *Governing the Commons*. Cambridge University Press.

Paavola, J. 2008. Science and social justice in the governance of adaptation to climate change. *Environmental Politics* 17: 644–659.

Plambeck, E.L., and C. Hope. 1996. An updated valuation of the impacts of global warming. *Energy Policy* 24: 783–793.

Raleigh, C. 2010. Political marginalization, climate change, and conflict in African Sahel states. *International Studies Review* 12: 69–86.

Raleigh, C., A. M. Linke, and J. O'Loughlin. 2014. Extreme temperatures and violence. *Nature Climate Change* 4: 76–77.

Rippy, B.R. 2015. The U.S. drought of 2012. *Weather and Climate Extremes* 10: 57–64.

Salifu, B. 2020. Climate change impacts and traditional coping mechanisms of Borana pastoralists in southern Ethiopia. Building adaptive capacity and resilience from an Indigenous people's perspective. Masters Thesis, University of Tromso, Norway. Available: https://munin.uit.no/bitstream/handle/10037/18690/thesis.pdf?sequence=2&isAllowed=y

Sarsons, H. 2015. Rainfall and conflict: A cautionary tale. *Journal of Development Economics* 115: 62–72.

Schilling, J., R. Locham, T. Weinzierl, J. Vivekananda, and J. Scheffran. 2015. The nexus of oil, conflict, and climate change vulnerability of pastoral communities in northwest Kenya. *Earth System Dynamics* 6: 703–717.

Schleussner, C. F., J. F. Donges, R. V. Donner, and H. J. Schellnhuber. 2016. Armed-conflict risks enhanced by climate- related disasters in ethnically fractionalized countries. *Proceedings of the National Academy of Sciences* 113: 9216–9221.

Schreurs, M. A., 2008. From the bottom up: Local and subnational climate change politics. *The Journal of Environment & Development* 17: 343–355.

Sharifi, A., D. Simangan, C. Y. Lee, S. R. C. Reyes, T. Katramiz, J. C. C. Josol, L. dos Muchangos, H. Virji, S. Kaneko, T. K. Tandog, and T. Tandog. 2021. Climate-induced stressors to peace: A review of recent literature. *Environmental Research Letters* 16: 073006.

Skidmore, M., and H. Toya. 2014. Do natural disasters enhance societal trust? *Kyklos* 67: 255–279.

Stringer, L. C., J. C. Dyer, M. S. Reed, A. J. Dougill, C. Twyman, and D. Mkwambisi. 2009. Adaptations to climate change, drought and desertification: Local insights to enhance policy in southern Africa. *Environmental Science & Policy* 12: 748–765.

Sultana, N., M. M. Rahman, and R. Khanam. 2022. Environmental kuznets curve and causal links between environmental degradation and selected socioeconomic indicators in Bangladesh. *Environment, Development and Sustainability* 24: 5426–5450.

Tache, B., and G. Oba. 2009. Policy-driven inter-ethnic conflicts in Southern Ethiopia. *Review of African Political Economy* 36: 409–426.

Theisen, O. M., H. Holtermann, and H. Buhaug. 2012. Climate wars? Assessing the claim that drought breeds conflict. *International Security* 36:79–106.

Thomas, D. S., C. Twyman, H. Osbahr, and B. Hewitson. 2007. Adaptation to climate change and variability: Farmer responses to intra-seasonal precipitation trends in South Africa. *Climatic Change* 83: 301–322.

Tilahun, M., A. Angassa, and A. Abebe. 2017. Community based knowledge towards rangeland condition, climate change, and adaptation strategies: The case of Afar pastoralists. *Ecology Process* 6: 29.

Tir, J. and J. T. Ackerman. 2009. Politics of formalized river cooperation. *Journal of Peace Research* 46: 623–640.

Tir, J., and D. M. Stinnett. 2011. The institutional design of riparian treaties: The role of river issues. *Journal of Conflict Resolution* 55: 606–631.

Tol, R. S., 2002. Estimates of the damage costs of climate change. Part 1: Benchmark estimates. *Environmental and Resource Economics* 21: 47–73.

Tramblay, Y., G. Villarini, and W. Zhang. 2020. Observed changes in flood hazard in Africa. *Environmental Research Letters* 15: 1–9.

Turner, M. D. 2004. Political ecology and the moral dimensions of 'resource conflicts': The case of farmer-herder conflict in the Sahel. *Political Geography* 23: 863–889.

Turner, M. D., A. A. Ayantunde, K. P. Patterson, and E. D. Patterson III. 2012. Conflict management, decentralization and agropastoralism in dryland West Africa. *World Development* 40: 745–757.

United Nations. 2020. *Climate Security Mechanism: Toolbox*. New York. Available: https://dppa.un.org/sites/default/files/csm_toolbox-1-briefing_note.pdf

Van Weezel, S. 2015. Economic shocks and civil conflict onset in sub-Saharan Africa, 1981–2010. *Defense and Peace* Economics 26:153–177.

von Uexkull, N., M. Coicu, H. Fjelde, and H. Buhaug. 2016. Civil conflict sensitivity to growing season drought. *Proceedings of the National Academy of Sciences* 113: 12391–12396.

von Uexkull, N., M. d'Errico, and J. Jackson. 2020. Drought, resilience, and support for violence: Household survey evidence from DR Congo. *Journal of Conflict Resolution* 64: 1994–2021.

Walch, C., 2018. Disaster risk reduction amidst armed conflict: Informal institutions, rebel groups, and wartime political orders. *Disasters* 42: S239–S264.

Watts, M. 1983. *Silent Violence: Food, Famine, and Peasantry in Northern Nigeria*. University of Georgia Press.

Weinthal, E., N. Zawahri, and J. Sowers. 2015. Securitizing water, climate, and migration in Israel, Jordan, and Syria. *International Environmental Agreements: Politics, Law and Economics* 15: 293–307.

Westerhoff, L. and B. Smit. 2009. The rains are disappointing us: Dynamic vulnerability and adaptation to multiple stressors in the Afram Plains, Ghana. *Mitigation and Adaptation Strategies for Global Change* 14: 317–337.

Wisner, B., P. Blaikie, T. Cannon, and I. Davis. 2004. *At Risk: Natural Hazards, People's Vulnerability and Disasters*. Routledge.

Wolf, A. T., 2007. Shared waters: Conflict and cooperation. *Annual Review of Environmental Resources* 32: 241–269.

Yin, Y., L. Wang, Z. Wang, Q. Tang, P. Shilong, C. Deliang, X. Jun, T. Conradt, L. Junguo, W. Yoshihide, C. Ximing, Z. Xie, Q. Duan, X. Li, J. Zhou, and J. Zhang. 2020. Quantifying water scarcity in northern China within the context of climatic and societal changes and south-to-North water diversion. *Earth's Future* 8: 1–17

15

PLACING PEACE

The pedagogies of positive peace and environmental justice

Mark Ortiz, María Belén Noroña, Lorraine Dowler and Joshua Inwood

Introduction

> Radical geographers and critical education scholars of various theoretical traditions have moved beyond such a treatment of the classroom as a mere stage on which teaching and learning take place. Instead, they insist that the classroom is embedded in and shaped by institutional, governmental, and interpersonal power geometries. It constitutes one site out of many where various systems of power—e.g., white supremacy, cisheteropatriarchy, settler colonialism, and capitalism—work in and through those who participate in the classroom. Radical approaches to the classroom must thus account for the socio-spatialities of power, violence, and inequality that characterize its political geographies.
>
> (Catungal 2019, 45)

As the opening quote suggests, the classroom is a space where power entangles with unique and distinctive ways of understanding social life. Political geographers engage with uneven power relationships when guiding students through actual threats to their survival, the survival of their communities, and their planet. As educators, we recognize that vulnerability is "shaped by the intersecting dynamics of racialization, gender, sexuality, ability, and other axes of difference, as well as by the topic-or issue-based content of our courses" (McLeod et al. 2021, 36). Unlearning racialized and gendered understandings of the world through the lens of power can be "unsettling" for our students. As a discipline, we must uniformly engage in anti-racist and decolonial pedagogies and not delegate this responsibility to "Indigenous, people of colour, women and queer faculty" (Daigle and Sundberg 2017, 340). Crucial to this approach

DOI: 10.4324/9781003345794-15

is the need to learn the "nature of our contemporary predicament and the means by which we might collectively engage in resistance that would transform our reality" (hooks 1994, 67).

As a point of entry to the discussion and study of justice, this chapter's intellectual and pedagogical contributions are located at the intersection of two essential and interrelated literatures. The first flows from work in environmental education and centers around the climate crisis that threatens the ability of the planet to continue to support and maintain conditions necessary for life (Harvester and Blenkinsop 2010). The second, driven by social, economic, and political conflict, focuses on the research and teaching of peace: Specifically how we might foster the conditions necessary for the de-escalation of conflict, a more just and sustainable sharing of the social surplus, and in creating the conditions necessary for humans to live in communities that sustain the conditions needed for life.

Each of these literatures contributes to a broad-based critique of our modern world and, perhaps most importantly, offers paths toward creating a more just society. Randall Amster notes that when examining the interrelationships between these two fields of study, there is a kind of "proactive sensibility" in each of these approaches that goes beyond simple negative approaches or critiques of existing structural conditions. Instead, they each draw from approaches focusing on existing structural inequalities and how those structural conditions might be supplanted through practices and values promoting peace and justice. Amster goes on to note, "[s]uch invocations of 'positive peace' have defined the field for decades, and perhaps nowhere are they more germane than around contemporary issues of climate change" and environmental justice (Amster 2013, 473).

As a starting point for this discussion, we consider how our work around environmentalism, activism, and peace informs our pedagogical and intellectual approaches in the classroom. Specifically, through the work of Amster (2013) and others, we build the concept of positive peace to advocate for an approach to geography education that focuses on a set of engagements that foster and promote positive peace. A critical intervention in the peace literature focuses on the distinction between negative peace, which often functions in the *prevention* of violence, and positive peace, which are actions that encourage the development of knowledge and institutions that address the structural conditions that make conflict likely (Inwood and Tyner 2011). A crucial site for taking up the challenge of positive peace work is in the classroom; specifically, pedagogy informed through an approach that contributes to conditions that positively promote peace.

This approach intersects with the conception of negative and positive peace and through a set of Black feminist practices and thoughts that have long animated our collective scholarship. For example, the tension between "thinking" and "doing" in peace education approaches aligns with the question of what steps and actions can be individually and collectively taken to build a more peaceful and just world (see Hill Collins 1986). The same questions align with work in environmental pedagogy that also asks how we can move toward a more proactive and engaged scholarship that provides students with skills to recognize their fundamental role in the broader struggles for justice.

Taking these realizations as a starting point, we have framed this essay around pedagogies of peace. We take our cue from peace geographers who caution against utopian notions of peace and instead understand peace as a process that "can also be deepened by looking at how it is, and is not, inclusive of difference" (Koopman 2011a, 193). Therefore, just as we understand peace as having different meanings for different groups in different places (see Koopman 2011b; Loyd 2012; Williams 2015), we also view environmental education as being similarly positioned (see Greenwood 2010).

Bajaj and Chiu (2009, 441) argue that there is a deep relationship between pedagogy in peace studies and environmental education, as a dialogue "toward greater equity and social justice." We suggest that this relationship goes beyond this exchange to also engage in a deep and sustained commitment to taking the placeness of peace and environmental education approaches seriously. As positive peace approaches are focused on divining "opportunities where crises persist" (Amster 2013, 476), we can use these moments of disruption to leverage pro-peace pedagogical engagements that offer ways to transform our current realities.

Within this context, it is increasingly important to focus on militarization and its connection to the broader global climate crisis. In doing so, we move from a negative approach focused on critiquing militarism to one grounded in an understanding of positive peace and the potential to transform these current realities into productive learning moments. The reality is that activities that reproduce militarized knowledge and violent ontologies are related to the very extractive technologies that have come to define the global climate crisis. Greenwood (2008) notes that this style of knowledge production found in many university classrooms and beyond focuses on fear and loss. This way of thinking connects with the presumption that the U.S. is losing its position in the global hierarchy, resulting from challenges posed by the rise of foreign adversaries that will eventually dominate the U.S. as a global hegemon. Positive approaches to peace stand in opposition to this thinking. It provides a robust counter to the vision of a world defined through global geopolitical competition. Instead, it asks the most fundamental of questions: *what world do we want to live in, and how can we collectively work to make it happen?* More simply stated, to supplant war and militarism while seriously addressing threats posed from extractive practices, we must address negative understandings of peace. For example, a broader understanding of war culture and militarization allows for a reimagination of the worlds we can create through pedagogies grounded in positive peace perspectives.

In the next section of this paper, we introduce a series of teaching vignettes to articulate the context for creative and critical thinking around positive peace, climate change, and militarization. In the first vignette, Joshua Inwood explores how militarization not only relates to the size and scale of arms and armies but also to a range of identity positions and ways we organize society around violence. An approach grounded in positive peace makes visible militarism's role in perpetuating global climate change and assists students in understanding how positive peace can be transformative to their daily engagements. In the second vignette, Mark Ortiz explores the production of knowledge in transnational youth climate movements, as they embody inspiring models of public pedagogy. Since these movements are founded and led by young people similar in age to students in our classrooms, and since activism takes place across many of the social media platforms that our students use daily, students often identify with these movements' concerns. Thus, diverse students see themselves reflected in the activism and leadership of these networks. In the third vignette María Belén Noroña, inspired by women of color feminisms and Indigenous and rural female epistemology in Latin America, explores the embodiment of peace and environmental pedagogies. Her thinking of positive peace pedagogies draws from women's epistemology to acknowledge interconnectedness through "acuerpar," or the collective embodiment of each other across difference.

Pedagogies of peace are environmental, Joshua Inwood

As we have discussed in our introduction, militarization permeates university campuses and is a critical piece of how U.S. universities orient themselves. Giroux (2008) notes that in the wake of the attacks on U.S. targets on September 11, 2001 (or "9/11") and the unfolding U.S.-led war on terrorism, militarization at U.S. universities is shaped by various aspects of university life and produces a set of militarized values within the university itself. The reality is that for most undergraduate students, by the time they enter the university, they have never had a formal class on peace, nor have they spent much time working to understand what it means to live in a peaceful community. On the other hand, they have spent considerable time learning about war and the way war is central to the unfolding of geography. As a result, adopting a position that promotes peace in the classroom is revelatory. It can open space for positive peace-based approaches that create conditions that promote peaceful communities. Because work in environmental education also focuses on creating conditions that promote life, engaging through environmental pedagogy is one way to address the realities of the unfolding climate crisis.

While students may not have spent much time thinking about peace, in my experience, they have spent a considerable amount of time thinking and worrying about the looming climate crisis. These students' feelings are shared by scientists who, in the recent Intergovernmental Panel on Climate Change (IPCC 2022) report on global warming, paint a dire picture of the planet's climate and our collective future The report notes that fossil fuels and other global warming gasses are fueling an acceleration in climate change, and fossil fuels are causing significant changes that are threatening biodiversity across the globe. In addition, a warming planet is causing more frequent and severe storms, wildfires, and extreme weather events like droughts and major hurricanes. Perhaps most dire of all—the IPCC (2022) declares that collectively we are falling critically short of attaining the reductions to our fossil fuel usage needed to slow the current crisis. This prediction suggests we will not be successful in limiting the worst aspects of global warming, including limiting global temperature rise to 1.5 degrees Celsius. The IPCC report also documents that global greenhouse emissions "are not evenly distributed—the wealthiest countries are responsible for disproportionately more emissions than developing countries" while "developing countries are experiencing more severe climate impacts" (2022, 15). This reality calls into question the ability of our planet and its ecosystems to support life. A fact that students are all too aware of.

As we noted, the work of climate and environmental educators placed in a context that promotes peace can challenge climate inaction and promote more peaceful communities. Andrezejewski et al. (2009, 11) note that foundational to environmental education is an understanding of the role of peace in building "a world of caring, respect, compassion" that puts into practice "values of human rights, liberty, justice, dignity, democracy, freedom of body, speech, and religion and access to adequate food, water, shelter and health care." This awareness aligns with the values in pedagogies of peace attentive to making the world more humane (Bar-Tal 2002). More to the point, Bar-Tal (2002) argues that for peace education to be positive, it must be relevant and engage with people in their everyday lives. In other words, peace education is context-dependent because it must consider the kinds of conditions within the classroom space and the perspectives and life within it. This approach means addressing militarism and its role in threatening the planet.

Turning specifically to the classroom experience, and from my perspective, students understand the basic science behind global climate change and remain hyper-focused on ways to reduce carbon emissions at the individual level. Students talk honestly about their carbon footprint and ways that they can work to reduce their carbon budget. Unsurprisingly, given how students come to understand their place in the world, this attention to individual solutions to problem-solving extends into every aspect of how we organize our society. This perspective ignores the role of large corporations, wealthy individuals, and international militaries in contributing to global climate gasses. Recent work by Belcher et al. (2020) documents the staggering sum of carbon emissions the U.S. military releases into the atmosphere annually. In 2019 the U.S. military released as much carbon as more than 140 countries. If the U.S. military were a nation-state, it would rank slightly ahead of countries like Denmark and Portugal for the amount of global carbon it releases into the atmosphere (Kitchlew 2022).

Belcher et al. (2020) document that the logistics and movement of materials and forces from bases worldwide are a massive driver of the U.S. military's global carbon emissions. A final cause of emissions is the shift to more advanced military technologies. Kitchlew (2022) documents how advanced technologies, such as the new F-35 fighter jet, burn more fuel than older models like the F-16. They note that the F-35 jet, developed to replace the F-16, requires about 5,600 liters per hour to fly compared with the 3,500 liters of fuel needed for the F-16. This difference indicates how new military technologies are often more carbon-intensive. Perhaps most problematic is how global military carbon use is not counted toward a nation's greenhouse gas emissions. The 1997 Kyoto Protocol—the first global climate treaty—specifically exempted a nation's military from counting toward global emissions. Apart from the difficulty in fully accounting for militarism's role in climate change, one result is that most of the effort to curb global climate change falls onto civilians and others not associated with the military-industrial complex who have, by comparison, a smaller share of the global climate impact. For this reason, making visible the role of militarization is critical to enacting a positive pedagogy of peace and environmental justice.

Returning to the idea that peace and environmental education must occur within a pro-peace pedagogy to inform how students see and understand their place within a contested global terrain, militarism still skews values and shifts responsibility if left out of the discussion. Understanding militarism's role in climate education necessarily means engaging in a broader critique of militarism to help students see their role in providing alternatives to a war culture. For many students, teachers, university administrators, and the public, militarism's ubiquity means its myriad manifestations are largely unremarked upon realities of our daily life. This understanding includes how we understand militarism's role in perpetuating global climate change. Borrowing from the anti–Vietnam War group *Another Mother for Peace* global climate change "is unhealthy for children and other living things" (see Loyd 2011, 403). This reality necessitates that peace education links to climate education, and by engaging in peace work, we are also concomitantly engaging in climate justice. Bajaj and Chiu (2009) note that over the last two decades, there has been increased awareness around the intersections of peace and environmental education and that the two approaches share a focus "of stopping violence" and sharing the social surplus across the community (Bajaj and Chiu 2009, 444). Such an approach strives to create a context in which students move from seeing themselves as individuals operating largely independently to instead concentrate on the innumerable connections and interdependencies that sustain our lives.

Peace pedagogies in youth climate movements, Mark Ortiz

As we have described, militarization works through proliferating armed forces and normalizing violent processes across civil society. At its core, the militarization of civil society relies on rendering invisible those most impacted by structural violence while obscuring our roles and responsibilities in violent processes at various scales. Structural violence refers to harm that is perpetrated by structural factors or social institutions, and it is generally made invisible by hegemonic cultural forces. In outlining pedagogies of peace, we must think about how our teaching can address what Edward Said called the "normalized quiet of unseen power" (Said 2001, as quoted in Nixon 2011, 6). Building on Said's work, Nixon's (2011) elaboration of "slow violence" draws attention to several difficulties associated with teaching about environmental and structural violence. In the case of forms of "slow violence" like climate change, the question for Nixon (2011) and others becomes, how do we represent and communicate about violence that entangles generations and geographies in relations of uneven and unclear responsibility?

The contemporary global university is simultaneously a site where militarized knowledge and violent epistemologies are recreated while also a place of potentially liberatory pedagogies and organizing. Pedagogies of peace must promote critical interrogation aimed at transforming practices and educational models that obfuscate and reproduce the environmental violence of the everyday. As one approach, Sultana (2022, 2) calls for centering stories and "lived experiences" for those most impacted by climate change as a corrective course for "climate coloniality," a phrase that describes the "uneven and unequal vulnerabilities and marginalizations of deaths and devastation taken for granted." Since slow and structural violence traffic in obfuscation, the power of pedagogy is in bringing what may appear invisible for some students into a different kind of visibility (Davies 2022). Pedagogy, in this form, becomes an exercise in political reimagination and a powerful conduit for social transformation.

To engage students on a deeper level, it is essential that they see themselves in our course content, and that they recognize their concerns as being reflected in what we teach. That is why we look to transnational youth climate movements as inspiring sites of peace pedagogy that strive toward positive peace. These movements encourage solutions addressing the root causes of climate change by rigorously describing its intersections with structural and interpersonal violence, such as militarization (see DiNucci, n.d.). These movements are led by global networks of young people similar in age to those we teach, and they take shape across many of the social media platforms our students use daily. For these reasons, our students can often resonate with the concerns of these youth networks, and diverse students see themselves reflected in their leadership.

One example is the global youth movement founded by Greta Thunberg, Fridays for Future (FFF). Fridays for Future is best known for coordinating the school strikes for climate in which children and young adults boycott schools to protest climate inaction. Since 2018, school strikes for climate have taken place in over 140 countries involving millions of young people from vastly different backgrounds (Milman 2019). The signs, protest chants, figures, and phrases associated with FFF constitute an "oppositional pedagogy of cultural production" positioned against formal schooling that takes root and travels across contexts (Giroux 2008, 75). FFF has played a central role in global climate politics in the last four years. As it has grown transnationally, the network elaborates and works toward a "radical trans-relational peace" that encompasses thinking about how "multiple 'locals' nurture or undermine ecological dignity in a globalizing world" (Courtheyn 2018, 743).

The Fridays for Future movement has emerged as a transnational network of youth climate activism models, a praxis of positive peace and climate justice. At a moment in which many countries are responding to climate change by militarizing borders and insisting on the nation-state as impermeable, the transnational mobilization of FFF is all the more critical. As more activists from what FFF terms MAPA—Most Affected Peoples and Areas—become involved in the movement, its messaging has evolved from an emphasis on the future impacts of climate change to a more intersectional climate justice framing focusing on the present and its uneven dynamics (see Ravi 2021). Activists are learning from one another and creating deliberative spaces of dialogue that acknowledge and seek to transform the unevenness of existing geopolitics by building communities of care and solidarity across boundaries. Activists share memes, hopes, fears, grief, and love across the internet and build communities of action each Friday by posting photos of themselves striking (see Tan et al. 2022). This type of engaged political praxis starkly contrasts learning about climate change impacts through school modules, instead creating intimate relations of solidarity, care, and learning across geographies.

FFF's political analysis locates and calls out "organized irresponsibility" (Beck 2009, 27) across institutional landscapes and the slow violence that results. When youth activist Greta Thunberg described the global climate negotiations in a protest as "blah blah blah," she articulated a critique that applies to a range of adult-dominated institutions from the UN to formal school systems. Such institutions reinforce the militarized status quo and the routinization of uneven ecological harm through technical language that works to neutralize urgency coupled with the dragging tempos of administration and bureaucracy (Nelson 2016). Schools, like the UN climate negotiations, often participate in a particular rendering of climate change that fails to acknowledge the historical roots of the climate crisis and the responsibilities that emerge from this acknowledgment (Kashwan and Ribot 2021).

By contrast, structural critique has been central to the framings of FFF, which have repeatedly questioned why school attendance should take priority over urgent political action (Kahn 2019). In their movement-building, young people situated within "imperial constellations that cross generations, borders, and oceans" are remapping the geographies of the present-future against the reproduction of militarized civil society and extractivism (Gergan et al. 2022, 1). In the protest spaces of affective resistance, young activists come together outside adult imprimatur, collectively embodying a complex tapestry of emotions from despair to "revolutionary potentiality" (Sultana 2022, 2). Climate crisis becomes perceived through the collective, embodied refusal of youth protestors pushing against the abstraction of climate and environmental violence present in schools, and dominant intergenerational relations which position young people as learners rather than agents.

FFF's work also includes an impressive array of digital content which can be used as pedagogical tools in the classroom. This digital content includes a podcast, newsletter, and other educational multimedia material on platforms like Twitter and Instagram (see Fridays for Future Digital, n.d.). These youth-created knowledge products are helpful companions for encouraging students to reflect on their embeddedness in intersecting systems of power and their potential roles in addressing slow violence. FFF's materials are created by and for young people and characterize climate change not as a distant process known through Western science but as intimately interwoven with our everyday practices, spaces, and discourses. Thus, climate change is a problem that must be approached with creativity, humor, and everyday grammars (Tan 2021). Following FFF activists, I incorporate creative course outputs in my classes, including podcasts, artwork, and music, to encourage students to reflect collectively on climate change across various registers.

The pedagogical approach of Fridays for Future works in the space of what my colleagues in the North Carolina Climate Justice Collective think about as "reimagining." Reimagination engages "how we think about ourselves with each other and the whole" (Lassetter, n.d., 1). It forms a critical element of social change and positive peace alongside practices of reform, resistance, and recreation. In their form and function, youth climate networks reimagine geopolitics through a historically grounded planetary lens, critique the institutional forms that have shaped climate deadlock and ongoing militarization, and seek to recreate intergenerational relations to honor and value the ingenuity of young people. This work of cultural and ecological reimagination is instructive as a critical example of young people exercising their capacities to "assist others in plotting a course from a current state to an imagined future," building intersectional critiques of where we are and at the same time building engaging imaginaries of where we may wish to go (Jeffrey and Dyson 2022, 1340).

"Acuerpando" or embodying conflict and peace pedagogies, María Belén Noroña

Peace geographers have defined peace as an engaged political practice confronting hegemonic power structures (Macaspac and Moore 2022). Positive peace, especially, is understood as situated, place-based, individual, and/or collective processes aimed at addressing underlying causes of violence while building transnational solidarities (Inwood and Tyner 2011; Loyd 2012; Koopman 2011b).

Inspired by women of color feminisms and Indigenous and rural female epistemology in Latin America, I argue for positive peace as an embodied pedagogy. This scholarship has a tradition of building networks and solidarities based on shared understandings of peace and oppression experienced in different ways by different bodies (Cabnal 2010, 2017; Moraga and Anzaldúa [1981] 2015; Rodriguez 2021). Still, producing common or shared understandings of conflict in academic spaces becomes problematic as divisive politics occur through spatial hierarchies and binaries that separate us through social difference such as race, class, and gender; not to mention metanarratives producing national borders, cultural belonging, politics, religion, etc. (Reynolds, Silvernell, and Mercer 2020). Indeed, Rivera-Cusicanqui (2010) and other Latin American Indigenous intellectuals (TERERBV 2016) denounce how most individuals tend to understand the effects of militarism, extractive processes, climate change, and ongoing forms of colonial oppression as structural processes existing at a distance, rather than processes deeply embedded within our communities, our relationships, and even our bodies.

As scholars and educators, our tendency to teach and research about conflict and oppression outside our communities without problematizing our interwoven realities is highly troubling. Such a practice not only reproduces the violence we try to address in our teaching but also continues to naturalize the spatiality of social difference, affecting our ability to see how we enable those processes and their effect. Feminist political ecologists such as Mollet and Faria (2013, 2016) and decolonial scholars such as Espinosa, Gómez, and Ochoa (2014) argue that we lack frameworks to communicate across differences and hierarchies. Indeed, we lack practical tools to enable productive conversations amid these conflicts within geographies of peace. This lack of communication further harms students' ability to connect with urgent social and climate realities, decreasing political engagement and opportunities for mobilization.

Still, for rural and Indigenous women in the global south who survive the struggles of protecting communal land from endless extraction, militarization, and biodiversity loss, our perceived social differences become Western tricks used to naturalize ongoing violence and dispossession (Silva Santisteban 2019). Indigenous women believe all bodies are connected and interdependent, including nature and land (Haesbaert 2020; Ulloa 2021; Echeverri 2004). Similarly, women of color feminisms have argued for understandings of social difference as imbricated processes (Davis 2008), stressing embodied aspects of survival. A survival that is fleshed through common struggles addressed through an intimacy of solidarity (Moraga and Anzaldúa [1981] 2015).

In thinking of positive peace pedagogies, I draw from women's epistemology to acknowledge such fleshed and embodied interconnection through "acuerpar" or embodying each other across skins and differences (Cabnal 2010). By incorporating acuerpamiento into pedagogy, students are encouraged to listen to marginalized voices and realities with both their minds and bodies (Cabnal 2017; Noroña, forthcoming). Acuerpar requires embodying and embracing another's reality to attempt to travel to each other's worlds (Lugones 1987; Rodriguez 2021; Noroña, forthcoming).

Traveling to and through each other's worlds requires understanding rural and marginalized realities beyond negative peace policies that call for preventive measures and affirmative climate policies to ameliorate the devastating consequences of land grabs, climate change, continued extraction, and military intervention. A positive peace pedagogy requires acknowledging that, in addition to preventive and affirmative climate and peace policies, we must reveal the connections between our skins and those of marginalized populations at the frontlines of extraction and warfare. Only then can we recognize ourselves as complicit and vulnerable to gradual violence, a deeply personal process that encourages solidarity and political engagement across geographies. Thus, acuerpar responds to peace geographers' calls for a peace agency built on strong interpersonal relations and shared understandings of vulnerability (Koopman 2011a, 2011b; Woon 2014).

Questions guiding acuerpamiento or an embodied pedagogy for positive peace include asking how we feel specific types of conflict and violence in our community, where do we feel conflict and violence in the body, and how do these processes materialize in our communities, among others (CMCTF 2017). Moreover, class discussions that reflect acuerpar open spaces for students to connect by building campus communities and, despite diverse subjectivities and positionalities, to produce common understandings of conflicts and violence discussed in class. Such an exercise becomes a prerequisite for shared accounts of how social justice and peace should look and feel and what we must do individually and collectively to induce change.

Other ways to have students connect their lives with the realities we study through acuerpamiento is to ask them to write assignments from the perspective of those more directly affected by violence. This writing can include human and nonhuman perspectives, such as those of animals, rivers, and forests. Indeed, acuerpamiento among Indigenous people, particularly in the Amazon rainforest, requires feeling and thinking from the bodies of nonhuman beings to redefine the conditions of what it means to be human (AMWAE 2019; Valdivia et al. 2021).

In general, female and southern epistemologies of acuerpar can be helpful to positive peace processes and peace pedagogies, as the politics of embodiment and emotion are essential for learning processes in ways that encourage theory and action (Wright 2010). When our students successfully connect the experiences of racialized and feminized others with their realities, they become aware that detachments from issues such as slow warfare and climate change are part of a structure that conceals our participation and our vulnerability in these processes. Indeed, these pedagogies help us delineate a political praxis to navigate across place, space, and difference by dwelling in the intimate ways we internalize geographies of conflict and difference. Such awareness encourages positive peace processes at the body and subjective levels in the classroom and the community. All these are spaces where we not only have the power to challenge and contest power structures but to transform them positively.

Conclusion

> We need an education that allows us to go beyond the space we already inhabit and accompanies us into the unknown.
>
> (UNESCO 2021, 2)

By working with the intersecting theories of positive peace and environmental education, we aim to enable learners to acquire knowledge grounded in an understanding of positive peace both in and across their worlds. Pedagogies grounded in positive peace and environmental justice encourage students and educators to reimagine their immediate and distant worlds in ways that enrich our shared humanity.

Despite the mythology of universities as peaceful sanctuaries of political expression, intersectional campus life can be isolating and violent and a place of devaluation. For students who are at risk for identity-based trauma such as LGBTQA+ and gender-nonconforming students, women, students of color, and as first-generation college students' their lives can hinge on the liberatory potential of education (Wood 2020). As the vignettes in this chapter illustrate, pedagogical experiences of positive peace are defined by working directly with difference in a context of interconnectedness to deepen our understanding of differing worldviews and foster attitudes of acceptance and respect. Not to do so supports militarized and violent outcomes.

The authors recently witnessed firsthand how a focus on negative peace and militarization can have traumatic outcomes. The conservative student group "Uncensored America" invited Gavin McInnes, founder of the Proud Boys, and Alex Stein, another far-right actor, to speak at Pennsylvania State University. Individuals who were assumed to be members of the Proud Boys, an organization identified as a hate group by the Southern Poverty Law Center, sprayed an irritant (bear spray) onto protesting students. In the days that followed, we heard student protestors' testimonies of the trauma that ensued, such as trying to use water bottles to help wash out their friends' eyes.

At the same time, mounted police officers watched on and did not intervene. The only arrest that evening was a Penn State student for disorderly conduct.

For the authors, this event crystallized the need for peace pedagogies that go beyond the classroom and create new ways for administrators to enact an ethic of care that recognizes that universities are places in crisis; where politics, education, and socialization processes remain deeply divided and polarized (see Dowler, Cuomo, and Laliberte 2014). As Ruth Wilson Gilmore advises, a crisis is a form of instability "that can be fixed only through radical measures which include developing new relationships and new or renovated institutions of what already exists" (Gilmore 2007, 26). This urgent call for understanding our planet through pedagogies that are nonviolent and environmentally aware is one path forward to a more peaceful future.

References

Amster, R. 2013. Toward a climate of peace. *Peace Review* 25: 473–479.
AMWAE (Asociación de Mujeres Waorani de la Amazonia Ecuatoriana). 2019. *Cuidanderas Documentary*. Urgent Action Fund for Latin America and the Caribean. Araguato Films. Accessed March 17, 2023. https://www.youtube.com/watch?v=Xzz05Se_mHU&ab_channel=FondodeAcci%C3%B3nUrgenteAmericaLatinayelCaribe
Andrzewewski, A., M. Baltodanao, and L. Symcox. 2009. *Social Justice, peace, and environmental education: transformative standards*. Routledge.
Bajaj, M., and B. Chiu. 2009. Education for sustainable development as peace education. *Peace & Change* 34: 441–455.
Bar-Tal, D. 2002. The elusive nature of peace education. In *Peace Education: The Concept, Principles, and Practices Around the World*, eds. G. Solomon and B. Nevo, 27–36. Psychology Press.
Beck, U. 2009. *World at Risk*. Polity.
Belcher, O., P. Bigger, B. Neimark, and C. Kennelly. 2020. Hidden carbon costs of the "everywhere war": Logistics, geopolitical ecology, and the carbon boot-print of the US military. *Transactions of the Institute of British Geographers* 45: 65–80.
Cabnal, L. 2010. Feminismos Diversos: El Feminismo Comunitario. *ACSUR, Las Segovias*. Accessed March 17, 2023. https://porunavidavivible.files.wordpress.com/2012/09/feminismos-comunitario-lorena-cabnal.pdf
Cabnal, L. 2017. Tzk'at, Red de Sanadoras Ancestrales Del Feminismo Comunitario Desde Iximulew-Guatemala. *Ecología Política* 54: 98–102.
Catungal, J.P. 2019. Classroom. In *Keywords in Radical Geography: Antipode at 50*, eds. J. Tariq, A. K. Kent, N. McKittrick, S. Theodore, P. Chari, V. Chatterton, and N. Gidwani, 45–49. Wiley.
CMCTF (Colectiva Miradas Críticas del Territorio desde el Feminismo). 2017. *Mapeando El Cuerpo-Territorio: Guía Metodologica Para Mujeres Que Defienden Sus Territorios [Mapping Body-Territory: Methodological Guide for Women Defending Their Territories]*. Ecuador. https://miradascriticasdelterritoriodesdeelfeminismo.files.wordpress.com/2017/11/mapeando-el-cuerpoterritorio.pdf
Courtheyn, C. 2018. Peace geographies: Expanding from modern-liberal peace to radical trans-relational peace. *Progress in Human Geography* 42: 741–758.

Daigle, M. and J. Sundberg. 2017. From where we stand: Unsettling geographical knowledges in the classroom. *Transactions of the Institute of British Geographers* 42: 338–341.

Davies, T. 2022. Slow violence and toxic geographies: 'Out of sight' to whom? *Environment and Planning C: Politics and Space* 40: 409–427.

Davis, K. 2008. Intersectionality as buzzword: A sociology of science perspective on what makes a feminist theory successful. *Feminist Theory* 9: 67–85.

DiNucci, O. n.d. Fridays for Future #UprootTheSystem Global Youth Climate Strike - Cut the Pentagon Day #13 *Codepink*. Accessed December 14, 2022. https://www.codepink.org/ctp-day-13

Dowler, L., D. Cuomo, and N. Laliberte. 2014. Challenging 'The Penn State Way': A feminist response to institutional violence in higher education. *Gender, Place & Culture* 2: 387–394.

Echeverri, A. 2004. Territorio Como Cuerpo y Territorio Como Naturaleza: Diálogo Intercultural. In *Tierra Adentro: Territorio Indígena y Percepción Del Entorno*, eds. A. Surralles and P. García Hierro, 259–275. Grupo Internacional de Trabajo sobre Asuntos Indígenas.

Espinosa, Y., D. Gómez, and K. Ochoa, eds. 2014. *Tejiendo de otro modo: Feminismo, epistemología, y apuestas decoloniales en Abya Yala*. Popayán. Universidad del Cauca.

Fridays for Future Digital. n.d. You Have a Voice. *FFF Digital*. Accessed December 14, 2022. https://fffdigital.carrd.co/

Gergan, M., S. Krishnan, S. Smith, and S. Young. 2022. Youth and decolonial politics in a relational context. *Antipode* https://doi.org/10.1111/anti.12906

Gilmore, R. W. 2007. *Golden Gulag: Prisons, Surplus, Crisis, and Opposition in Globalizing California*. University of California Press.

Giroux, H. A. 2008. The militarization of US higher education after 9/11. *Theory, Culture & Society* 25: 56–82.

Greenwood, D. A. 2008. A critical pedagogy of place: From gridlock to parallax. *Environmental Education Research* 14: 336–348.

Greenwood, D. A. 2010. Education in a culture of violence: A critical pedagogy of place in wartime. *Cultural Studies of Science Education* 5: 351–359.

Haesbaert, R. 2020. Del cuerpo-territorio al territorio-cuerpo (De la tierra): contribuciones Decoloniales. *Cultura y representaciones sociales* 15: 267–301.

Harvester, L. and S. Blenkinsop. 2010. Environmental education and ecofeminist pedagogy: Bridging the environmental and the social. *Canadian Journal of Environmental Education* 15: 120–134.

Hill Collins, P. 1986. Learning from the outsider within: The sociological significance of black feminist thought. *Social Problems* 33: 14–32.

hooks, b. 1994. *Teaching to Transgress: Education as the Practice of Freedom*. Routledge.

Inwood, J. and J. Tyner. 2011. Geography's pro-peace agenda: An unfinished project. *ACME: An International Journal for Critical Geographies* 10: 442–457.

IPCC. 2022. *Climate Change 2022: Impacts, Adaptation, and Vulnerability*. Contribution of Working Group II to the Sixth Assessment Report of the Intergovernmental Panel on Climate Change, H.-O. Pörtner, D.C. Roberts, M. Tignor, E.S. Poloczanska, K. Mintenbeck, A. Alegría, M. Craig, S. Langsdorf, S. Löschke, V. Möller, A. Okem, B. Rama eds. Cambridge University Press.

Jeffrey, C., & Dyson, J. 2022. Viable geographies. *Progress in Human Geography* 46: 1331–1348.

Kahn, B. 2019. Why go to school when the world is burning?' We asked students why they're striking for climate change. *Gizmodo*. March 13. Accessed December 14, 2022. https://gizmodo.com/why-go-to-school-when-the-world-is-burning-we-asked-1833264147

Kashwan, P. and J. Ribot. 2021. Violent silence: The erasure of history and justice in global climate policy. *Current History* 120: 326–331.

Kitchlew, I. 2022. Is super polluting Pentagon climate plan a military sized greenwash? *The Guardian*, March 10.

Koopman, S. 2011a. Let's take peace to pieces. *Political Geography* 30: 193–194.

Koopman, S. 2011b. Alter-geopolitics: Other securities are happening. *Geoforum* 42: 274–284.

Lassetter, J. n.d. The 4 Rs of social transformation. *North Carolina Climate Justice Collective*. Accessed December 14, 2022. https://www.ncclimatejustice.info/_files/ugd/393ec8_8d337ff39a0a4d7b9b4f4cf71ddf5a5e.pdf

Loyd, J. 2011. War is not healthy for children and other living things. *Environment and Planning D: Society and Space* 27:403–424.

Loyd, J. 2012. Geographies of peace and antiviolence: Peace and antiviolence. *Geography Compass* 6: 477–489.

Lugones, M. 1987. Playfulness,"world"-travelling, and loving perception. *Hypatia* 2: 3–19.

Macaspac, N. V., and A. Moore. 2022. Peace geographies and the spatial turn in peace and conflict studies: Integrating parallel conversations through spatial practices. *Geography Compass* 16. https://doi.org/10.1111/gec3.12614

McLeod, D., W. Wright, N. Laliberté, and A. Bain. 2021. Embodying controversy through feminist pedagogy. *Acme: An International Journal for Critical Geographies* 20: 479–490.

Milman, O. 2019. US to stage its largest ever climate strike: 'Somebody must sound the alarm', *The Guardian*. September 20. Accessed December 14, 2022. https://www.theguardian.com/world/2019/sep/20/climate-strikes-us-students-greta-thunberg

Mollett, S. and C. Faria. 2013. Messing with gender in feminist political ecology. *Geoforum* 45: 116–125.

Mollett. S. and C. Faria. 2016. Critical feminist reflexivity and the politics of whiteness in the field. *Gender, Place & Culture* 23: 79–93.

Moraga, C. and G. Anzaldúa. 2015 [1981]. *This Bridge Called My Back: Writings by Radical Women of Color*. 4th Edition. State University of New York.

Nelson, S. 2016. The slow violence of climate change. *Jacobin*. February 17. Accessed December 4, 2022. https://jacobin.com/2016/02/cop-21-united-nations-paris-climate-change/

Nixon, R. 2011. *Slow Violence and the Environmentalism of the Poor*. Harvard University Press.

Noroña, M. B. Forthcoming. Visceral empathy: Communicating across difference in the Amazon of Ecuador. *Antipode*.

Ravi, D. 2021. The climate crisis is about the Global South's present. *Al Jazeera* November 12. Accessed December 14, 2022. https://www.aljazeera.com/opinions/2021/11/12/climate-crisis-is-global-south-today-not-global-norths-tomorrow

Reynolds, H., D. Silvernell, and F. Mercer. 2020. Teaching in an era of political divisiveness: An exploration of strategies for discussing controversial issues. *The*

Clearing House: A Journal of Educational Strategies, Issues and Ideas 93. https://www.tandfonline.com/doi/abs/10.1080/00098655.2020.1762063

Rivera-Cusicanqui, S. 2010. *Violencias (Re) Encubiertas En Bolivia*. Piedra Rota.

Rodriguez Castro, L. 2021. *Decolonial Feminisms, Power and Place: Sentirpensando with Rural Women in Colombia*. Palgrave Macmillan.

Said, E. 2001. The public role of writers and intellectuals. *The Nation*, September 17.

Silva Santisteban, R., ed. 2019. *Mujeres Indígenas Frente al Cambio Climático*. IWGIA, SERVINDI, ONAMIAP, COHARYIMA.

Sultana, F. 2022. The unbearable heaviness of climate coloniality. *Political Geography*, Pre-Print. https://doi.org/10.1016/j.polgeo.2022.102638

Tan, M.J. 2021. The mainstream climate change movement needs to get more creative. *Teen Vogue*. June 1. Accessed December 4, 2022. https://www.teenvogue.com/story/climate-change-movement-creativity

Tan, M.J., Y. Baluch, A. Calderon Hernandez, F. Faruk Jhumu, and M. Ortiz. 2022. Fighting climate change in the Global South. *The Forge*. April 28. Accessed December 14, 2022. https://forgeorganizing.org/article/fighting-climate-change-global-south

TERERBV (Tercer Encuentro Regional para el Buen Vivir desde las Mujeres y los Pueblos). 2016. *Sanando Nuestro Territorio Cuerpo-Tierra*. Documentary, Universidad de Costa Rica, Quince UCR. https://www.youtube.com/watch?v=fwba3zTJxvw&t=1433s&ab_channel=Quince-UCR

Ulloa, A. 2021. Repolitizar la vida, defender los cuerpos-territorios y colectivizar las acciones desee los feminismos indígenas. *Ecología Política* 61: 38–48.

UNESCO, 2021. *Futures of Education, A New Social Contract*. 2021. Accessed March 17, 2023. https://en.unesco.org/futuresofeducation

Valdivia, G., F. Lu, A. Weay Cawiya, and M. Omari Ima Omene. 2021. Real and mythical bodies weaving social skin: Two Waorani women disruption genres of Amazonial humanity. In *Feminist geography unbound: Discomfort, bodies, and prefigured futures*, eds. S. Smith, C. Neubert, M. Hawkins, and B. Gokariksel, 95–115. West Virginia University Press.

Williams, J. 2015. Women, weapons, peace and security. *SUR* 12: 31–37.

Wood, J. 2020. Teaching students at the margins: A feminist trauma-informed care pedagogy. In *Lessons from the Pandemic: Trauma-Informed Approaches to College, Crisis*, eds. P. Thompson and J. Carello, 23–38. Palgrave Macmillan.

Woon, C. Y. 2014. Precarious geopolitics and the possibilities of nonviolence. *Progress in Human Geography* 38: 654–670.

Wright, M. W. 2010. Geography and gender: Feminism and a feeling of justice. *Progress in Human Geography*, 34: 818–827.

INDEX

Printed in the United States
by Baker & Taylor Publisher Services